Lecture Notes in Computer Science 15251

Founding Editors

Gerhard Goos
Juris Hartmanis

Editorial Board Members

Elisa Bertino, *Purdue University, West Lafayette, IN, USA*
Wen Gao, *Peking University, Beijing, China*
Bernhard Steffen, *TU Dortmund University, Dortmund, Germany*
Moti Yung, *Columbia University, New York, NY, USA*

The series Lecture Notes in Computer Science (LNCS), including its subseries Lecture Notes in Artificial Intelligence (LNAI) and Lecture Notes in Bioinformatics (LNBI), has established itself as a medium for the publication of new developments in computer science and information technology research, teaching, and education.

LNCS enjoys close cooperation with the computer science R & D community, the series counts many renowned academics among its volume editors and paper authors, and collaborates with prestigious societies. Its mission is to serve this international community by providing an invaluable service, mainly focused on the publication of conference and workshop proceedings and postproceedings. LNCS commenced publication in 1973.

Tianqing Zhu · Jin Li · Aniello Castiglione
Editors

Algorithms and Architectures for Parallel Processing

24th International Conference, ICA3PP 2024
Macau, China, October 29–31, 2024
Proceedings, Part I

 Springer

Editors
Tianqing Zhu 🆔
City University of Macau
Macau, China

Jin Li 🆔
Guangzhou University
Guangzhou, China

Aniello Castiglione 🆔
University of Salerno
Fisciano, Italy

ISSN 0302-9743 ISSN 1611-3349 (electronic)
Lecture Notes in Computer Science
ISBN 978-981-96-1524-7 ISBN 978-981-96-1525-4 (eBook)
https://doi.org/10.1007/978-981-96-1525-4

© The Editor(s) (if applicable) and The Author(s), under exclusive license
to Springer Nature Singapore Pte Ltd. 2025

This work is subject to copyright. All rights are solely and exclusively licensed by the Publisher, whether the whole or part of the material is concerned, specifically the rights of translation, reprinting, reuse of illustrations, recitation, broadcasting, reproduction on microfilms or in any other physical way, and transmission or information storage and retrieval, electronic adaptation, computer software, or by similar or dissimilar methodology now known or hereafter developed.
The use of general descriptive names, registered names, trademarks, service marks, etc. in this publication does not imply, even in the absence of a specific statement, that such names are exempt from the relevant protective laws and regulations and therefore free for general use.
The publisher, the authors and the editors are safe to assume that the advice and information in this book are believed to be true and accurate at the date of publication. Neither the publisher nor the authors or the editors give a warranty, expressed or implied, with respect to the material contained herein or for any errors or omissions that may have been made. The publisher remains neutral with regard to jurisdictional claims in published maps and institutional affiliations.

This Springer imprint is published by the registered company Springer Nature Singapore Pte Ltd.
The registered company address is: 152 Beach Road, #21-01/04 Gateway East, Singapore 189721, Singapore

If disposing of this product, please recycle the paper.

Preface

On behalf of the Conference Committee, we welcome you to the proceedings of the 2024 International Conference on Algorithms and Architectures for Parallel Processing (ICA3PP 2024), which was held in Macau Special Administrative Region, China from October 29–31, 2024. ICA3PP 2024 was the 24th in this series of conferences (started in 1995) that are devoted to algorithms and architectures for parallel processing. ICA3PP is now recognized as the main regular international event that covers the many dimensions of parallel algorithms and architectures, encompassing fundamental theoretical approaches, practical experimental projects, and commercial components and systems. This conference provides a forum for academics and practitioners from countries around the world to exchange ideas for improving the efficiency, performance, reliability, security, and interoperability of computing systems and applications.

A successful conference would not be possible without the high-quality contributions made by the authors. This year, ICA3PP received a total of 265 submissions from authors in various countries and regions. Based on rigorous peer reviews by the Program Committee members and reviewers, 131 high-quality papers were accepted to be included in the conference proceedings and submitted for EI indexing. In addition to the contributed papers, distinguished scholars were invited to give keynote lectures, providing us with the recent developments in diversified areas in algorithms and architectures for parallel processing and applications.

Among the accepted papers, the papers with the highest weighted review mark in each round received the Best Paper Award. The Best Papers were *Updates Leakage Attack against Private Graph Split Learning,* Hao Yang, Zhuo Ma, Yang Liu, Xinjing Liu, Beiwei Yang, and Jianfeng Ma, *Data-Free Encoder Stealing Attack in Self-supervised Learning,* Chuan Zhang, Xuhao Ren, Haotian Liang, Qing Fan, Xiangyun Tang, Chunhai Li, Liehuang Zhu, and Yajie Wang, and *Federated Meta Continual Learning for Efficient and Autonomous Edge Inference* Bingze Li, Stella Ho, Youyang Qu, Chenhao Xu, Tom H. Luan, and Longxiang Gao. Best Student paper was *DIsFU: Protecting Innocent Clients in Federated Unlearning* Fanyu Kong, Xiangyun Tang, Yu Weng, Tao Zhang, Hongyang Du, Jiawen Kang, and Chi Liu.

We would like to take this opportunity to express our sincere gratitude to the 262 Program Committee members and reviewers for their dedicated and professional service. We highly appreciate the track chairs for their hard work in promoting this conference and organizing the reviews for the papers submitted to their tracks. We are so grateful to the publication chairs, Zuobin Ying and Sheng Wen, for their meticulous work in editing the conference proceedings. We must also say "thank you" to all the volunteers who helped us at various stages of this conference.

Moreover, we would like to extend our appreciation to the following chairs for their invaluable contributions:

Local Chairs
Wenjian Liu, Chris Chu, Max Kuok

Workshop Chairs
Jia Gu, Chaofeng Zhang, Fengshi Jin, Chi Liu

Publicity Chairs
Gengshen Wu, Lefeng Zhang

Registration Chairs
Kaiyao Jiang, Congcong Zhu

We were so honored to have many renowned scholars be part of this conference. Finally, we would like to thank all speakers, authors, and participants for their great contribution to and support for the success of ICA3PP 2024.

October 2024

Wanlei Zhou
Paulo Quaresma
Albert Zomaya
Willy Susilo
Tianqing Zhu
Jin Li
Aniello Castiglione

Organization

General Chairs

Wanlei Zhou   City University of Macau, China
Paulo Quaresma   University of Évora, Portugal
Albert Zomaya   University of Sydney, Australia
Willy Susilo   University of Wollongong, Australia

Program Chairs

Tianqing Zhu   City University of Macau, China
Jin Li   Guangzhou University, China
Aniello Castiglione   University of Salerno, Italy

Local Chairs

Wenjian Liu   City University of Macau, China
Chris Chu   City University of Macau, China
Max Kuok   City University of Macau, China

Publication Chairs

Zuobin Ying   City University of Macau, China
Sheng Wen   Swinburne University of Technology, Australia

Workshop Chairs

Jia Gu   City University of Macau, China
Chaofeng Zhang   Advanced Institute of Industrial Technology, Japan
Fengshi Jin   City University of Macau, China
Chi Liu   City University of Macau, China

Publicity Chairs

Gengshen Wu	City University of Macau, China
Lefeng Zhang	City University of Macau, China

Registration Chairs

Kaiyao Jiang	City University of Macau, China
Congcong Zhu	City University of Macau, China

Program Committee

Aniello Castiglione	Department of Management and Innovation Systems
Bangbang Ren	National University of Defense Technology
Bin Wu	Chinese Academy of Sciences
Bo Li	Swinbourn University of Technology
Bo Liu	University of Technology Sydney
Bowen Liu	Nanjing University
Chao Li	Beijing Jiaotong University
Chao Wang	University of Science and Technology of China
Chaokun Zhang	Tianjin University
Chen Zhang	City University of Hong Kong
Chentao Wu	Shanghai Jiao Tong University
Chi Liu	City University of Macau
Chris Chu	City University of Macau
Chuan Zhang	Beijing Institute of Technology
Chuang Hu	Wuhan University
Congcong Zhu	City University of Macau
Daniel Andresen	Kansas State University
Dayong Ye	University of Wollongong
Deze Zeng	China University of Geosciences
Dezun Dong	National University of Defense Technology
En Shao	Institute of Computing Technology
Faqian Guan	China University of Geosciences
Fei Lei	NUDT
Fuliang Li	Northeastern University
Fuyuan Song	Nanjing University of Information Science and Technology
Geng Sun	Jilin University

Gongming Zhao	University of Science and Technology of China
Guang Wang	Florida State University
Guangwu Hu	Shenzhen Institute of Information Technology
Guo Chen	Hunan University
Guozhu Meng	Institute of Information Engineering
Hai Xue	University of Shanghai for Science and Technology
Haikun Liu	Huazhong University of Science and Technology
Hailong Yang	Beihang University
Haipeng Dai	Nanjing University
Haiping Huang	College of Computer
Haonan Lu	University at Buffalo
Haozhe Wang	University of Exeter
Heng Qi	Dalian University of Technology
Hongwei Zhang	Tianjin University of Technology
Hua Huang	University of California
Hui Sun	City University of Macau
Humayun Kabir	Microsoft
Ioanna Kantzavelou	University of West Attica
Jaya Prakash	Champati University of Victoria
Jiahui Li	Jilin University
Jin Li	Guangzhou University, China
Jinbin Hu	Hong Kong University of Science & Technology
Jing Gong	KTH Royal Institute of Technology
Jinguang Han	Southeast University
Jingwen Leng	Shanghai Jiao Tong University
Jinwen Xi	Beijing Zhongguancun Laboratory
Jordan Samhi	CISPA – Helmholtz Center for Information Security
Jun Shao	School of Computer and Information Engineering
Kaiping Xue	University of Science and Technology of China
Kejiang Ye	SIAT
Ladjel Bellatreche	LIAS/ENSMA
Lanju Kong	Shandong University
Laurent Lefevre	Inria
Lefeng Zhang	City University of Macau
Lei Wang	Soochow University
Letian Zhang	Middle Tennessee State University
Li Duan	Beijing Jiaotong University
Li Ma	ShangHai Jiao Tong University
Lijie Xu	Nanjing University of Posts and Telecommunications

Lin He	THU
Lingjun Pu	Nankai University
Liu Yuling	Institute of Information Engineering
Lizhao You	Xiamen University
Longxiang Gao	Qilu University of Technology
Lu Zhao	Nanjing University of Posts and Telecommunications
Mahbubur Rahman	City University of New York
Marc Frincu	West University of Timisoara
Massimo Cafaro	University of Salento
Massimo Torquati	University of Pisa
Max Kuok	City University of Macau
Meixuan Ren	Sichuan Normal University
Meng Li	Hefei University of Technology
Meng Li	Nanjing University
Meng Shen	Beijing Institute of Technology
Mengying Zhao	Shandong University
Mi Zhang	ICT
Minfeng Qi	City University of Macau
Minghao Zhao	East China Normal University
Minghui Xu	Shandong University
Mingwu Zhang	Hubei University of Technology
Minyu Feng	Southwest University
Mirazul Haque	Research Scientist
Peter Kropf	University of Neuchâtel
Philip Brown	University of Colorado Colorado Springs
Qianhong Wu	Beihang University
Qing Fan	BIT
Qiong Huang	South China Agricultural University
Radu Prodan	University of Klagenfurt
Ravishka Rathnasuriya	University of Texas at Dallas
Roman Wyrzykowski	Czestochowa University of Technology
Rongxing Lu	University of New Brunswick
Sa Wang	Institute of Computing Chinese Academy of Sciences
Shaojing Fu	National University of Defense Technology
Shen Dian	Southeast University
Sheng Ma	NUDT
Shenglin Zhang	Nankai University
Shuai Gao	Beijing Jiaotong University
Shuai Xu	Nanjing University of Aeronautics and Astronautics

Shuai Zhou	City University of Macau
Shuang Chen	Huawei Cloud
Shujie Han	Peking University
Shuxin Zhong	Rutgers University
Simin Chen	UTD
Songwen Pei	Dept. of Computer Science and Engineering
Su Yao	Tsinghua University
Susumu Matsumae	Saga University
Tao Wu	National University of Defense Technology
Tianqing Zhu	University of Technology
Tianyi Liu	HUAWEI
Tie Qiu	Tianjin University
Tingwen Liu	Institute of Information Engineering
Vladimir Voevodin	RCC MSU
Wei Bao	The University of Sydney
Wei Wang	Central South University
Weibei Fan	Nanjng University of Posts and Telecommunications
Weihua Zhang	Fudan University
Weitian Tong	Georgia Southern University
Weixing Ji	Beijing Normal University
Weizhi Meng	Lancaster University
Wenjuan Li	The Education University of Hong Kong
Wenxin Li	Tianjin University
Wenzheng Xu	Sichuan University
Xiang Zhang	Nanjing University of Information Science and Technology
Xiangyu Kong	Dalian University of Technology
Xiangyun Tang	BIT
Xiangyun Tang	Minzu University of China
Xiaojie Zhang	Hunan First Normal University
Xiaoli Gong	Nankai University
Xiaolu Li	Huazhong University of Science and Technology
Xiaoyang Xie	Rutgers University
Xiaoyi Tao	Dalian University of Technology
Xiaoyong Tang	School of Computer and Communication Engineering
Xiaoyu Wang	Soochow University
Xin He	Nanjing University of Posts and Telecommunications
Xin Xie	The Hong Kong Polytechnic University
Xuan Liu	Hunan University

Xueqin Liang	Xidian University
Yajie Wang	BIT
Yajie Wang	Beijing Institute of Technology
Yang Du	University of Science and Technology of China
Yanyan Wang	Nanjing University
Yi Ding	University of Texas at Dallas
Yi Zhao	Beijing Institute of Technology
Yifei Zhu	Shanghai Jiao Tong University
Yitao Hu	Tianjin University
Yizhi Zhou	Dalian University of Technology
Yongkun Li	University of Science and Technology of China
Yongqian Sun	Nankai University
Youyang Qu	Qilu University of Technology
Youyou Lu	Tsinghua University
Yu Zhang	Huazhong University of Science and Technology
Yuan Cao	Ocean University of China
Yuben Qu	Shanghai Jiao Tong University
Yuchao Zhang	Beijing University of Posts and Telecommunications
Yueming Wu	Nanyang Technological University
Yukun Yuan	University of Tennessee at Chattanooga
Yunxia Lin	Yangzhou University
Yutong Gao	Minzu University of China
Ze Zhang	University of Michigan/Cruise
Zhaoyan Shen	Shandong University
Zhen Ling	Southeast University
Zhengkai Wu	Citadel Securities
Zhengxiong Li	The University of Colorado Denver
Zhenlin An	The Hong Kong Polytechnic University
Zhiqiang Li	University of Nebraska
Zhiquan Liu	Jinan University
Zhou Qin	Amazon
Zhuoxuan Du	Ant Group
Zichen Xu	Nanchang University
Zongheng Wei	School of Computer Science
Zonghua Gu	UMU
Zuobin Ying	City University of Macau

Contents – Part I

Fake News Detection Across Multiple Domains Using Fuzzy Association Rules

Xiaofeng Xu[1,2,3], Weisha Zhang[4], Yuanyuan Huang[1,2,3],
Hongyu Lu[1,2,3], and Jiazhong Lu[1,2,3(✉)]

[1] School of Cybersecurity, Xin Gu Industrial College, Chengdu University of
Information Technology, Chengdu 610225, China
`{3220809015,3220811059}@stu.cuit.edu.cn, hy@cuit.edu.cn`
[2] Advanced Cryptography and System Security Key Laboratory of Sichuan Province,
Chengdu 610225, China
`ljz@cuit.edu.cn`
[3] SUGON Industrial Control and Security Center, Chengdu 610225, China
[4] School of Foreign Languages, University of Electronic Science and Technology of
China, Chengdu 611731, China
`zhangweisha@uestc.edu.cn`

Abstract. The detection of fake news is not confined to a single domain, and due to domain transfer, the reliance on detection methods specific to a single domain becomes inadequate when dealing with multi-domain issues. When a multi-domain detection occurs, labels from one domain focus exclusively on the news characteristics of that particular domain, ignoring the possibility that part of the news may include functions from more than one domain. To address these challenges, we propose a multi-domain fake news detection model based on fuzzy rules, incorporating a fuzzy mechanism and an optimization module for fuzzy association rules. In this model, we first introduce a fuzzy mechanism that applies fuzzy processing to domains through neural networks, constructing fuzzy domain labels for each news domain. Subsequently, we incorporate fuzzy association rules to extract a broader range of sample features, thereby enhancing detection effectiveness. Experiments conducted on a fake news dataset demonstrate that our model achieves commendable detection performance with a relatively low training cost, reaching an F_1 score of over 97.13% for sample detection. This model successfully overcomes the limitations of single-domain detection in fake news identification and improves the precision of feature extraction through fuzzy rules, showing superior performance in handling multi-domain fake news detection.

Keywords: fake news detection · social media · multi-domain · fuzzy rules

1 Introduction

In the Internet age, fake news is ubiquitously distributed across various domains. To date, the majority of detection methods [1,2] focus on processing data from

ⓒ The Author(s), under exclusive license to Springer Nature Singapore Pte Ltd. 2025
T. Zhu et al. (Eds.): ICA3PP 2024, LNCS 15251, pp. 1–20, 2025.
https://doi.org/10.1007/978-981-96-1525-4_1

a single domain, termed Single-Domain Fake News Detection (SFND). Taking politics and economics as examples, SFND allows for modular data processing, targeted analysis of domain characteristics, and prevalent rumors, thereby enhancing detection precision and efficiency. However, SFND still faces challenges. While it plays a crucial role in certain domains, its heavy reliance on the sufficiency of specific domain data limits its applicability in news domains with scarce data.

In early studies, scholars often aggregated data from various domains without distinguishing between domains for fake news detection. These approaches did not fully leverage domain labels nor methods for modeling the relationships between domains. Differences in data distribution, word frequency, and dissemination patterns across domains give rise to what is known as domain transfer phenomena [3]. In certain domains, the quantity of fake news might be quite limited. Nonetheless, real-world news platforms publish news from a wide array of domains daily. Hence, to tackle the problem of sparse data and improve performance across diverse domains, Multi-Domain Fake News Detection emerges as a promising strategy. MFND leverages multi-domain data to effectively tackle challenges posed by uneven data distribution across different domains. Significant differences in content and vocabulary exist across news domains, along with substantial variations in dissemination patterns [4]. For instance, tech news might generate more buzz and shares on social media due to its relevance to daily life, while medical news may attract more attention from professionals and industry insiders. Financial news, in contrast, might interest investors and economists more because it directly affects personal and business financial states and interests. The scarcity of labeled data in some domains can hinder the training of accurate models. Thus, Nan et al. [5] proposed a Multi-domain Fake News Detection Model that utilizes single-domain labels to address this issue. These labels, which signify a news item's domain, allow the model to extract shared characteristics of fake news across various domains and distinguish unique features of particular domains. This ability to detect fake news across various domains ensures accurate information transmission and effective audience understanding, underscoring the significance of MFND in enhancing performance across domains.

However, single-domain labels have limitations. A news item may span multiple domains with varying weights, so categorizing it under a specific domain does not fully capture its content. In such cases, single-domain labels fall short in helping models accurately extract multi-domain features. To improve the model, we introduce a fuzzy mechanism and fuzzy association rules. The fuzzy mechanism constructs membership fuzzy labels for each domain, better representing the news domain features and aiding the model in extracting multi-domain content. Experimental results from the weibo21 multi-domain dataset show that our model is superior to others.

The main contributions of this paper are as follows:

(1) Addressing the deficiencies of current single-domain label-based fake news detection when faced with multi-domain information by designing a model that captures multi-domain fake news features.
(2) Addressing the inability of current single-domain label-based approaches to extract news with multi-domain features by employing a fuzzy mechanism for domain fuzzification and fuzzy label extraction. Expanding extraction of news features and establishing fuzzy association rules to improve detection effectiveness.
(3) Our model demonstrates certain advantages compared to other models, with an overall good detection performance, and the highest F_1 score reaching over 97.13%.

2   Related Work

Current fake news detection methods primarily diverge into two directions: based on social context and based on news content. This section briefly introduces the research status of these two directions and the progress in multi-domain learning.

Social Context-Based Detection Methods: These methods focus on the study of news dissemination patterns, the interactions between publishers and other users in the news, and the social network relationships such as the social topology of publishers. Kwon et al. [6] studied a set of associated features to investigate the cumulative propagation patterns of misinformation over time. This approach allows tracking the changing characteristics of false information, thereby enhancing prediction accuracy. These studies elucidate the theoretical mechanisms behind the spread of misinformation and provide new insights for early detection. Guo et al. [7] introduced a Hierarchical Attention-based Social Bi-LSTM network (HAS-BLSTM) for detecting fake news. This approach represents a fake news event as a hierarchical time series comprising information at various semantic levels. Specifically, an event is segmented into several sub-events containing multiple posts, each of which is then broken down into words. These segments are fed into a hierarchical Bi-LSTM network, utilizing social features as an additional indicator to identify the crucial aspects of fake news. This method outperforms others in early detection scenarios. Song et al. [8] introduced the concept of "trustworthy detection points," which dynamically acquires the minimum forwarding information needed for reliable detection of each news item during the detection process. The Credible Early Detection (CED) considers all forwards of a piece of news content as a sequence, finding a "trustworthy detection point" for each piece of news to make an early and reliable prediction. Results show that this model significantly reduces the prediction time span. Most detection algorithms focus solely on finding clues from the content of the news, where fake news often misleads users by mimicking real news. Therefore, exploring auxiliary information to improve detection performance is necessary. Shu et al. [9] studied three internal connections formed in the process of social media news transmission: the relationships between the publisher, the content of the news, and the users. These relationships have the potential to enhance fake news detection. An embedded framework for the triadic relationship was proposed, modeling the interactions among them simultaneously for the classification of fake news content, improving detection effectiveness across different dimensions. Many propagation techniques depend on neural networks to understand how individual news items spread, but this approach is inadequate for capturing the diverse dissemination capabilities of news and ignores important global connections between news and users, which restricts detection accuracy. Therefore, Sun et al. [10] introduced a joint learning model. Without knowing the content of the news and the identity of the user, the model analyzes the behavior of other users and local users to identify the difference between real and fake news when the news begins to spread. An interactive learning module using hypergraphs was designed to obtain users' global preferences from their common propagation relationships, and node centrality encoding was introduced to augment the influence of users in hypergraph learning. Additionally, a local context learning module based on self-attention mechanism was designed, initially introducing the propagation state to highlight the dissemination capabilities of news and users, thereby providing additional signals for verifying the authenticity of news.

Detection Methods Based on News Content: Saqib et al. [11] constructed an integrated classification model comprising decision trees, random forests, and extra-trees classifiers to categorize features extracted from news content. This approach addressed issues present in some studies, such as improper feature extraction, low parameter tuning rate, and dataset imbalance, achieving higher accuracy rates in training and testing on the Liar and ISOT datasets. Qi et al. [12] proposed a novel visual neural network framework that integrates visual information from both frequency and pixel domains, performing multimodal modeling. This framework extracts visual characteristics at various semantic levels along with textual content, utilizing an attention mechanism to dynamically merge feature representations of the two domains. This method has significantly improved its performance in detecting fake news in multiple modes, increasing it by more than 5.2%. Relying on neural networks for feature extraction of fake news content presents certain challenges. To mitigate issues related to noise and redundant data, Ma et al. [13] presented a Dual Channel Convolutional Neural Network with an attention mechanism, incorporating Skip-Gram and Fast-Text for improved performance. This approach effectively filters out noisy and improves the model's capability to learn words with low frequency. A novel parallel dual-channel pooling layer was suggested to substitute the standard CNN pooling layer in DC-CNN. The max pooling layer, serving as one channel, maintains the capacity to capture local information between adjacent words. The attention pooling layer, featuring a multi-head attention mechanism and acting as another pooling channel, improves the comprehension of contextual semantics and global dependencies. This model maximizes the learning benefits of both channels, mitigating the risk of losing local-global feature relevance in pooling layers. Mouratidis et al. [14] developed a novel deep neural network architecture that enables adaptable input fusion at various network layers. Using paired text inputs, combined with word embeddings as well as linguistic and social account features, news is divided into headlines and body text, extensively using both for classification tests. This method differentiates between fake and real news by analyzing linguistic and related features. Rawat et al. [15] introduced a novel approach to enhance existing automatic fake news detection methods by automatically gathering evidence for each statement. Supportive evidence is extracted from web articles, and appropriate texts are selected as evidence sets. Pre-trained summarization models are subsequently applied to these sets of evidence, and the extracted summaries act as supporting evidence to assist in the classification task.

The core idea of multi-domain learning is to simultaneously learn multiple related but not necessarily identical domains or tasks within a single model. Qin et al. [16] investigated the multitasking of user activity streams by modeling users' sequential behavior in a neural multitask learning environment, proposing a new framework, the Mixed Sequence Expert Model (MOSE), which utilizes Long Short-Term Memory networks to model sequential users explicitly. Zhao et al. [17] introduced a multi-domain task learning framework combined with Graph Attention Networks to obtain fine-grained representations of conversational actions. Multi-domain learning has been demonstrated to be effective in numerous applications [18–24], with an emphasis on capturing the relationships among different tasks that share numerous features. Each task is strengthened through its connections, which also encompass the variability in correlations between tasks.

Social context-based detection methods utilize a wide range of external data, providing more comprehensive information for accurately assessing the credibility of specific publishers or platforms. However, due to limitations in data availability and reliability, this approach may not timely capture emerging fake news creators or patterns. News content-based detection methods directly extract features from the content of news, enabling quick response to new fake news creators and patterns. However, for some technically sophisticated fake news, feature extraction alone may not make accurate judgments. To address these issues and enhance detection effectiveness, we introduce fuzzy mechanisms and fuzzy association rules in multi-domain content detection. These methods can effectively deal with the complexity of data, improving the capability to detect multi-domain fake news.

3   Model

This article introduces a multi-domain fake news detection model based on fuzzy rules, termed as FRMFND. Similar to traditional approaches, this detection method is also treated as a classification problem. We utilize the Roberta model to extract content from news and introduce a fuzzy mechanism to process each news domain fuzzily, constructing fuzzy domain labels. Fuzzy domain tags are used to extract the multi-domain features of news and generate the comprehensive feature representation. Subsequently, we employ a discriminator module to utilize these comprehensive feature representations for news detection. The structure of the entire framework is illustrated in Fig. 1.

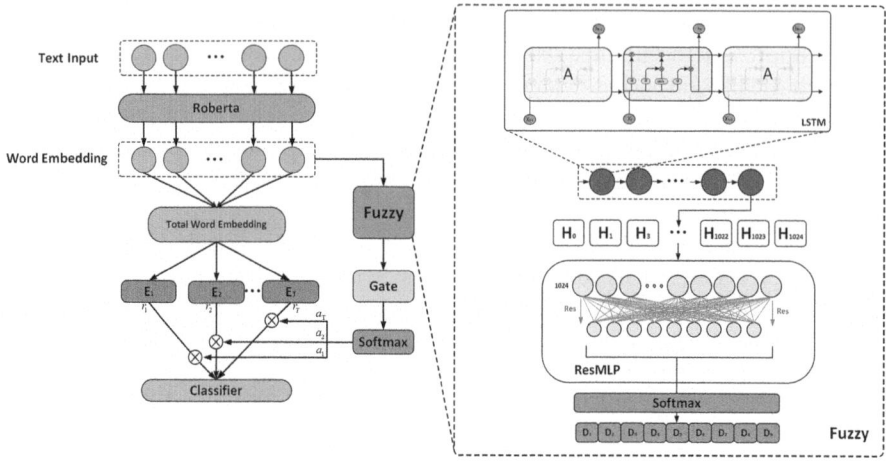

Fig. 1. Overall framework of Model

3.1   Representation Extraction

To represent news content, we use Roberta for feature extraction. For a piece of news, the text is first preprocessed by removing stopwords and punctuation, followed by tokenization. The text is divided into a series of word segments, and each segment is mapped to a corresponding token number in the vocabulary. The content is tokenized using BertTokenizer [25], with special CLS and SEP symbols added to denote the beginning and separation of the text sequences, respectively. Enter these tokens into Roberta to get word embeddings $W = \left[w_{[cls]}, w_1, w_2, \cdots\cdots w_n, w_{[sep]}\right]$, where all word embeddings are handled through the mask attention mechanism. Feature extraction from the news is performed using techniques such as Mixture-of-Expert [26]. Taking advantage of the Mixture-of-Experts model, which consists of multiple expert models each responsible for processing different parts of the input data, their outputs are dynamically combined through a gating mechanism specialized for different domains. For each news domain, the corresponding domain expert extracts the domain-specific features.

In our model, each expert network is a TextCNN [27]. An expert network model can be denoted as $E_i\left(W; \theta_i\right) (1 \leq i \leq T)$, W representing a set of word embeddings input to the expert network, θ_i representing learnable and trainable parameters, and T being a hyperparameter that represents the number of expert networks. r_i is used to denote the final input of each expert network, allowing us to obtain the feature representation extracted by the expert network corresponding to each domain:

$$r_i = E_i\left(W; \theta_i\right) \tag{1}$$

This approach enables targeted feature extraction from different news domains, leveraging the specificity of each domain expert network to enhance the overall detection capability of the model.

3.2 Fuzzy: Membership Function

First, the fuzzy set is represented by a membership function, assigning a nine-dimensional membership degree to each news domain. A fuzzy mechanism is used for fuzzy domain processing of news domains, constructing fuzzy domain labels. The fuzzy module mainly involves three parts: LSTM, ResMLP, and Softmax, which together constitute the membership function. The word vectors output by Roberta are input into the fuzzy module. Compared to the Bert model, the Roberta model increases the dimensionality of each output embedding vector. In deep neural networks, we add residual analysis on top of MLP to mitigate the vanishing gradient problem. Residual connections help in building the model at deeper levels, making the network easier to train. Throughout the training process, we continually train the entire module to function as a classifier for news domains.

For a piece of news, the LSTM transforms and outputs the input word embeddings to the ResMLP layer, which then transforms and outputs a nine-dimensional vector. Finally, the Softmax function normalizes the output of the ResMLP, resulting in a vector D, representing different degrees of membership across nine domains. Therefore, there are nine dimensions in total, with the sum of all dimensions equaling 1. We denote the fuzzy membership function as F, and $\theta_1, \theta_2, \theta_3$ representing the parameters of the fuzzy membership function, LSTM, and ResMLP, respectively. D represents the overall fuzzy domain label for each piece of news:

$$D = F\left(W; \theta_1\right) = Softmax\left(ResMLP\left(LSTM\left(W; \theta_2\right); \theta_3\right)\right) \tag{2}$$

Following the detailed application of fuzzy membership functions to a news dataset, this study successfully mapped the distribution of information across specific domains by categorizing news content and allocating corresponding weights, as shown in Fig. 2. The analysis distinctly highlights that the processed news content primarily focuses on the critical areas of cultural entertainment and social life. This process underscores the significant capability of fuzzy logic techniques in managing ambiguity and uncertain information, thereby effectively enhancing the accuracy and applicability of news content classification. This work not only demonstrates the practical value of fuzzy logic in the analysis of news content but also provides a precise methodological reference for subsequent research in this field.

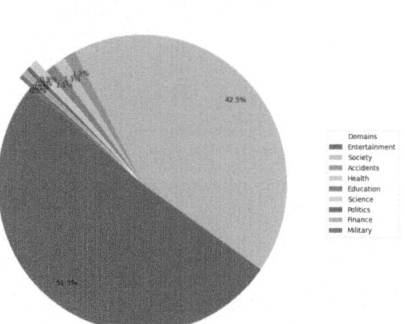

Fig. 2. Percentage of fuzzy domains

To enhance the detection results, fuzzy association relationships are utilized on the detected data, employing fuzzy association rules for re-evaluation. Due to the limited features in the original dataset, we have expanded upon it. By crawling based on each news item's original ID, additional features were gathered, including poster account information, verification status, number of followers, level, date and time of posting, as well as the number of shares, comments, and likes. Define the feature dimension of news content as n and the dimension of class labels as m, then:

$$X = \{y_1, \cdots, y_n\}, Y = \{y_{n+1}, \cdots, y_{n+m}\} \tag{3}$$

Let X be the feature set and Y be the label set. In the fuzzy mean processing, the dataset and class labels can be represented as:

$$
\begin{aligned}
S = \{S_i = (X, Y)_i &= \{y_1, \cdots, y_n; y_{n+1}, \cdots, y_{n+m}\}_i\} \\
&= \{(y_{i1}, \cdots, y_{in}; y_{i(n+1)}, \cdots, y_{i(n+m)}) \,|\, i = 1, \cdots, k\}
\end{aligned} \tag{4}
$$

The fuzzy sets for the dataset and the class labels are as follows:

$$MF = \{A_a^b(y_{ia}) \,|\, i = 1, \cdots, k; a = 1, \cdots, n + m; b = 1, \cdots, b_{ia}\} \tag{5}$$

then:

$$F\sup(X) = \sum_{i=1}^{k} \prod_{a=1}^{n} t_i(y_a) = \sum_{i=1}^{k} \prod_{a=1, b=1, \cdots, b_{ia}}^{p} \max\{A_a^b(y_{ia})\} \tag{6}$$

The fuzzy support of the fuzzy association rule "$X \rightarrow Y$" is:

$$F\sup(X) = \sum_{i=1}^{k} \prod_{a=1}^{n+m} t_i(y_a) = \sum_{i=1}^{k} \prod_{a=1, b=1, \cdots, b_{ia}}^{p+m} \max\{A_a^b(y_{ia})\} \tag{7}$$

The fuzzy confidence of the fuzzy association rule "$X \rightarrow Y$" is:

$$FConf = \frac{FSup(X \cup Y)}{FSup(X)} \tag{8}$$

3.3 Domain Gate

Given that news from different domains is characterized by features extracted by their respective expert network models, achieving optimal model performance requires more than simply averaging the feature extractions of all expert models. To adaptively select features from different domains, we input the fuzzy domain labels into a gating mechanism to obtain weighting scores. These scores consist of T numbers, corresponding to the T experts. And then output a vector a, representing the weight ratio for each expert. We denote the gating mechanism as $G(\cdot; \phi)$, and ϕ as the trainable parameters of the gating mechanism:

$$a = (a_1, a_2, ..., a_T) = G(D; \phi) = Softmax(ResMLP(D; \phi)) \tag{9}$$

The domain gate $G(\cdot; \phi)$ is a feedforward neural network that utilizes weight scores to aggregate the domain features, resulting in the final feature vector of the news:

$$v = \sum_{i=1}^{T} a_i r_i \tag{10}$$

3.4 Discriminator Prediction

The final feature vector is fed into a Classifier, which is an MLP network with a Softmax output layer. This enables us to obtain the model's predicted value $\hat{p}$, used for fake news detection:

$$\hat{p} = Softmax(MLP(v)) \tag{11}$$

Where p^i represents the actual value and $\hat{p}^i$ represents the predicted value, training is conducted using binary cross-entropy loss:

$$L = -\sum_{i=1}^{N} \left(p^i \log \hat{p}^i + \left(1 - p^i\right) \log \left(1 - \hat{p}^i\right)\right) \tag{12}$$

4   Experiment

4.1 Datasets

We use the dataset, Weibo21 [5]. This dataset comprises 4,488 fake news items and 4,640 real news items, encompassing 9 distinct news domains. The distribution of fake and real news items across these domains is outlined in Table 1. To illustrate domain-specific vocabulary usage, we generate word clouds for different domains based on word frequency in Fig. 3.

Table 1. Data Statistics of Weibo21

domain	Science	Military	Education	Accidents	Politics
real	143	121	243	185	306
fake	93	222	248	591	546
all	236	343	491	776	852
domain	Health	Finance	Entertainment	Society	All
real	485	959	1000	1198	4640
fake	515	362	440	1471	4488
all	1000	1321	1440	2669	9128

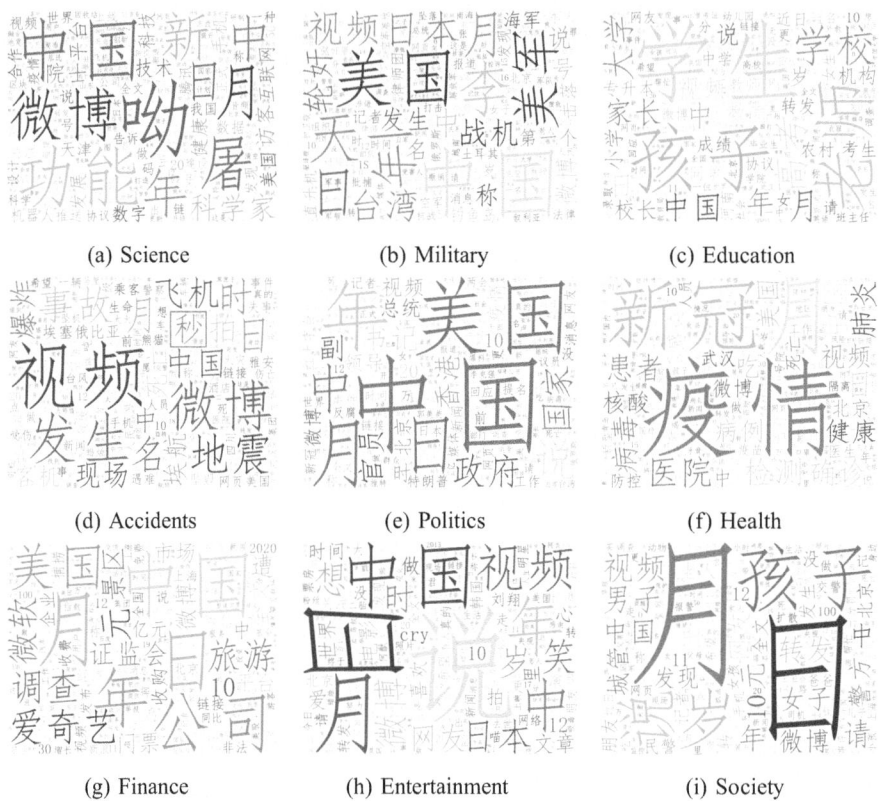

(a) Science   (b) Military   (c) Education

(d) Accidents   (e) Politics   (f) Health

(g) Finance   (h) Entertainment   (i) Society

Fig. 3. Different fields of word cloud

4.2 Baseline Methods

In the experimental comparisons, four types of baselines are used: (1) Single-domain baselines: TextCNN_single [27], BiGRU_single [2], and BERT_single [25]; (2) Mixed-domain baselines: TextCNN_all [27], BiGRU_all [2], and

BERT_all [25]; (3) Multi-domain baselines: EANN [28], MMOE [29], MOSE [16], and EDDFN [4]; (4) Multi-domain models: MDFEND [5] and FuDFEND [30]. TextCNN is a convolutional neural network-based text classification model that uses one-dimensional convolutional operations to extract local features from text, followed by pooling operations and fully connected layers for classification. BiGRU is a text classification model based on bidirectional gated recurrent units that can effectively capture contextual information in text, with a final fully connected layer for classification. EANN is a multi-domain model for event-related tasks that utilizes attention mechanisms to address the correlations between different domains. MMOE is a multi-domain model that handles input features from different domains through a multi-gate mechanism and expert models. MOSE is a multi-domain model for multi-objective learning tasks, employing self-supervised embedding techniques to process input data from different domains. EDDFN is an embedding-based domain-aware dual fusion network for multi-domain text classification tasks. MDFEND is a multi-domain detection model that enhances feature representation using single-domain labels. FuDFEND is an improved version of MDFEND, adopting multi-domain feature representations and introducing a fuzzy domain mechanism. In single-domain baselines, experiments are conducted in one domain at a time for different domains, and the average of the results across all domains is taken as the final column result. In mixed-domain baselines, a model is trained on all domains together, and the performance is calculated for each domain individually, with the final column result being the aggregate of all domain data. Multi-domain baseline models organize data according to different domain structures, whereas the final multi-domain models can extract common features of news across various domains, then differentiate specific features of specific domains through domain labels.

4.3   Experiment Setting

During training, the parameters of Roberta are fixed, with the dimensionality of the input embedding vectors set to 1024, and the maximum length of a sentence capped at 300. The learning rate is set at 0.0005, and the MLP network consists of 1024 layers. The regularization strength parameter, gamma, is set at 0.98. The training process employs the Adam [31] optimizer for optimization, with a batch size of 64 for all experiments. Model evaluation is conducted using Accuracy (Acc), Precision (Pr), Recall (Re), and the F_1 score.

Table 2. Confusion matrix

Actual	Predicted: Real news	Predicted: Fake news
Real news	TP	FN
Fake news	FP	TN

Based on Table 2, we calculated the following metrics, represented in the mathematical formulas below.

$$Accuracy = \frac{TP + TN}{TP + TN + FP + FN} \tag{13}$$

Accuracy (Acc) measures the proportion of correctly predicted observations to the total observations.

$$Recall = \frac{TP}{TP + FN} \tag{14}$$

Recall (Re), also known as sensitivity, measures the ability of the classifier to identify all positive samples.

$$Precesion = \frac{TN}{TN + FP} \tag{15}$$

Precision (Pr) measures the accuracy of positive predictions.

$$F_1 = 2 \cdot \frac{Recall \cdot Precesion}{Recall + Precesion} \tag{16}$$

The F_1 is a weighted average of model accuracy and recall, with a maximum value of 1 and a minimum value of 0.

It balances the trade-offs between precision and recall, providing a single metric to assess the performance of a model where both the precision and recall are important. Setting a minimum support of 0.1 and a minimum confidence of 0.2 defines meaningful association rules. Let $i_1, i_2, i_3, \cdots, i_{n-1}, i_n$ represent the n features of each news item, the result is i_{n+1}, use $true$ and $false$ to respectively indicate if the news is true or fake. Some association rules are shown in Table 3, where i_1, i_2, i_3, i_4 represents whether the news publisher is verified, and the amounts of shares, comments, and likes a news post receives. A value of $i_1 = 1$ indicates a verified user, while $i_1 = 0$ indicates an unverified user. Three membership intervals, denoted as i_2, i_3, i_4 are established to represent low, medium, and high levels, respectively. Through the application of fuzzy mean clustering to quantities, these quantities are categorized into three classes, corresponding to low, medium, and high amounts. In total, 55 association rules have been defined, with the table below displaying a subset of these rules. When the support is less than 0.1 and the confidence is less than 0.2, the rule is considered meaningless. As can be seen from Table 3 below, it can be seen that the third and eighth rules yield meaningless results. Overall, whether the news publisher is a verified user is a key point in determining the authenticity of the news.

4.4   Ablation Study

To confirm the significance of the key elements in our proposed approach, we conducted ablation experiments. The experimental results indicate that, when

Table 3. Partial Association Rules

Rules
$\{i_1 = 1, i_2 = medium, i_3 = medium\} \Rightarrow i_5 = true$
$\{i_1 = 1, i_2 = medium, i_4 = medium\} \Rightarrow i_5 = true$
$\{i_1 = 1, i_3 = medium, i_4 = high\} \Rightarrow i_5 = true$
$\{i_1 = 1, i_2 = low, i_3 = low\} \Rightarrow i_5 = true$
$\{i_1 = 1, i_3 = low, i_4 = low\} \Rightarrow i_5 = true$
$\{i_1 = 1, i_2 = low, i_4 = low\} \Rightarrow i_5 = true$
$\{i_1 = 1, i_2 = low, i_3 = low, i_4 = low\} \Rightarrow i_5 = true$
$\{i_1 = 0, i_2 = low, i_3 = low, i_4 = low\} \Rightarrow i_5 = false$
$\{i_2 = low, i_3 = low, i_4 = medium\} \Rightarrow i_5 = false$
$\{i_1 = 0, i_3 = low, i_4 = low\} \Rightarrow i_5 = false$
$\{i_1 = 0, i_2 = low, i_3 = low\} \Rightarrow i_5 = false$
$\{i_1 = 0, i_2 = low, i_4 = low\} \Rightarrow i_5 = false$

dealing with the experimental data of this paper, the LSTM module is more suitable for processing complex sequential data, especially long text data, compared to the GRU module. Both LeakyReLU and PReLU are commonly used methods to address the vanishing gradient problem, with LeakyReLU solving the neuron dying issue of basic ReLU. The choice of which activation function to use generally depends on the specific data and task requirements. In this experiment, LeakyReLU performed better than PReLU. These results are instructive for the improvement and optimization of our method and offer insights for related work in the research field. Here, we use F_1 score as the evaluation metric and use D_1–D_9 to represent nine different domains, testing the performance of our modules across these domains.

Based on comparative experimental data, we replaced the modules for processing long texts and the activation function modules in the domain gate, comparing them under the same parameters. As shown in Fig. 4, it can be observed that in most cases, the LSTM module outperforms the GRU module when dealing with the long text data of this experiment. When using ReLU-type activation functions in the domain gate, the LeakyReLU module outperforms the PReLU module in comprehensive situations. Considering all experimental data, we selected the LSTM module and LeakyReLU module as our text sequence processing module and activation function module in the domain gate, respectively.

The essence of ablation experiments is to conduct controlled variable studies on the effects of individual variables. For those variables that cannot be removed, such as the choice of hyperparameters, one should control other aspects of the system to remain unchanged and test a set of hyperparameters to study their impact on the system. Thus, hyperparameter tuning is essentially an ablation experiment. They explore the changes in system performance when only one parameter is changed, thereby demonstrating the impact of the parameter on

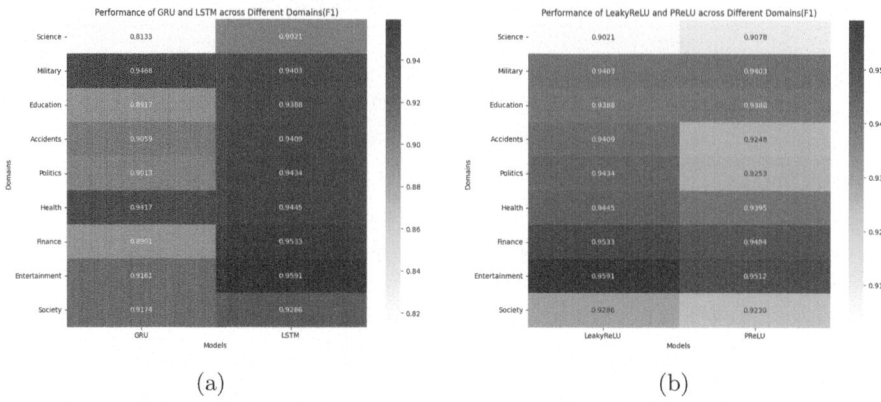

(a)   (b)

Fig. 4. Influence of Module Selection on Model Performance: (a) Different text sequence processing modules; (b) Different activation functions handle modules

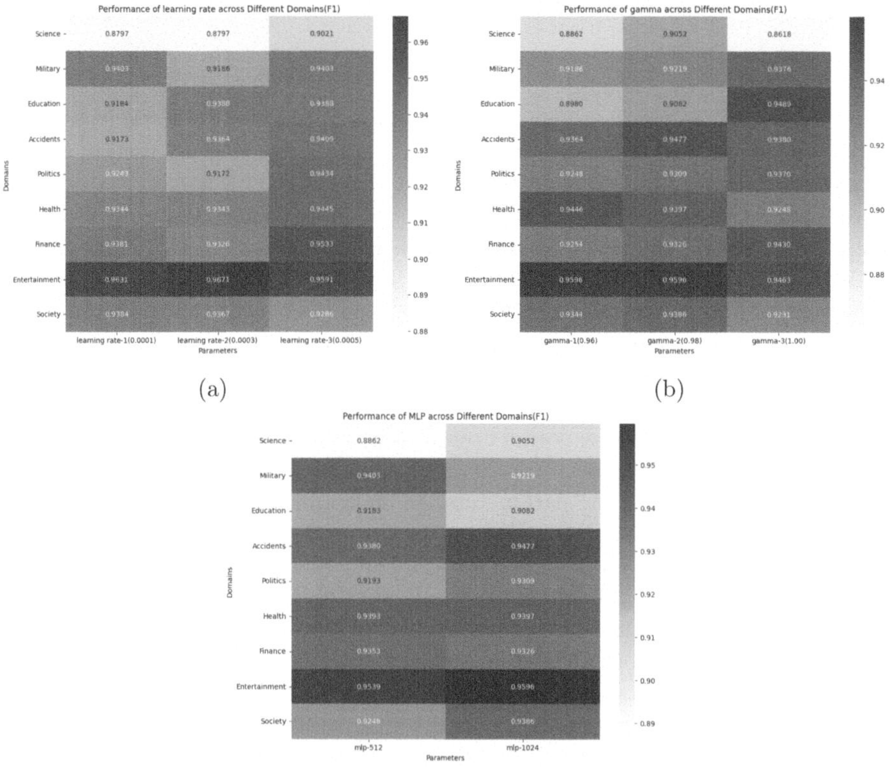

(a)   (b)

(c)

Fig. 5. Influence of hyperparameter selection on model performance: (a) Different learning rate parameters; (b) Different regularization intensity parameters; (c) Different MLP network layers

the system. Here, we continue to use the F_1 score as the evaluation metric and conduct tests in nine different domains on the learning rate(lr), MLP layers, and the parameter gamma, which represents the model's regularization strength. In the experimentation detailed within Fig. 5(a), learning rates denoted as lr_1, lr_2 and lr_3 are established at 0.0001, 0.0003, and 0.0005, respectively. Concurrently, Fig. 5(b) delineates the model's regularization strength parameters g_1, g_2 and g_3 as 0.96, 0.98, and 1.00 respectively. Furthermore, Fig. 5(c) contrasts the Multi-Layer Perceptron (MLP) configurations, with $mlp1$ and $mlp2$ representing architectures of 512 and 1024 layers accordingly. Upon a comprehensive analysis of the experimental data, we have adjudicated the selection of a learning rate of 0.0005, a gamma regularization coefficient of 0.98, and an MLP architecture encompassing 1024 layers as the optimal hyperparameters for our experimental model, thereby facilitating enhanced model performance and robustness in application.

4.5   Results

The proposed model is compared with other methods, and the effectiveness of the model is verified. The results are shown in the following figure. The figure demonstrates the detection effects of multi-domain experimental data on the model.

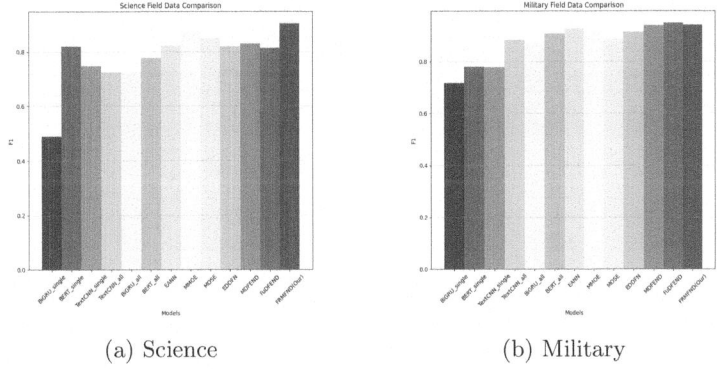

(a) Science   (b) Military

Fig. 6. The experimental results of the model in Science and Military domains

From Fig. 6, it can be seen that in the fields of science and military, hybrid and multi-domain models are generally superior to single-domain models. The advantage is more significant in the military domain. This is because the news data in the military field is relatively sparse compared to other fields, while the scientific field involves a wide range of cross-disciplinary areas, making multi-domain models generally superior.

Figure 7 shows the situation in the fields of education and disaster accidents. In the field of education, single-domain models even show a significant advantage,

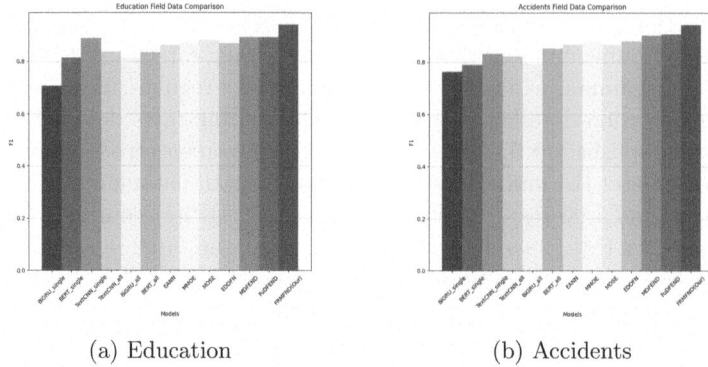

(a) Education

(b) Accidents

Fig. 7. The experimental results of the model in Education and Accidents domains

basically exceeding hybrid models and some multi-domain models. This indicates that in certain fields, it is not possible to improve detection efficiency through simple mixed data, and the redundancy of data may even reduce the detection effect in that field. The overall data volume in the disaster accident domain is moderate, and the detection efficiency shows a gradual improvement overall.

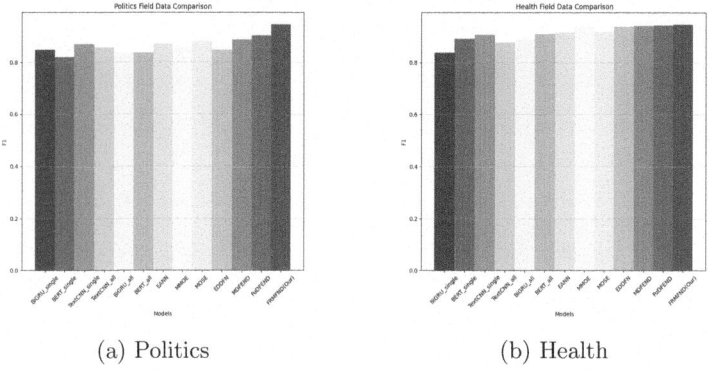

(a) Politics

(b) Health

Fig. 8. The experimental results of the model in Politics and Health domains

As shown in Fig. 8, the overall performance in the fields of politics and health is relatively average, with multi-domain models showing a superior performance. Additionally, the data volume proportion in the fields of politics and health is moderate, indicating that these two fields involve a rich variety of multi-domain content, and therefore, the effect of detection through multi-domain models is better than other models.

As shown in Fig. 9, the news content in the fields of business and finance, and entertainment and lifestyle occupies a relatively large proportion, with a clear

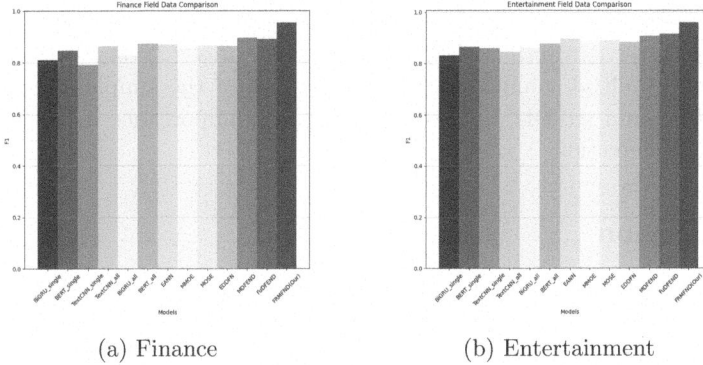

(a) Finance   (b) Entertainment

Fig. 9. The experimental results of the model in Finance and Entertainment domains

overall trend. Multi-domain models and others perform relatively well, resulting in an overall average detection level.

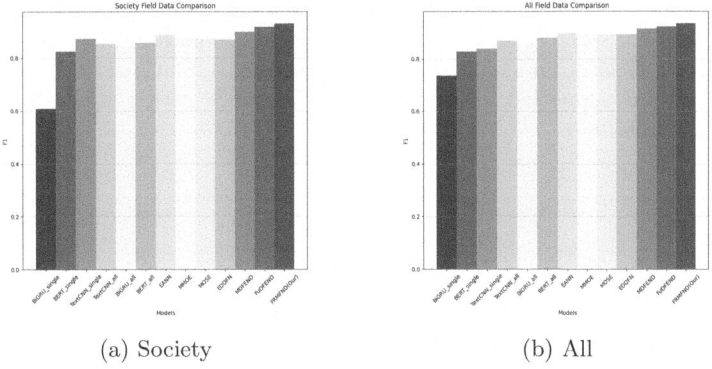

(a) Society   (b) All

Fig. 10. The experimental results of the model in Society domains

Social life has the largest proportion in the news, as shown in Fig. 10(a). In domains with large amounts of data, multi-domain models show greater advantages. Figure 10(b) shows the overall average detection level. Overall, the multi-domain model outperforms the single-domain model, highlighting the importance of data volume on model performance. At the same time, the performance of the multi-domain model is also superior to that of the simple mixed-domain model, indicating that the process of multi-domain learning is effective. However, in some domains with less data volume, the performance of the single-domain model may surpass that of the simple mixed model, suggesting that simple mixing of domain data might have a negative impact on model performance.

Our designed model was compared with previous models to verify its effectiveness. The experimental results further revealed the following conclusions: (1)

Fuzzy labels can enhance the detection effect of news under multiple domains, facilitating better extraction of multi-domain features. (2) The overall performance of the multi-domain model surpasses that of the simple single-domain and mixed-domain models, further confirming the importance and necessity of multi-domain detection.

5   Conclusion

This paper explores the challenge of detecting fake news across multiple domains, employing the MFND dataset Weibo21. By incorporating a fuzzy mechanism and association rules into the base model, this method extracts multi-domain fuzzy labels of news through the fuzzy mechanism, and then leverages association rules to extract more features, thereby strengthening the detection effect. We compared our model's performance against various types of previous models, and the experiments have proven the effectiveness of our model.

Acknowledgments. The authors would like to acknowledge the National Natural Science Foundation of China (Grant No. 62102049) and"Computational Communication Strategy for Chinese Technology Image on Twitter" (No. 20CXW016) which is supported by National Social Science Fund of China. The authors also extend their appreciation to Natural Science Foundation of Sichuan Province (Grant No. 2022NSFSC0557). This research was supported by the National Natural Science Foundation of China (Grant No. 62102379).

References

1. Ma, B., Lin, D., Cao, D.: Content representation for microblog rumor detection. In: Angelov, P., Gegov, A., Jayne, C., Shen, Q. (eds.) Advances in Computational Intelligence Systems. AISC, vol. 513, pp. 245–251. Springer, Cham (2017). https://doi.org/10.1007/978-3-319-46562-3_16
2. Ma, J., et al.: Detecting rumors from microblogs with recurrent neural networks (2016)
3. Zhuang, F., et al.: A comprehensive survey on transfer learning. Proc. IEEE **109**(1), 43–76 (2020)
4. Silva, A., Luo, L., Karunasekera, S., Leckie, C.: Embracing domain differences in fake news: cross-domain fake news detection using multi-modal data. Proc. AAAI Conf. Artif. Intell. **35**(1), 557–565 (2021)
5. Nan, Q., Cao, J., Zhu, Y., Wang, Y., Li, J.: MDFEND: multi-domain fake news detection. In: Proceedings of the 30th ACM International Conference on Information & Knowledge Management, pp. 3343–3347 (2021)
6. Kwon, S., Cha, M., Jung, K.: Rumor detection over varying time windows. PLoS ONE **12**(1), e0168344 (2017)
7. Guo, H., Cao, J., Zhang, Y., Guo, J., Li, J.: Rumor detection with hierarchical social attention network. In: Proceedings of the 27th ACM International Conference on Information and Knowledge Management, pp. 943–951 (2018)
8. Song, C., Yang, C., Chen, H., Tu, C., Liu, Z., Sun, M.: CED: credible early detection of social media rumors. IEEE Trans. Knowl. Data Eng. **33**(8), 3035–3047 (2019)

9. Shu, K., Wang, S., Liu, H.: Beyond news contents: the role of social context for fake news detection. In: Proceedings of the Twelfth ACM International Conference on Web Search and Data Mining, pp. 312–320 (2019)

10. Sun, L., Rao, Y., Lan, Y., Xia, B., Li, Y.: HG-SL: jointly learning of global and local user spreading behavior for fake news early detection (2023)

11. Hakak, S., Alazab, M., Khan, S., Gadekallu, T.R., Maddikunta, P.K.R., Khan, W.Z.: An ensemble machine learning approach through effective feature extraction to classify fake news. Fut. Gener. Comput. Syst. **117**, 47–58 (2021)

12. Qi, P., Cao, J., Yang, T., Guo, J., Li, J.: Exploiting multi-domain visual information for fake news detection. In: 2019 IEEE International Conference on Data Mining (ICDM), pp. 518–527. IEEE (2019)

13. Ma, K., et al.: DC-CNN: dual-channel convolutional neural networks with attention-pooling for fake news detection. Appl. Intell. **53**(7), 8354–8369 (2023)

14. Mouratidis, D., Nikiforos, M.N., Kermanidis, K.L.: Deep learning for fake news detection in a pairwise textual input schema. Computation **9**(2), 20 (2021)

15. Rawat, M., Kanojia, D.: Automated evidence collection for fake news detection. In: Bandyopadhyay, S., Devi, S.L., Bhattacharyya, P. (eds.) Proceedings of the 18th International Conference on Natural Language Processing (ICON), National Institute of Technology Silchar, Silchar, India, December 2021, pp. 456–464. NLP Association of India (NLPAI) (2021). https://aclanthology.org/2021.icon-main.55

16. Qin, Z., Cheng, Y., Zhao, Z., Chen, Z., Metzler, D., Qin, J.: Multitask mixture of sequential experts for user activity streams. In: Proceedings of the 26th ACM SIGKDD International Conference on Knowledge Discovery & Data Mining, pp. 3083–3091 (2020)

17. Zhao, M., Wang, L., Jiang, Z., Li, R., Lu, X., Hu, Z.: Multi-task learning with graph attention networks for multi-domain task-oriented dialogue systems. Knowl. Based Syst. **259**, 110069 (2023)

18. Wang, Z., Shen, L., Duan, T., Zhan, D., Fang, L., Gao, M.: Learning to learn and remember super long multi-domain task sequence. In: Proceedings of the IEEE/CVF Conference on Computer Vision and Pattern Recognition, pp. 7982–7992 (2022)

19. Park, C., Jeong, S., Kim, J.: ADMit: improving NER in automotive domain with domain adversarial training and multi-task learning. Exp. Syst. Appl. **225**, 120007 (2023)

20. Yan, S., Zhao, J., Xu, L.: Adaptive multi-task learning for cross domain and modal person re-identification. Neurocomputing **486**, 123–134 (2022)

21. Tomczak, A., et al.: Multi-task multi-domain learning for digital staining and classification of leukocytes. IEEE Trans. Med. Imaging **40**(10), 2897–2910 (2020)

22. Lyu, F., Wang, S., Feng, W., Ye, Z., Hu, F., Wang, S.: Multi-domain multi-task rehearsal for lifelong learning. Proc. AAAI Conf. Artif. Intell. **35**(10), 8819–8827 (2021)

23. Zhang, J., Li, W., Ogunbona, P.: Unsupervised domain adaptation: a multi-task learning-based method. Knowl. Based Syst. **186**, 104975 (2019)

24. Liu, A.-A., Xu, N., Nie, W.-Z., Su, Y.-T., Zhang, Y.-D.: Multi-domain and multi-task learning for human action recognition. IEEE Trans. Image Process. **28**(2), 853–867 (2018)

25. Devlin, J., Chang, M.-W., Lee, K., Toutanova, K.: BERT: pre-training of deep bidirectional transformers for language understanding. In: Burstein, J., Doran, C., Solorio, T. (eds.) Proceedings of the 2019 Conference of the North American Chapter of the Association for Computational Linguistics: Human Language Technologies, Volume 1 (Long and Short Papers), June 2019, pp. 4171–4186, Minneapolis, Minnesota. Association for Computational Linguistics (2019). https://aclanthology.org/N19-1423

26. Zhu, Y., et al.: Learning to expand audience via meta hybrid experts and critics for recommendation and advertising. In: Proceedings of the 27th ACM SIGKDD Conference on Knowledge Discovery & Data Mining, pp. 4005–4013 (2021)

27. Kim, Y.: Convolutional neural networks for sentence classification. In: Moschitti, A., Pang, B., Daelemans, W. (eds.) Proceedings of the 2014 Conference on Empirical Methods in Natural Language Processing (EMNLP), Doha, Qatar, October 2014, pp. 1746–1751. Association for Computational Linguistics (2014). https://aclanthology.org/D14-1181

28. Wang, Y., et al.: EANN: event adversarial neural networks for multi-modal fake news detection. In: Proceedings of the 24th ACM SIGKDD International Conference on Knowledge Discovery & Data Mining, pp. 849–857 (2018)

29. Ma, J., Zhao, Z., Yi, X., Chen, J., Hong, L., Chi, E.H.: Modeling task relationships in multi-task learning with multi-gate mixture-of-experts. In: Proceedings of the 24th ACM SIGKDD International Conference on Knowledge Discovery & Data Mining, pp. 1930–1939 (2018)

30. Liang, C., Zhang, Yu., Li, X., Zhang, J., Yu, Y.: FuDFEND: fuzzy-domain for multi-domain fake news detection. In: Lu, W., Huang, S., Hong, Yu., Zhou, X. (eds.) Natural Language Processing and Chinese Computing: 11th CCF International Conference, NLPCC 2022, Guilin, China, September 24–25, 2022, Proceedings, Part II, pp. 45–57. Springer, Cham (2022). https://doi.org/10.1007/978-3-031-17189-5_4

31. Kingma, D.P., Ba, J.: Adam: a method for stochastic optimization. CoRR abs/1412.6980 (2014). https://api.semanticscholar.org/CorpusID:6628106

A Scheme of Dynamic Location Privacy-Preserving with Blockchain in Intelligent Transportation System

Xuhan Zuo[1][ID], Minghao Wang[2][ID], Dayong Ye[1][ID], and Shui Yu[1(✉)][ID]

[1] University of Technology Sydney, Ultimo 2007, Australia
Xuhan.Zuo-1@student.uts.edu.au, {Dayong.Ye,Shui.Yu}@uts.edu.au
[2] City University of Macau, Macao, Macao, Special Administrative Region of China
mhwang@cityu.edu.mo

Abstract. In the promotion process of the Internet of Vehicles (IoV), location privacy issues arise when large-scale transmissions are applied. The current solutions for addressing location privacy issues in IoV are hindered by several pain points, including limited effectiveness, scalability challenges, and reliance on a single privacy mechanism. While a blockchain-based method for preserving location privacy has been proposed, there remains an academic concern regarding the suitability of such a scheme. In this paper, we propose a Dirichlet-based location privacy-preserving scheme for IoV, oriented towards vehicle density awareness and adjustable sensitivity by a certified organization. This scheme is applicable to intelligent transportation systems and addresses the limitations of current blockchain-based methods for preserving location privacy. Finally, we implement the proposed scheme in an industrial blockchain, evaluating its performance and conducting a theoretical analysis from security and privacy perspectives.

Keywords: Blockchain · Privacy-Preserving · VANET

1 Introduction

The Internet of Vehicles (IoV) is a prominent and integral part of the Internet of Things (IoT). IoV represents a network of automobiles equipped with sensors, software, and technologies that facilitate communication between them. The foundation of the IoV scenario is established by Vehicular Ad-hoc Networks (VANETs), which leverage a variety of protocols (e.g., IEEE 802.11P, IEEE 1609) [21] and communication modalities, including Vehicle to Vehicle (V2V) and Vehicle to Infrastructure (V2I). IoV harnesses the capabilities of IoT to maximize utility, enabling innovative applications aimed at reducing accidents, alleviating traffic congestion, and providing other information services.

However, similar to other industries implementing big data, the IoV encounters significant challenges in data security and privacy during its advancement [30]. A paramount concern in IoV networks is the preservation of location privacy.

© The Author(s), under exclusive license to Springer Nature Singapore Pte Ltd. 2025
T. Zhu et al. (Eds.): ICA3PP 2024, LNCS 15251, pp. 21–39, 2025.
https://doi.org/10.1007/978-981-96-1525-4_2

The disclosure of a vehicle's precise position poses considerable risks to the traffic system and high-mobility vehicles on freeways. For instance, if an attacker compromises IoV systems in a high-speed vehicle, even minor disruptions can lead to significant traffic congestion. To mitigate these risks, Location Privacy Preserving Methods (LPPMs) have been developed to secure sensitive vehicle data, such as real-time location and trajectories. Current solutions for location privacy in IoV can be categorized into several techniques: anonymity-based methods, Multi-agent Reinforcement Learning (MARL) methods [27], encryption-related methods, and Blockchain-based Location Privacy Preserving Methods (BCLPPMs).

Anonymity-based LPPMs in IoV aim to protect users' trajectory data over time. Event-level privacy, on the other hand, is concerned with safeguarding the user's position at each timestamp [14]. A limitation of these methods is their focus on mitigating data mining risks after user data has been compromised. MARL solutions have addressed some of these issues in implementing location privacy-preserving approaches in IoV. Federated learning perspectives are also being explored on third-party platforms [18,28]. Location privacy in trajectory data is crucial for intelligent transportation systems, reflecting users' movements, such as travel paths. However, these solutions often prove less feasible in actual industrial applications. Encryption-related solutions face challenges in demonstrating general effectiveness, especially when integrated into current business platforms. They struggle to maintain consensus among all platform users within a reasonable timeframe. Blockchain-based LPPMs, often combined with trust management systems [22], such as [12] and [19], typically focus on two main features: traceability and anonymity. However, these solutions frequently overlook the mobility aspect of vehicles in IoV, as highlighted in studies like [4]. The most effective method among the four categories mentioned earlier is the Blockchain-based Location Privacy Preserving Method (BCLPPM) [4].

Several platforms, such as those mentioned in [30], and [24], currently employ blockchain's distributed ledger and smart contract technology to ensure data integrity on the chain, while utilizing desensitization algorithms for privacy protection. However, there are three main limitations at the current stage:

- Centralized while they simulation: In BCLPPM, a decentralized system is employed for managing reputation, a feature recognized as advantageous in IoV for location-based crowdsourcing tasks. Roadside Units (RSUs) are configured to work in conjunction with vehicles within the IoV [13]. Each block within this system is capable of containing a set number of vehicle ratings. However, there is a concern that these ratings in the IoV might inadvertently disclose users' private information.
- Evaluation metrics is insufficient: From a performance perspective, Li et al. [10] consider the simulation results to demonstrate the feasibility and effectiveness of Hyperledger, despite the lack of latency and throughput metrics in the proposed scheme.
- Experiments are not comprehensive: Further experiments are essential to identify the optimal settings for selecting and creating blocks, as well as to

assess the scalability of the proposed plans. Once the feasibility and appropriate settings are established, it becomes crucial to ascertain whether they align with the demands of real-world Intelligent Transportation Systems (ITS).

To address the limitations previously discussed, we introduce a decentralized scheme for preserving the privacy of vehicle density location data. This work integrates the Dirichlet differential privacy mechanism with the capabilities of Vehicle to Infrastructure (V2I) communications. When vehicles register with Roadside Units (RSUs), the timestamp can signify a specific location at a given time, given that the coordinates of the RSUs are known to the intelligent transportation systems, as illustrated in Fig. 1.

This scheme challenges conventional design principles by focusing not only on recognizing vehicle density but also on incorporating a flexible mechanism suitable for Blockchain-based Location Privacy Preserving Methods (BCLPPM) in practical Intelligent Transportation Systems (ITS). Moreover, the method allows for privacy sensitivity adjustments by a certified organization. A key challenge is to verify if the scheme meets the practical requirements for decentralized features. Specifically, the proposed privacy mechanism could be implemented within a smart contract. Once in place, assessing the performance of internal systems presents a significant challenge.

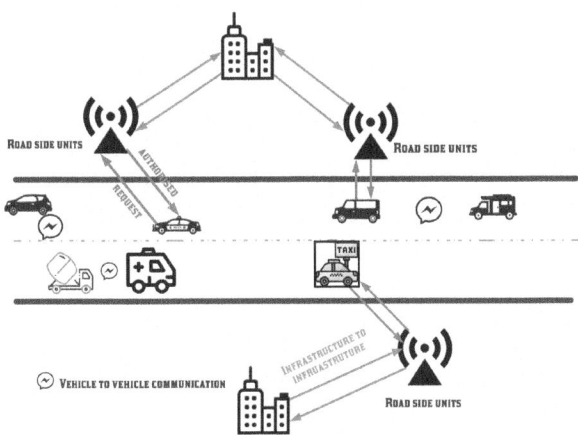

Fig. 1. Overview scenario for Road Side Units Communication. The main components in this figure is road side units and vehicles, and how they communicate with each other.

Based on the blockchain, this paper has proposed and implemented a more effective and privacy focused scheme:

- This paper addresses the location privacy problem in awareness of vehicle density via blockchain. The vehicle density definition comes from V2I communication in an intelligent transportation system, and the sensing network in IoV enable the V2I communication.

– This paper raises various issues regarding the density of vehicles in intelligent transportation and the effect of these factors on the privacy of the vehicular location.
– This paper analyzes the security properties and privacy. We evaluate the performance of the proposed scheme. The simulation results show that the vehicle density awareness method with multiple ratings proposed in this paper can significantly validate the feasibility and efficiency.

2   Related Work

Current solutions for location privacy issues in the Internet of Vehicles (IoV) can be divided into two categories: level-based classification (event-level and user-level) and technique-based classification.

2.1   Level-Based Classification

Location can indicate health, personal habits, and professional commitments by linking locations or paths visited. The Internet of Things (IoT) includes many technologies that enable innovative services across applications. Vehicles have become active and efficient network nodes due to the car industry's extensive integration with IoT technologies and strong research base. Event-level scenarios require excellent time efficiency and reliability due to their complexity and variety. To simplify method analysis, prototypes always use event-level methods. The Internet of Vehicles event-level privacy protection using blockchain technology must not compromise system effectiveness. However, privacy protection can significantly impair vehicle operation management application operations. Data analysis quality, timeliness, depth, and trustworthiness are essential for traffic congestion forecast, decision-making, and worldwide perception positioning.

2.2   Technique-Based Classification

In the Internet of Vehicles (IoV) domain, Crowd sensing LPPM is pivotal for user participation in mobile crowd-sensing, yet existing strategies often overlook the varied quality of task execution among participants [17,31]. Legislation enforcement on data submission architecture is essential for data privacy, especially on edge servers [23], while k-anonymity and blockchain integration offer a solution for secure data contribution and reward receipt [15]. Singh and Kim's BCLPPM system, integrating a local dynamic blockchain with the main blockchain and employing the Intelligent Vehicle Trust Point (IVTP), ensures reliable vehicle-to-vehicle interactions [7,29]. Additionally, the PBFT-based PoR (PPoR) consensus method addresses the challenges of V2V energy trading in vehicular networks [2,3]. The integration of IoV with AI is revolutionizing intelligent transportation systems and enhancing VANET communication protocols [11,20]. Blockchain architectures prove effective in detecting security attacks within IoV, offering enhanced privacy and efficiency [23] (Table 1).

Table 1. Summary of Key Aspects in IoV Research

Aspect	Details
Crowd Sensing LPPM	Importance in mobile crowd-sensing with varied task execution quality [17,31]
Data Privacy	Legislation enforcement on data submission and edge server protection [23]
Blockchain	Integration for secure data contribution and rewards [15]
BCLPPM System	IVTP for reliable V2V interactions [7]
PPoR Consensus	Addressing V2V energy trading challenges in vehicular networks [2,3]
AI and IoV	Advancing intelligent transportation and VANET protocols [11,20]
Security in IoV	Blockchain architectures for attack detection and privacy [1,23]

3 Preliminary

3.1 Differential Privacy

Differential privacy, as introduced by Dwork [6], is a robust notion of data privacy, also known as the trusted curator model. A differential privacy (DP) mechanism protects sensitive data by introducing distortion, typically by adding noise to numeric or non-numeric query results. The objective of DP is to minimize privacy leakage while maximizing data utility. Specifically, the probability that a query to one database (D) will yield the same result as a query to another (D'), where the two databases differ by at most one record, should be constrained by an exponentiated function of ϵ, reflecting a bounded increase in the likelihood of any outcome. Commonly, to achieve this level of privacy, controlled random noise—often drawn from a Laplace distribution—is added to the query output.

In the context of differential privacy, two datasets D and D' are considered adjacent if they differ by only one record are considered adjacent if they differ by only one record. A function f, mapping a dataset D to an abstract range R, is associated with a sensitivity S, which measures the greatest change in the function's output between two adjacent datasets. Sensitivity dictates the magnitude of noise that needs to be added to the query response to ensure an individual's privacy.

Definition 1 (ε-Differential Privacy)

Let D and D' are neighboring data sets, which means at most one record is different between D and D'. We claim random algorithm M holds ε-Differential Privacy when M satisfies: $\mathrm{Prob}[M(D) \in \mathrm{Out}\,] \leq \exp(\varepsilon) \cdot \mathrm{Prob}\,[M\,(D') \in \mathrm{Out}\,]$

Where Out is the subset of $M\varepsilon$, the term privacy budget is used. Significantly, the protection level ought to be raised the lower ε's privacy budget is. Innately, differential privacy ensures that adjacent sensitive data items will have statistically comparable privatised values.

3.2 Dirichlet Mechanism

A Dirichlet mechanism without any need of projection, maps

$$\mathbb{P}[\mathcal{M}_D^{(k)}(p) = x] = \frac{1}{\mathrm{B}(kp)} \prod_{i=1}^{n-1} x_i^{kp_i - 1} \left(1 - \sum_{i=1}^{n-1} x_i \right)^{kp_n - 1}, \tag{1}$$

where

$$\mathrm{B}(kp) := \frac{\prod_{i=1}^{n} \Gamma(kp_i)}{\Gamma\left(k \sum_{i=1}^{n} p_i \right)} \tag{2}$$

The multi-variable beta function serves as a Dirichlet mechanism, noteworthy for not requiring projection mappings. This mechanism is denoted by Dir and is parameterized by k. The multi-variable beta function is succinctly completed by the Dirichlet mechanism, which operates under an assumption regarding the parameter k. This parameter k is pivotal for adjusting the balance between the accuracy of the Dirichlet mechanism and the level of privacy it provides. The privacy guarantees of the Dirichlet mechanism can be achieved through this methodical adjustment.

Initially, the proposed approaches employ the Dirichlet mechanism to assess identity queries. A sensitive vector p becomes practically indistinguishable from its neighboring sensitive vectors due to the application of the Dirichlet process. Here, $\triangle k$ signifies the sensitive data's storage region.

As mentioned, when considering the VANET's characteristics, it is practical to assume privacy over a set of locations T, which may include customized elements. The cardinality of these elements can be denoted by K, and the set itself by H.

3.3   Problem Definition

To formalize the problem within the Internet of Vehicles (IoV), let each vehicle have a unique identifier V_{id}. Correspondingly, for each vehicle V_i, there exists a key key_v associated with V_{id}. The RSUs are capable of validating V_{id} when vehicles pass by an RSU identified by RSU_{id}. Each RSU_{id} is associated with a significant and stable location within the RSU pool in the IoV network.

The objective of the proposed method is to preserve the location of any vehicle V_i at any given time $Time_i$. Hence, the timestamp $Time_i$ for vehicle V_i can represent the relative location in the IoV, effectively replacing the need for explicit location coordinates.

To ensure the safety requirements for IoV are met, and to avoid conflicts in special cases, such as different vehicles registering with the same RSU_j at the exact same time, each vehicle's transaction is uniquely identified by its transaction id in our proposed scheme.

Once a sufficient number of transactions have been established, our privacy mechanism becomes operational. The RSU, denoted as RSU_j, will transmit privacy-related and vehicle density-related parameters to the smart contract, ensuring the integrity of the privacy preservation process.

4   Proposed Approach

Differential privacy technologies have the potential for widespread application across large institutions through strategic integration. Within such institutions, which often consist of numerous departments, mutually trusted entities can expedite the process of certification exchange. This approach allows for efficient information sharing within a decentralized framework. Thus, while institutions may not share the most precise data on-chain, the utility of the data remains uncompromised.

The method we propose focuses on integrating the concepts of density and the standard deviation associated with Laplace noise. This has a direct relevance to urban planning and is strongly correlated with the natural development of public transportation in urbanization. By offering viable solutions for the Internet of Vehicles (IoV), our method fosters an amalgamation of digitization, flexibility, and traffic management. Some necessary notations are summarized in Table 2

4.1   System Modelling

To deploy hash-linked data storage, Hyperledger Fabric utilizes a specialized infrastructure type. Hyperledger Fabric is a permissioned blockchain platform that enables the development of distributed applications and networks. It follows a modular architecture, which allows for customization and flexibility in designing blockchain solutions. The blocks are assembled such that the data is stored in a database-like structure, featuring multiple distinct hash instances, as illustrated in Fig. 2. Hyperledger Fabric's architecture consists of several key components:

Peers: Peers are the fundamental elements of the network, responsible for maintaining the ledger and participating in the consensus process. They execute smart contracts (known as chaincode) and validate transactions.
Orderers: Orderers are responsible for establishing the order of transactions and creating blocks. They ensure the consistency and reliability of the transaction sequence across the network. **Chaincode**: Chaincode is the smart contract in Hyperledger Fabric. It defines the business logic and rules for transaction processing and ledger updates. Chaincode is executed within a secure container environment on the peers.
Channels: Channels provide a mechanism for data isolation and confidentiality. They allow subsets of network participants to communicate and transact privately, without exposing their data to the entire network.

Hyperledger Fabric's modular design and permissioned nature enable fine-grained access control, privacy, and confidentiality. It supports the use of different consensus algorithms. By leveraging these architectural components and the hash-linked data storage mechanism, Hyperledger Fabric provides a robust and flexible platform for building blockchain-based applications in various domains. The overview of how proposed approach effects in blockchain is shown in Fig. 2.

Table 2. Notation Table

Notation	Description
(V_{id})	Vehicle ID
(RSU_{id})	Roadside Unit ID
(key_v)	Token for vehicle (V_i)
(key_{RSU})	Token for RSU
(Pool)	Pool of registered vehicle and RSU IDs
(Time)	Timestamp
(Pos)	Position
(λ)	Traffic flow
(ε_j)	Traffic density at location (j)
(λ_j)	Traffic volume at location (j)
(δ_j)	Average speed at location (j)
(k)	Traffic flow control constant
(D)	Vehicle density
(N)	Number of vehicles
(P)	Dirichlet parameter
(Dir())	Dirichlet distribution

4.2   Security and Privacy Requirements

To ensure the integrity and confidentiality of the system, several security and privacy requirements must be met by the key participants in the process: the Certificate Authority (CA), vehicles, and roadside units (RSUs).

- **Certificate Authority:** The CA, whether a trustworthy organization or certified third-party corporation, plays a crucial role in maintaining the security of the system. It is responsible for implementing secure chain code, properly identifying and authenticating roadside units, and configuring the system with robust security measures. The CA must possess strong communication and computational capabilities to prevent unauthorized access and protect sensitive data.
- **Vehicles:** Vehicles, as participants in the system, must ensure secure interactions with roadside units and maintain the privacy of their location data when cooperating with Location Based Services (LBS). The on-board components of the Internet of Vehicles (IoV) should provide secure Vehicle-to-Infrastructure (V2I) connectivity for transmitting traffic information updates. To protect user privacy, the vehicle density D used for traffic forecasting in the intelligent transportation system should be aggregated and anonymized, preventing the identification of individual vehicles or users. Secure communication protocols and encryption mechanisms should be employed to safeguard the transmitted data.

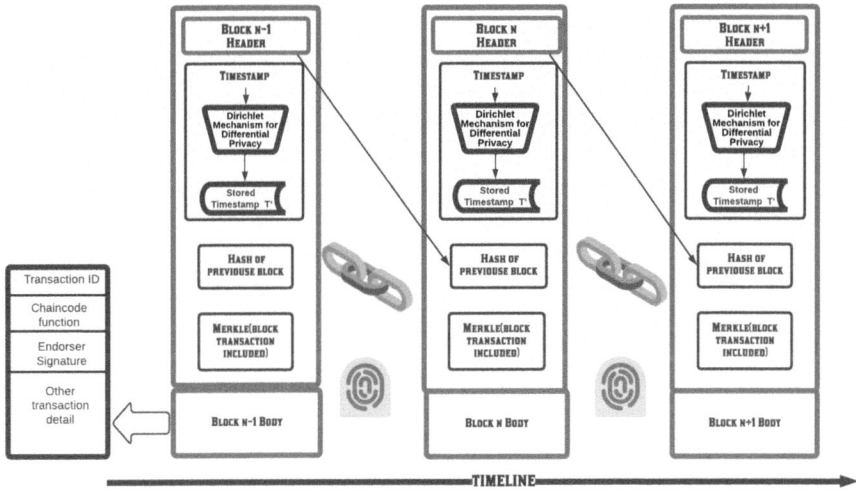

Fig. 2. Overview proposed approach effects in blockchain

- **Roadside Units:** RSUs play a critical role in collecting real-time vehicle position data and must be properly certified and validated by authorized entities on the blockchain. The certification process ensures that only legitimate and trusted RSUs are part of the system, reducing the risk of rogue or compromised units. RSUs should securely store and transmit the collected vehicle position data, protecting it from unauthorized access or tampering. Secure communication channels and data encryption should be used when RSUs interact with vehicles and other system components. The strategic positioning of designated RSUs should consider both optimal data collection and the security of the infrastructure.

By addressing these security and privacy requirements for the CA, vehicles, and RSUs, the system can establish a robust and trustworthy framework for secure data collection, transmission, and storage. Implementing strong authentication mechanisms, secure communication protocols, and data encryption techniques will help protect sensitive information and maintain user privacy. Regular security audits and updates should be conducted to identify and address any potential vulnerabilities in the system. By prioritizing security and privacy, the intelligent transportation system can ensure the integrity of the collected data and maintain the trust of its users.

4.3 Initialization

According to Algorithm 1, the initial phase of our method is outlined. This approach's major objective is to guarantee that the registration procedure is successful. The initialization process comprises issuing IDs, which are designated as V_{id} and RSU_{id}, to unauthenticated RSUs and vehicles. The above step serves the improvement the Hyperledger authentication method that was initially developed. The V_{id} should be checked to see whether it is duplicated on the blockchain in the event that the registration process is unsuccessful. whether it is found to be duplicated, a new V_{id} should be provided. Afterwards, it has been established the V_{id}, the proposed mechanism will proceed to generate the token key_v for the vehicle V_i, as seen in lines 5 and 6. The process of verifying RSU_{id} is carried out in a manner that is comparable. During the registration process, the procedure checks that all V_{id} and RSU_{id} have been successfully merged into the pool. This prevents any ID clashes from occurring. The method is located on line 12. Furthermore, this phase acts as a pre-measure in Algorithm 2, to avoid a possible excessive amount of SHA-256 duplication. RSU_{id} is the identifier that is used to record subsequent transactions on the blockchain.

Algorithm 1. Privacy Trajectory building

Require: V_{id}, RSU_{id}
Ensure: $RegisterSucess, key$
 1: $Register\ success = False$;
 2: **if** $V_{id} \in V_{pool}$ **then**
 3:    **return** Change V_{id}
 4: **end if**
 5: $key_v \leftarrow Blockchain$;
 6: $V_{id} \leftarrow key_v$;
 7: **if** $RSU_{id} \in RSU_{pool}$ **then**
 8:    **return** Change RSU_{id}
 9: **end if**
10: $key_{RSU} \leftarrow Blockchain$;
11: $RSU_{id} \leftarrow key_{RSU}$;
12: $Pool \leftarrow Pool \cup V_{id} \cup RSU_{id}$;
13: $Register\ success = True$;
14: **return** $RegisterSuccess, key$

4.4   Vehicle Sign up with RSU

This procedure focuses on the vehicle registering its data with roadside units. Algorithm 2 aims to phrase the important message while the vehicle communicates with the roadside units. Due to the nature of the roadside device, its position and the vehicle's coordinates at the moment of registration are steady. Consequently the timestamp $Time$ and the RSU_{id} mean the position for the V_{id}. Furthermore, there is a function in blockchain to track the existing transactions, which is straightforward to collect the trajectory of the particular vehicles. A complete verification is supposed to serve for this algorithm in Line 2. This method also includes the generation vehicle condition, and the time is appended to the block header whenever a new block is created. Again, this algorithm is commonly repeated. Contracting state Algorithm 2 with another algorithm would not decrease latency.

Algorithm 2. Vehicle Sign-Up with RSU

Require: $Time$, Position Pos, V_{id}, RSU_{id}
Ensure: Sign-Up Successful
1: Vehicle passes the RSU
2: RSU validates the vehicle identity V_{id}
3: **if** $V_{id} \in V_{pool}$ **then**
4:     RSU generates a record $\{V_{id}, Time, Pos, RSU_{id}\}$
5:     RSU stores the record $\{V_{id}, Time, Pos, RSU_{id}\}$ in the transaction database
6:     **return** Sign-Up Successful
7: **else**
8:     **return** Invalid V_{id}, Sign-Up Unsuccessful
9: **end if**

Vehicle Density. The proposed method integrates design, simulation, verification, deployment, and operation across systems. It facilitates safe and reliable collaboration and intractability between heterogeneous information systems and physical systems. The goal is to achieve reliable, efficient, real-time perception and decision-making control for intelligent networked vehicles, thereby enhancing driving comfort.

$$\lambda \cdot \varepsilon_j = f\left(\frac{\lambda_j}{\delta_j}\right) + k$$

– k is the traffic flow control constant.

This method categorizes the existing research on measuring location privacy into two broad groups, based on distinct definitions of location privacy for different vehicle densities and the associated projected inference error. The concept of rural areas provides a significant conceptual basis for location privacy solutions that aim to maintain consistent traffic flow control. We acknowledge the varied density ranges and propose that mobility and speed cannot always be clearly distinguished from density in current traffic density studies. These studies provide users with high mobility in manipulating vehicle traffic indicators (speed, volume, and density).

The scenario can be summarized from three distinct perspectives. Moreover, it is both feasible and desirable to develop a location obfuscation mechanism that effectively combines these two unique privacy concepts. Our scheme is designed to be advantageous and practicable. Lastly, incorporating user-defined constraints enhances usability—a key factor in privacy protection models—and supports adaptive noise adjustment for distinguishability. It also meets mobile users' customized privacy/utility requirements, assuming that each driver reacts to stimuli from other vehicles ahead or behind in some manner.

4.5 Data Submission

As outlined in Algorithm 3, the most significant feature is the data processor equipped with a differential privacy mechanism. This allows us to treat the specific timestamp $Time_i$ as private. The rationale is to adhere to a fundamental blockchain rule: the timestamp of the last block must precede that of the next block. To maintain the utility or readability of the blockchain after applying the privacy mechanism, $Time$ is adjusted using the Dirichlet mechanism for the Query model. The final adjusted result is then recorded in the chain code. The complexity of ensuring high-level security requirements is notable; for example, RSU_{id} must be verified as part of the certificated RSU_{pool}. The input vehicle density D, the Dirichlet parameter P, and the output of Algorithm 3 are central to this algorithm. In case we consider the additional new RSUs, we would like to start the initialization step, which is similar to the Algorithm 1.

Algorithm 3. Data Submission Process

Require: Density D, Number of Vehicles N, Dirichlet parameter P, RSU_{id}
Ensure: Submit Successful
 1: Ensure a sufficient number of vehicles pass the RSU
 2: RSU computes the Dirichlet distribution Dir() based on P, D, and N
 3: RSU augments Dir() to the timestamp of the record to generate $\{V_{id}, \text{Dir}(Time), D, N\}$
 4: RSU sends the record $\{V_{id}, \text{Dir}(Time), D, N\}$ to the Smart Contract
 5: **if** $RSU_{id} \in RSU_{pool}$ **then**
 6:     Smart Contract commits the record to the blockchain
 7:     RSU clears the record from the transaction database
 8:     **return** Submit Successful
 9: **else**
10:     **return** Invalid RSU_{id}, Submit Unsuccessful
11: **end if**

---

5   Security and Privacy Analysis

The privacy analysis is based on the analysis of the query part in the Dirichlet mechanism. Based on the linear query collection feature, it is necessary to acknowledge the qualification criteria of the privacy guarantees afforded via the Dirichlet mechanism. As mentioned in the preliminary [16], privacy guarantees would satisfy the requirement of the bounding ratios of the Dirichlet distributions. Moreover, a part of the function should include *gamma* functions. Assumption [8].

Assumption 1. *In $\Delta_{n,U}^{(\eta,\bar{\eta})}$, $\eta > 0$, $\bar{\eta} > 0$, and $\eta + \bar{\eta} < \frac{1}{2}$.*
  Allowing d to be a vector in $\mathbb{R}^n$, [9] use the notation $d_{(i,j)}$ to indicate the vector $(d_i, d_j)^T \in \mathbb{R}^2$, where $(\cdot)^T$ is a vector's transposition, and $d_{(i,j)} \in \mathbb{R}^{n-2}$ to denote the vector d with i^{th} and j^{th} entries removed. $\mathbb{D}[\cdot]$ denotes the probability of an event. For a random variable, $\mathbb{E}[\cdot]$ denotes its expectation and Var $[\cdot]$ denotes its variance. We use the notation $|\cdot|$ for the cardinality of a finite

set. $\| \cdot \|_1$ *represents a vector's 1-norm We also make advantage of the specific functions, which are the gamma, beta, digamma functions, respectively.*

$$\Gamma(x) = \int_0^\infty z^{x-1} \exp(-z)dz, \quad x \in \mathbb{R}_+$$

$$beta(a,b) = \int_0^1 t^{a-1}(1-t)^{b-1}dt = \frac{\Gamma(a)\Gamma(b)}{\Gamma(a+b)}, \quad a,b \in \mathbb{R}_+$$

$$\psi^{(0)}(x) = \frac{d}{dx}\log(\Gamma(x)), \quad \psi^{(1)}(x) = \frac{d^2}{dx^2}\log(\Gamma(x)),$$

$$x \in \mathbb{R}_+$$

Assumption 2. *For the Dirichlet mechanism by* [9] $\mathcal{M}_T^{(k)}$, *the parameter k satisfies*

$$k \geq \max\left\{\frac{1}{\eta}, \frac{1}{1 - \eta - \bar{\eta}}\right\}.$$

We later employ the texting option to fine-tune the compromise we have made between the Dirichlet mechanism's precision and privacy. After that, we prove that the Dirichlet mechanism is indeed secure.

Lemma 1. *Let Assumptions 1 and 2 hold. Let W be a given set of indices which is used to construct $\Delta_{n,U}^{(\eta,\bar{\eta})}$ and let p, d be any b-adjacent vectors in $\Delta_{n,U}^{(\eta,\bar{\eta})}$ with their i^{th} and j^{th} entries different. Then, for a constant $k \in \mathbb{R}_+$, we have that*

$$\frac{beta(kd_i, kd_j)}{beta(kp_i, kp_j)} \leq \frac{beta(kd_i, k(1 - \bar{\eta} - d_i))}{beta(kp_i, k(1 - \bar{\eta} - p_i))}.$$

where the parameter $\gamma \in (0,1)$ defines the set Ω_1.

As previously stated, we present a simplified constraint on *epsilon* that offers a direct dependency of *epsilon* on other factors in the issue.

6   Performance Analysis

6.1   Set up and Results

Hyperledger Fabric 2.1 uses Docker 3.52.18, Docker Compose 1.29.2, and Calliper 0.4.1. The network has a CA, administrator, client, and peer node. Client nodes interact via command-line and install smart contracts on users' machines. Private collections and binary transitory data improve data privacy in version 2.1. Initialization batches transactions into 100-transaction blocks. Testing involves two peers per isolated organisation in three organisations. This section analyses Calliper results on VANET-based applications to improve Intelligent Transportation Systems. A single organisation with one peer to three organisations with two peers are simulated in the study (Table 3).

Table 3. Summary of Hyperledger Fabric 2.1 Implementation

Component	Description
Tools	Docker 3.52.18, Docker Compose 1.29.2, Caliper 0.4.1
Network Configuration	CA node, administrator node, two client nodes, two peer nodes
Interface	Command-line for client nodes
Smart Contracts	Installed on users' computers
Privacy Features	Private collections, transient data in binary
Transaction Batching	Default block size of 100 transactions
Testing	Two peers per organization in three organizations

- The preparation period involves several tasks, such as initializing the entire blockchain network, reading the configuration file, deploying smart contracts, and initiating monitoring setup.
- The testing phase commences with the client leveraging the benchmark configuration. Subsequent to conducting the statistical results are typically returned.
- The "reporting" stage entails the analysis of statistical data and the generation of HTML reports.

6.2   Reporting and Analysis

The data write throughput is substantially higher than the average volume and the federated chain's consensus structure will employ a Byzantine fault-tolerant mechanism that is both more efficient and less stressful on the environment [5] distributed. The provided approach has no significant influence on the reaction time or throughput of the blocks, and it preserves all of the federated chain's features. There are the results from mechanism based on the hyper ledger we designed, which we called the baseline result. As for the most existing proposed mechanism is based on the permissionless blockchain.

Experiments with fixed block sizes: The Figs. 3, 4 and 5 represent the outcomes of the validation and commitment of anywhere from 100 to 200 transactions for a single run, with this process being repeated 10 times. Each block of 100 transactions contains a collection of previous transactions. The topic of latency will be covered first, followed by throughput. For both delay and throughput, the rate is measured in transactions per second (TPS), and it varies depending on the size of the network. Due to batching, we display the latency for each individual transaction rather than the latency for each individual block.

The proposed framework's efficiency is reduced by blockchain latency, which may impose delay during the training phase. There is also involved in the throughput and send rate.

Data was separated into three pieces for box graphing. The first segment contains the median, while the second and third segments show the upper and lower half of the median for the baseline [26], D $=$ 10 D $=$ 100, with latency of 0.83,

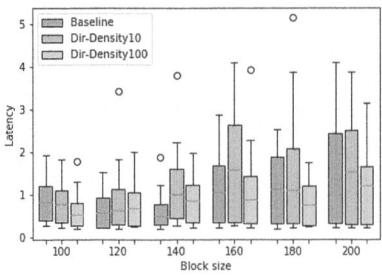
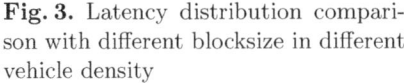
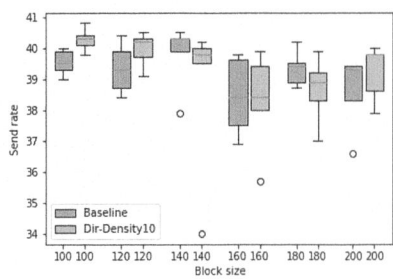

Fig. 3. Latency distribution comparison with different blocksize in different vehicle density

Fig. 4. Send rate distribution comparison with different blocksize in different vehicle density

0.79, and 0.70, respectively. The box plot is a classic way to show data distribution using five numbers: minimum, first quartile, median, third quartile, and maximum. Figures 3, 4, and 5 show the medians as segments within rectangles and the minimum and maximum values as boxes.

The line graph in Fig. 6 shows the average block construction time for add-on vehicle density of 10 and 100, respectively. Statistics show average delay from 200 to 1400 and for each round of that block size. 10 rounds of transactions would be built in a large block. As block size rises, vehicle density will rise, hence each block's latency will primarily rise. Baseline time consumption shares the shortest delay of 28 ms per block at 1400 block size. Given the time efficiency of the proposed scheme, our results align with the self-imposed goal of increasing throughput without incurring additional latency. In fact, our enhancements have slightly reduced peer latency, accompanied by an upward trend in the transmission rate. While the pipelining in scenarios with a vehicle density equal to 10

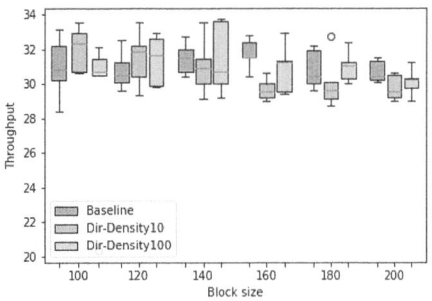
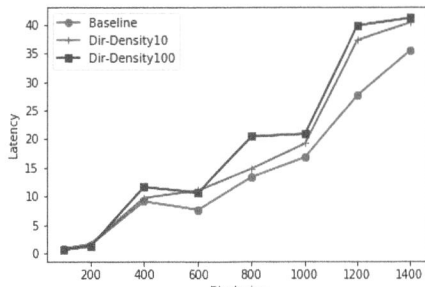

Fig. 5. Throughput distribution comparison with different block size in different vehicle density

Fig. 6. Latency comparison with different block size in different vehicle density

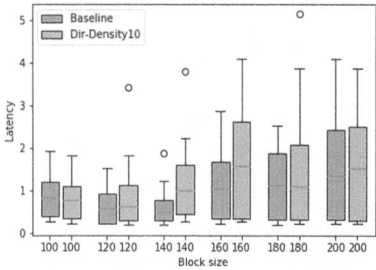
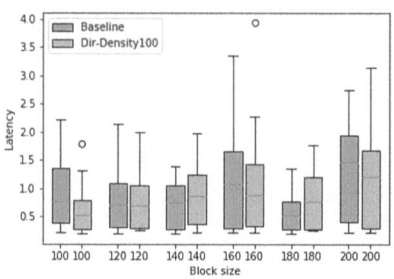

Fig. 7. Latency distribution with different block size in vehicle density $= 10$

Fig. 8. Latency distribution with different block size in vehicle density $= 100$

introduces some delay, the other benefits observed more than compensate for this increase.

The throughput of the proposed scheme, measured in transactions per second, reflects the total load managed by the scheme each second. Figure 5 presents the throughput for block sizes ranging from 100 to 200, showcasing a stable trend with slight fluctuations within an acceptable range. As the block size increases, there is a gradual decline in the median throughput. Moreover, the reliability of transactions per second is less affected by changes in density. In summary, as the target block size grows, the throughput experiences a marginal decline. In private blockchains, these factors may be adjusted as a design choice. Conversely, a permissioned blockchain permits an approximate target inter-block duration. To assess the efficacy of our simulation against various transaction execution times on the blockchain, a broad spectrum of transaction inclusion times was tested. The outcomes, predicated on our hypothesis, are presented in Figs. 7 and 8 below, depicted as box plots.

The box plots in Fig. 10 effectively illustrate the dispersion characteristics of the dataset. The most efficient use of box plots is to compare across different block sizes and qualitative data of the privacy budget to obtain grouped box plots. Outliers are values that lie outside the typical range of the data set. The distribution exhibits a rightward skew, with outliers clustering around the value state transition when the vehicle density equals 10 for block sizes of 400.

By displaying the latency results of Fabric with different densities, Figs. 7 and 8 are created. The grouped box plot of Dirichlet differential privacy, which, in comparison to the baseline experiment, seems to be considerably taller, may be compared with these. The experiment may be distinguished from other similar experiments by the significant changes that can be seen throughout the box plots, which highlight these distinctions. As vehicle density increases, the difference in latency between the DP add-on and the baseline method exhibits fluctuations of approximately 15% points. Considering that the block-created failure rate is 0.5 per cent, the solution provided in this study is for Fabric. The

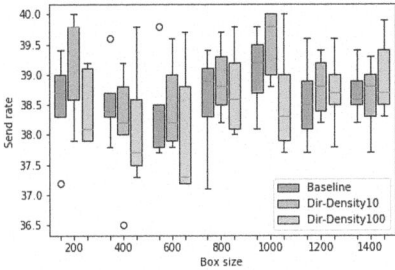

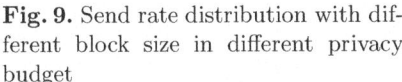

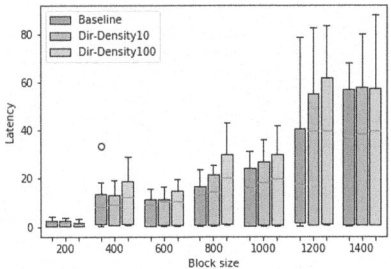

Fig. 9. Send rate distribution with different block size in different privacy budget

Fig. 10. Latency results of Fabric with varying different block-size, with high volume transportation location update requirements such as in rush hour

latency outcomes illustrated in Fig. 10 are applicable to every block, factoring in the rise in vehicle density and the blockchain's vulnerability to potential attacks, as well as resource awareness. The network successfully processes nearly all submitted transactions, indicating that the performance results are not exclusively contingent upon our proposed scheme. Moreover, vehicle density has a minimal impact on the transmission rate, as depicted in Fig. 9.

7 Conclusion

This research introduces a blockchain-based Dirichlet differential mechanism solution for intelligent transportation system privacy. When automobiles collide with roadside units, our system uses blockchain block timestamps to estimate geographic coordinates. Our trials show improved latency, send rate, and throughput. In addition, the Dirichlet mechanism handles vehicle density as a critical privacy concern using probabilistic mapping via Dirichlet distribution. Future research could examine blockchains in public or organisational settings. Regulation of the blockchain is necessary to provide strong security and privacy requirements for ordinary activities [25]. Most measurement tools fall short of these strict requirements, indicating a critical development area.

References

1. Rathore, S., Wook Kwon, B., Park, J.H.: BlockSecIoTNet: blockchain-based decentralized security architecture for IoT network. J. Netw. Comput. Appl. **143**, 167–177 (2019). https://doi.org/10.1016/j.jnca.2019.06.019
2. Abishu, H.N., Seid, A.M., Yacob, Y.H., Ayall, T., Sun, G., Liu, G.: Consensus mechanism for blockchain-enabled vehicle-to-vehicle energy trading in the internet of electric vehicles. IEEE Trans. Veh. Technol. **71**(1), 946–960 (2021)

3. Aggarwal, S., Kumar, N.: A consortium blockchain-based energy trading for demand response management in vehicle-to-grid. IEEE Trans. Veh. Technol. **70**(9), 9480–9494 (2021)

4. Alladi, T., Chamola, V., Sahu, N., Venkatesh, V., Goyal, A., Guizani, M.: A comprehensive survey on the applications of blockchain for securing vehicular networks. IEEE Commun. Surv. Tut. **24**(2), 1212–1239 (2022). https://doi.org/10.1109/COMST.2022.3160925

5. Buzachis, A., Filocamo, B., Fazio, M., Ruiz, J.A., Sotelo, M.Á., Villari, M.: Distributed priority based management of road intersections using blockchain. In: 2019 IEEE Symposium on Computers and Communications (ISCC), pp. 1159–1164. IEEE (2019)

6. Dwork, C., Rothblum, G.N.: Concentrated differential privacy. CoRR abs/1603.01887 (2016). http://arxiv.org/abs/1603.01887

7. Farooq, U., Javaid, N.: Blockchain based decentralized vehicular communication system and smart payment method. Management **6**(7), 8

8. Gohari, P., Wu, B., Hale, M., Topcu, U.: The dirichlet mechanism for differential privacy on the unit simplex. In: 2020 American Control Conference (ACC), pp. 1253–1258. IEEE (2020)

9. Gohari, P., Wu, B., Hawkins, C., Hale, M., Topcu, U.: Differential privacy on the unit simplex via the dirichlet mechanism. IEEE Trans. Inf. Forensics Secur. **16**, 2326–2340 (2021)

10. Li, B., Liang, R., Zhu, D., Chen, W., Lin, Q.: Blockchain-based trust management model for location privacy preserving in VANET. IEEE Trans. Intell. Transp. Syst. **22**(6), 3765–3775 (2020)

11. Liang, L., Ye, H., Li, G.Y.: Toward intelligent vehicular networks: a machine learning framework. IEEE Internet Things J. **6**(1), 124–135 (2019)

12. Liang, R., Li, B., Song, X.: Blockchain-based privacy preserving trust management model in VANET. In: Yang, X., Wang, C.-D., Islam, M.S., Zhang, Z. (eds.) ADMA 2020. LNCS (LNAI), vol. 12447, pp. 465–479. Springer, Cham (2020). https://doi.org/10.1007/978-3-030-65390-3_36

13. Liu, Y., et al.: VRepChain: a decentralized and privacy-preserving reputation system for social internet of vehicles based on blockchain. IEEE Trans. Veh. Technol. (2022)

14. Ma, Z., Zhang, T., Liu, X., Li, X., Ren, K.: Real-time privacy-preserving data release over vehicle trajectory. IEEE Trans. Veh. Technol. **68**(8), 8091–8102 (2019)

15. Peng, T., Liu, J., Chen, J., Wang, G.: A privacy-preserving crowdsensing system with muti-blockchain. In: 2020 IEEE 19th International Conference on Trust, Security and Privacy in Computing and Communications (TrustCom), pp. 1944–1949 (2020). https://doi.org/10.1109/TrustCom50675.2020.00265

16. Prince, P., Lovesum, S.: Privacy enforced access control model for secured data handling in cloud-based pervasive health care system. SN Comput. Sci. **1**(5), 1–8 (2020)

17. Qian, Y., Ma, Y., Chen, J., Wu, D., Tian, D., Hwang, K.: Optimal location privacy preserving and service quality guaranteed task allocation in vehicle-based crowdsensing networks. IEEE Trans. Intell. Transp. Syst. **22**(7), 4367–4375 (2021). https://doi.org/10.1109/TITS.2021.3086837

18. Qiu, D., Wang, Y., Zhang, T., Sun, M., Strbac, G.: Hybrid multi-agent reinforcement learning for electric vehicle resilience control towards a low-carbon transition. IEEE Trans. Ind. Inf. (2022)

19. Singh, P.K., Singh, R., Nandi, S.K., Ghafoor, K.Z., Rawat, D.B., Nandi, S.: Blockchain-based adaptive trust management in internet of vehicles using smart contract. IEEE Trans. Intell. Transp. Syst. **22**(6), 3616–3630 (2020)

20. Storck, C.R., Duarte-Figueiredo, F.: A survey of 5G technology evolution, standards, and infrastructure associated with vehicle-to-everything communications by internet of vehicles. IEEE Access **8**, 117593–117614 (2020)

21. Szott, S., et al.: Wi-Fi meets ML: a survey on improving IEEE 802.11 performance with machine learning. IEEE Commun. Surv. Tut. **24**(3), 1843–1893 (2022). https://doi.org/10.1109/COMST.2022.3179242

22. Wang, M., Zhu, T., Zhang, T., Zhang, J., Yu, S., Zhou, W.: Security and privacy in 6G networks: new areas and new challenges. Digit. Commun. Netw. **6**(3), 281–291 (2020)

23. Wang, M., Zhu, T., Zuo, X., Yang, M., Yu, S., Zhou, W.: Differentially private crowdsourcing with the public and private blockchain. IEEE Internet Things J. (2023)

24. Wang, M., Zhu, T., Zuo, X., Ye, D., Yu, S., Zhou, W.: Blockchain empowered multi-agent systems: advancing IoT security and transaction efficiency. IEEE Internet Things J. (2023)

25. Xu, H., Zhu, T., Zhang, L., Zhou, W., Yu, P.S.: Machine unlearning: a survey. ACM Comput. Surv. **56**(1) (2023). https://doi.org/10.1145/3603620

26. Xu, X., Sun, G., Luo, L., Cao, H., Yu, H., Vasilakos, A.V.: Latency performance modeling and analysis for hyperledger fabric blockchain network. Inf. Process. Manage. **58**(1), 102436 (2021)

27. Ye, D., Zhu, T., Cheng, Z., Zhou, W., Philip, S.Y.: Differential advising in multi-agent reinforcement learning. IEEE Trans. Cybern. **52**(6), 5508–5521 (2020)

28. Ye, D., Zhu, T., Zhu, C., Zhou, W., Philip, S.Y.: Model-based self-advising for multi-agent learning. IEEE Trans. Neural Netw. Learn. Syst. (2022)

29. Zhang, L., Zhu, T., Hussain, F.K., Ye, D., Zhou, W.: A game-theoretic method for defending against advanced persistent threats in cyber systems. IEEE Trans. Inf. Forensics Secur. **18**, 1349–1364 (2022)

30. Zhou, S., Liu, C., Ye, D., Zhu, T., Zhou, W., Yu, P.S.: Adversarial attacks and defenses in deep learning: from a perspective of cybersecurity. ACM Comput. Surv. **55**(8), 1–39 (2022)

31. Zou, S., Xi, J., Wang, H., Xu, G.: CrowdBLPS: a blockchain-based location-privacy-preserving mobile crowdsensing system. IEEE Trans. Ind. Inf. **16**(6), 4206–4218 (2019)

DBFIA: Diffusion-Based Face Image Anonymization

Hanyu Xue[1], Xin Yuan[2], Bo Liu[1]($\boxtimes$), and Ming Ding[2]

[1] Australian Artificial Intelligence Institute, School of Computer Science,
University of Technology Sydney, Ultimo, Australia
{hanyu.xue,bo.liu,tianqing.zhu}@uts.edu.au
[2] Data61, CSIRO, Eveleigh, Australia
{xin.yuan,ming.ding}@data61.csiro.au

Abstract. This paper presents a novel approach to address the issue of identity protection in facial image datasets. Our goal is to prevent any violation of privacy for the individuals depicted in the dataset while ensuring that the dataset is still useful for downstream training tasks and maintains the quality of the images. Previous methods have predominantly used Generative Adversarial Network (GAN)-based models to anonymize facial datasets. However, such models suffer from distortion and loss of detail when processing real-life images.

To overcome these challenges, we propose a method for anonymizing faces using a diffusion model. Our approach retains the original facial attributes and produces anonymized images with high-quality image details. These attributes are essential when using anonymized images for downstream tasks. Our proposed framework optimizes the latent space vectors of the conditional Denoising diffusion probabilistic model (DDPM) to maintain a certain distance between the identity features (in the feature space of ArcFace) of the anonymized images and the original identity features while preserving the facial attributes. Through qualitative and quantitative experiments, we demonstrate that our method can effectively anonymize the identity of facial images while preserving image details and facial attributes.

Keywords: diffusion autoencoder · dataset anonymization · attribute preserving · latent optimization

1 Introduction

In the era of deep learning research, the demand for image data has witnessed a remarkable surge, driven by the advancement of cutting-edge technologies and applications. As researchers delve into the potential of artificial intelligence and machine learning, the significance of data sharing and its associated legal and ethical considerations have come to the forefront. Ensuring data privacy and protection has become a paramount concern, with the General Data Protection

© The Author(s), under exclusive license to Springer Nature Singapore Pte Ltd. 2025
T. Zhu et al. (Eds.): ICA3PP 2024, LNCS 15251, pp. 40–59, 2025.
https://doi.org/10.1007/978-981-96-1525-4_3

Regulation (GDPR) [28] enacted by the European Union serving as a prominent example of comprehensive data protection legislation.

Facial images hold a unique position within the realm of image data due to their rich and sensitive personal information. Facial image datasets frequently contain identifiable information, both direct and indirect, that can lead to the re-identification of individuals, thereby exposing them to privacy breaches and unauthorized use of their data. In light of these risks, the anonymization of facial image datasets has emerged as a critical strategy to mitigate potential privacy threats while preserving the utility and value of the data for research and analysis [38].

The core objective of this paper is to address the crucial need for anonymizing facial image datasets and explore effective techniques for safeguarding individual privacy while retaining essential facial attributes and image details. By effectively anonymizing the data, we aim to minimize the risk of re-identification, enabling secure data sharing and fostering a culture of responsible data handling within the research community.

Several methods are commonly employed to anonymize face images. One prevalent technique is face blurring [6], which uses blurring algorithms or filters to obscure facial features. Another approach, pixelation [8], replaces the facial region with large pixels to conceal details. Face detection and masking techniques [43,46] utilize face detection algorithms to locate faces within an image and then overlay them with solid colours or patterned masks. Facial landmark removal [7] involves distorting or removing specific points on the face, such as eyes, nose, and mouth, altering the overall facial structure while retaining some contextual information, making recognising individuals challenging. However, these methods sacrifice facial attributes and image quality in exchange for privacy protection. Additionally, recent studies [21,23] have shown that while blurring and pixelation may be effective to human observers, deep learning algorithms can circumvent their de-identification effects, reducing their efficacy in preserving privacy.

Prominent approaches to address this issue primarily rely on Generative Adversarial Networks (GANs) [9]. GAN-based techniques generate synthetic faces that maintain anonymity while preserving visual quality. GAN-based face manipulation [33] involves altering attributes, including identities, of the original face, transforming the individual's appearance while retaining the image's structure. GAN-based methodologies are categorized into image inpainting [13], latent optimization [40], and attribute manipulation [17], each with varying effectiveness in anonymizing faces. Despite efforts, generating high-quality anonymous face images remains challenging. Notably, GAN-based Autoencoder [34], utilized in recent face anonymization work [1], extracts latent codes from real-world face images using a StyleGAN2-based encoder. This encoder maps images into the latent space of a pre-trained StyleGAN2 generator, showing promising outcomes in image manipulation. However, recent studies [26,40] have highlighted limitations of GAN-Based Autoencoders in manipulating facial images, attributed to GANs' reconstructive capabilities that fail to capture intricate details (Fig. 1).

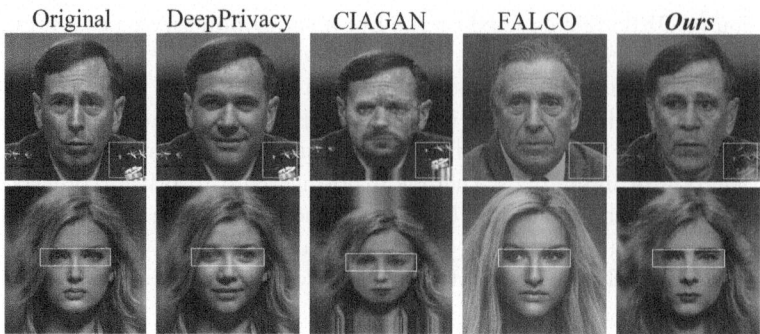

Fig. 1. Visual quality compared to DeepPrivacy [13], CIAGAN [20], FALCO [1]. **Ours**, along with the FALCO approach, exhibits the capability of concealing sensitive information within the background. Notably, **Ours** also demonstrates the ability to retain a richer set of attributes and intricate details.

Motivated by the aforementioned technological advancements, this paper pioneers the application of diffusion models [11] to address face anonymization, marking a significant departure from traditional GAN-based approaches. The diffusion model was chosen due to its innovative nature and potential to overcome the limitations associated with existing methods. Our proposed approach uses the DDIM decoder as the primary generator for synthesizing anonymized facial images. To ensure the faithful preservation of various facial attributes from the original dataset, we employ an attribute optimization technique for the generated images. This meticulous preservation of comprehensive facial attributes facilitates the training of diverse facial models, making our approach highly versatile for various applications. The main contributions of this paper are summarized as follows:

- We propose a novel face dataset anonymization framework based on the diffusion model, surpassing other state-of-the-art methods by delivering enhanced identity anonymization performance without compromising the preservation of crucial facial attributes.
- We propose a novel loss function that combines identity and attribute loss. This innovative approach, combined with a Denoising Diffusion Implicit Model (DDIM) decoder and attribute optimization technique, empowers the generation of an anonymized dataset that exhibits enhanced effectiveness and usability.
- Through evaluation and comparative analysis of the datasets such as Celeba-HQ and FFHQ, we demonstrate the superior efficacy of our approach in preserving facial attributes compared to other state-of-the-art methods. Moreover, our approach stands out by enabling the training of robust facial models that maintain the integrity of essential attributes, which are crucial for a wide range of practical applications.

2   Related Work

In recent image privacy researches [44,47,48], considerable attention has been paid to modifying identity-related information in images using various techniques, such as obfuscation, anonymization [4], GAN-based inpainting [32], feature space perturbation [2,3,35,36,50], differential privacy [18,37], and adversarial examples (AEs) [29,39]. It has been demonstrated that basic obfuscation techniques are ineffective when it comes to countering deep neural network (DNN)-based recognition systems [21,23].

GAN-Based Face Inpainting. The concept of GAN-based inpainting was introduced to generate content that effectively conceals sensitive information or the identity of an image while preserving the quality of the original image [41]. To enhance the training process, a conditional GAN (CGAN)-based approach was developed by incorporating labels into both the generator and discriminator networks [22]. Building upon the CGAN framework, two notable methods, namely Conditional Identity Anonymization Generative Adversarial Network (CIAGAN) [20] and DeepPrivacy [13], introduced the integration of autoencoder within the feature space of images. CIAGAN demonstrated the ability to anonymize faces and bodies, generating high-quality images and videos. While DeepPrivacy took into account factors such as pose and background to generate images with improved realism and privacy preservation.

Diffusion Models. Denoising diffusion probabilistic models (DDPMs) [11] are generative models that link image generation with the sequential isotropic Gaussian noise denoising process. Unlike other generative models such as GANs and most traditional-style VAEs that encode input data into a low-dimensional space, diffusion models maintain a latent space of the same size as the input. In the forward process, DDPMs progressively add noise to the image until it is completely degraded into pure Gaussian noise. Assuming an ideal forward process, where a real input image $\mathbf{x}_0$ undergoes T rounds of Gaussian noise addition, resulting in a purely Gaussian noise image $\mathbf{x}_T \sim \mathcal{N}(\mathbf{0}, \mathbf{I})$. Consequently, each step of noise addition can be formally expressed as the following probability function: $q(\mathbf{x}_t|\mathbf{x}_{t-1}) \sim \mathcal{N}(\mathbf{x}_t; \sqrt{1-\beta_t}\mathbf{x}_{t-1}, \beta_t\mathbf{I})$, where β_t is the coefficient associated with the noise. As a result, the cumulative noise in the image $\mathbf{x}_0$ after t processing steps can be represented as another Gaussian noise $q(\mathbf{x}_t|\mathbf{x}_0) \sim \mathcal{N}(\mathbf{x}_t; \sqrt{\alpha_t}\mathbf{x}_0, (1-\alpha_t)\mathbf{I})$, where $\alpha_t = \prod_{i=1}^{t}(1-\beta_i)$. In the reverse process, i.e., learning the distribution $p(\mathbf{x}_{t-1}|\mathbf{x}_t)$, the noise is gradually removed to generate a realistic image. To train a DDPM network, Ho et al. [11] give the distribution of $p_\theta(\mathbf{x}_{t-1}|\mathbf{x}_t) \sim \mathcal{N}(\mathbf{x}_{t-1}; \mu_\theta(\mathbf{x}_t, t), \sigma_t)$, and propose a learnable function $\epsilon_\theta(\mathbf{x}_t, t)$ to predict the noise added in each step. Despite requiring many noise injection and denoising steps to generate samples, DDPMs exhibit superior image fidelity and diversity compared to other generative models.

A notable limitation of diffusion models is their reliance on an extended sequence of diffusion steps to achieve desirable outcomes, resulting in sluggish generation speed. In contrast, Denoising Diffusion Implicit Models (DDIMs) [31], compared to DDPMs, overcome this constraint by relaxing the requirement for the diffusion process to conform to a Markov chain. As a result, DDIMs can employ fewer sampling steps, effectively expediting the generation process. Furthermore, DDIMs possess a distinctive characteristic in that generating samples from random noise is deterministic, obviating the need for intermittent random noise injections. This streamlined approach further accelerates the relatively computationally demanding sampling procedure inherent to DDPMs. By leveraging a deterministic forward-backward process, DDIMs demonstrate a near-perfect reconstruction capability. In this study, we employ conditional DDIMs as the generator to anonymize human facial images.

3   Preliminaries

3.1   Conditional Semantic Encoder

Our paper uses a conditional DDIM with an additional latent code, denoted as $\bar{\mathbf{z}}$. In contrast to certain other conditional DPMs [12,16,30] that employ spatial 2-D latent maps, the latent code we employ is a non-spatial code with a dimension of 512. The primary objective of the Conditional Semantic Encoder is to encode all the semantic information in an image into a high-level semantic space. Consequently, the latent code $\bar{\mathbf{z}} \in \mathbb{R}^{512}$ encompasses global semantics that is not specific to any spatial regions within the image, facilitating smooth optimization.

3.2   Diffusion-Based Decoder

The primary objective of the Diffusion-based Decoder is to model the distribution of the target dataset by training a network in the reverse process, denoted as $p(\mathbf{x}_{t-1}|\mathbf{x}_t)$. A successfully trained network can generate a realistic image with a Gaussian noise map. There are various methods to model this distribution, and one of them is through the DDIM introduced by Song et al. [31]:

$$p_\theta(\mathbf{x}_{t-1}|\mathbf{x}_t) \sim \mathcal{N}(\sqrt{\alpha_{t-1}}\mathbf{x}_0 + \sqrt{1 - \alpha_{t-1}}\frac{\mathbf{x}_t - \sqrt{\alpha_t}\mathbf{x}_0}{\sqrt{1 - \alpha_t}}, 0) \approx q(\mathbf{x}_{t-1}|\mathbf{x}_t, \mathbf{x}_0). \tag{1}$$

In this paper, we employ a conditional DDIM [25] for generating images, where the input is represented by a latent variable, $\mathbf{z} = (\bar{\mathbf{z}}, \mathbf{x}_T)$. This variable consists of two components: the high-level semantic latent code, $\bar{\mathbf{z}}$, generated by the conditional semantic encoder, and the low-level semantic latent maps, $\mathbf{x}_T$, generated through the forward process of conditional DDIM. During the image generation (reverse) process, the probability function is then written as $p_\theta(\mathbf{x}_{t-1}|\mathbf{x}_t, \bar{\mathbf{z}})$.

Drawing from Eq. (1), we define the generative process using the following equations:

$$p_\theta(\mathbf{x}_{0:T}|\bar{\mathbf{z}}) = p(\mathbf{x}_T) \prod_{t=1}^{T} p_\theta(\mathbf{x}_{t-1}|\mathbf{x}_t, \bar{\mathbf{z}}). \tag{2}$$

$$p_\theta(\mathbf{x}_{t-1}|\mathbf{x}_t,\bar{\mathbf{z}}) = \begin{cases} \mathcal{N}(\mathbf{f}_\theta^{(1)}(\mathbf{x}_1,\bar{\mathbf{z}}),\mathbf{0}) & \text{if } t=1 \\ q(\mathbf{x}_{t-1}|\mathbf{x}_t,\mathbf{f}_\theta^{(t)}(\mathbf{x}_t,\bar{\mathbf{z}})) & \text{otherwise} \end{cases}, \tag{3}$$

where

$$\mathbf{f}_\theta^{(t)}(\mathbf{x}_t,\bar{\mathbf{z}}) = \frac{1}{\sqrt{\alpha_t}}(\mathbf{x}_t - \sqrt{1-\alpha_t}\epsilon_\theta(\mathbf{x}_t,t,\bar{\mathbf{z}})). \tag{4}$$

where $\epsilon_\theta(\mathbf{x}_t,t,\bar{\mathbf{z}})$ is a pre-trained UNet model in [25]. Like the DDIM proposed by Song *et al.* [31], conditional DDIM follows a similar deterministic generative process:

$$\mathbf{x}_{t-1} = \sqrt{\alpha_{t-1}}\left(\frac{\mathbf{x}_t - \sqrt{1-\alpha_t}\epsilon_\theta(\mathbf{x}_t,t,\bar{\mathbf{z}})}{\sqrt{\alpha_t}}\right) + \sqrt{1-\alpha_{t-1}}\epsilon_\theta(\mathbf{x}_t,t,\bar{\mathbf{z}}). \tag{5}$$

3.3 Diffusion-Based Encoder

The diffusion-based encoder is the deterministic generative process of DDIM, which is used to encode an input image $\mathbf{x}_0$ into a lower semantic subcode $\mathbf{x}_T$. The following equation can represent the generative process:

$$\mathbf{x}_t = \sqrt{\alpha_t}\mathbf{f}_\theta(\mathbf{x}_{t-1},t-1,\bar{\mathbf{z}}) + \sqrt{1-\alpha_t}\epsilon_\theta(\mathbf{x}_{t-1},t-1,\bar{\mathbf{z}}). \tag{6}$$

Due to its restricted semantic information, $\mathbf{x}_T$ remains unchanged and does not participate in the optimization process for anonymizing image identities.

4 Methodology

This paper proposes a novel method for anonymizing faces in a given real-face dataset based on the diffusion model to gain better image utility. The proposed method optimizes the latent codes of images in the dataset within the latent space of a pre-trained conditional diffusion model [25]. The method involves creating a fake dataset $\mathcal{X}_F$ by randomly generating a large set of fake images such that $|\mathcal{X}_F| > |\mathcal{X}_R|$, where $\mathcal{X}_R$ is the real dataset. To obtain meaningful initial values for the latent codes that will be optimized to create the anonymized version of the real dataset $\mathcal{X}_A$, the real images from $\mathcal{X}_R$ are paired with the fake ones from $\mathcal{X}_F$ in the feature space of the ViT-based FaRL [49] image attribute encoder. The method then optimizes the successfully paired latent codes using two loss functions: (1) the identity loss, denoted by $\mathcal{L}_{id}$, ensures that the fake images remain a certain distance away from the real ones in terms of identity, and (2) the attribute preservation loss, denoted by $\mathcal{L}_{att}$, pulls the fake images closer to the real ones in the feature space of the FaRL [49] image encoder. In this way, the anonymized images inherit the attribute information of the real ones while possessing a different identity (Fig. 2).

The rest of the section is organized as follows: Sect. 4.1 briefly introduces the modules used in our framework. Section 4.2 shows the details of the fake dataset generation, Sect. 4.3 discusses the pairing and Sect. 4.4 presents the anonymization process and the losses.

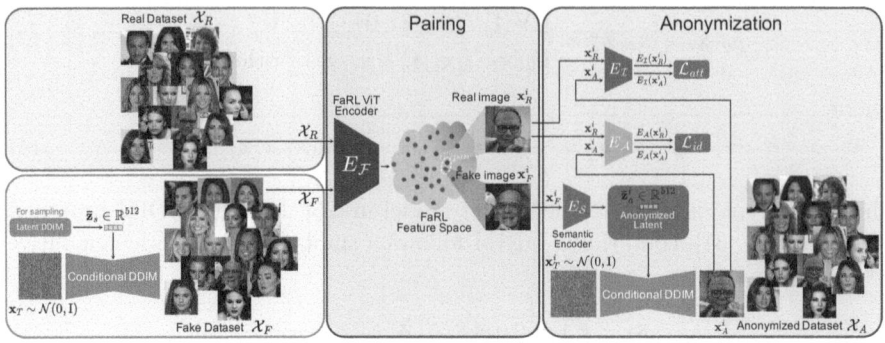

Fig. 2. The overview of face image privacy protection framework. Our protection framework optimizing the trainable anonymized latent vector $z_a^i \in \mathbb{R}^{512}$ using two loss functions, $\mathcal{L}_{id}$ and $\mathcal{L}_{att}$. This optimization aims to obfuscate the identity of the synthetic image x_A^i while maintaining the facial attributes.

4.1 Modules

Conditional DDIM. The conditional DDIM [25] image decoder takes a latent variable $z = (\bar{z}, x_T)$ as input, where $\bar{z}$ represents the high-level "semantic" subcode and x_T is the low-level "stochastic" subcode inferred by reversing the generative process of DDIM. Here, we keep the low-level stochastic subcode unchanged and optimize the image on the high-level semantic subcode.

ArcFace. To measure the similarity between the identities of two face images, we employ ArcFace [5], a method that maps images to a 512-dimensional feature space related to identity. Using the identity features, we optimize the semantic latent codes $\bar{z}$ of the conditional DDIM [25] to generate images that minimize the cosine similarity of the image identity features between real and fake images.

FaRL. FaRL [49] is a facial representation network trained in a contrastive manner on 20 million face image-text pairs for representing images in a meaningful and rich semantic feature space. The ViT image encoder of the FaRL framework is used to represent images in a 512-dimensional feature space and to find meaningful initial values for the latent codes that will be optimized to anonymize the real dataset. This approach provides a solid foundation for the anonymization of real datasets by capturing the underlying features of the input data and representing images in a high-dimensional feature space.

4.2 Fake Dataset Generation

Given a real face dataset $\mathcal{X}_R$, to generate a fake face dataset $\mathcal{X}_F$ such that $\mathcal{X}_F > \mathcal{X}_R$, we utilize the decoder of a conditional DDIM [25] as the generator. To sample $\bar{z}$ from the latent distribution and generate a synthetic dataset using diffusion autoencoders, we employ a pre-trained latent DDIM to approximate the latent distribution of $\bar{z}$. The latent distribution modelled by the latent DDIM is initially normalized to have a mean of zero and a variance of the unit. To perform unconditional sampling from a diffusion autoencoder, the process involves sampling from the latent DDIM and then unnormalizing it as $\bar{z}$. Next, $\mathbf{x}_T$ is drawn from a normal distribution with a mean of zero and a covariance matrix of $\mathbf{I}$, $\mathbf{x}_T \sim \mathcal{N}(\mathbf{0}, \mathbf{I})$. Finally, the fake image synthesis process involves using the decoder to decode $\mathbf{z} = (\bar{z}, \mathbf{x}_T)$.

4.3 Pairing

To pair real images in dataset $\mathcal{X}_R$ with fake images in dataset $\mathcal{X}_F$, we employ the pre-trained FaRL [49] ViT-based encoder $E_{\mathcal{F}}$. All images from both datasets are represented in a 512-dimensional feature space using the class token representation. This approach yields a robust feature representation of both datasets, which we use to train a kNN classifier. Using this classifier, we can identify the fake image in $\mathcal{X}_F$ that is closest to each real image in $\mathcal{X}_R$ based on the Euclidean distance. After generating and pairing the images, the real dataset $\mathcal{X}_R$ and the fake dataset $\mathcal{X}_F$ are paired, creating a set of pairs consisting of images from the real dataset $\mathcal{X}_R$ and images from the fake dataset $\mathcal{X}_F$.

4.4 Anonymization

To create an anonymized version $\mathcal{X}_A$ of the real dataset $\mathcal{X}_R$, we utilize the pairs of real-fake images obtained from the pairing section. These pairs consist of real and fake images that are semantically similar based on the FaRL image representation features. Specifically, for each pair, the real image $\mathbf{x}_R^i$ and its corresponding anonymized image $\mathbf{x}_A^i$, generated using the anonymized latent code $\mathbf{z} = (\bar{z}_A^i, \mathbf{x}_T^i)$, are used to calculate the proposed losses.

The identity loss $\mathcal{L}_{id}(\mathbf{x}_A^i, \mathbf{x}_R^i)$ ensures that the identity of $\mathbf{x}_A^i$ maintains a desired distance from the identity of $\mathbf{x}_R^i$. The attribute preservation loss $\mathcal{L}_{att}(\mathbf{x}_A^i, \mathbf{x}_R^i)$ enforces the preservation of facial attributes from the original image in the anonymized image.

For each pair consisting of a real image $\mathbf{x}_R^i$ and its anonymized version $\mathbf{x}_A^i$, i ranges from 1 to the total number of images in $\mathcal{X}_R$. The anonymization is achieved by optimizing the following losses:

Identity Loss. The identity loss is defined as

$$\mathcal{L}_{id} = \left| \frac{E_{\mathcal{A}}(\mathbf{x}_A^i) \cdot E_{\mathcal{A}}(\mathbf{x}_R^i)}{\left\| E_{\mathcal{A}}(\mathbf{x}_A^i) \right\|_2 \cdot \left\| E_{\mathcal{A}}(\mathbf{x}_R^i) \right\|_2} \right| = \left| \cos(E_{\mathcal{A}}(\mathbf{x}_A^i), E_{\mathcal{A}}(\mathbf{x}_R^i)) \right|, \tag{7}$$

where $E_{\mathcal{A}}$ is a pre-trained ArcFace [5] identity encoder used to extract identity features from facial images. $\cos(\cdot, \cdot)$ denotes the cosine similarity, which is a measure of similarity that calculates the cosine of the angle between two vectors. The loss yields a value ranging from 0 to 1. A value of 1 indicates that the vectors are identical, while a value of 0 signifies no similarity.

Attribute Loss. The attribute loss is defined as

$$\mathcal{L}_{att}(\mathbf{x}_A^i, \mathbf{x}_R^i) = \left\| E_{\mathcal{I}}(\mathbf{x}_A^i) - E_{\mathcal{I}}(\mathbf{x}_R^i) \right\|_1, \tag{8}$$

where $E_{\mathcal{I}}$ is the patch-level features, specifically the 14×14 768-dimensional features, obtained from the ViT-based image encoder introduced in the FaRL [49] paper. These 14×14 768-dimensional features are subsequently flattened into $14 \times 14 \times 768$-dimensional vectors for the purpose of loss calculation.

5   Experiments

In this section, we aim to investigate and evaluate four crucial dimensions of our framework: face detection accuracy, identity anonymity, image quality, and facial attribute preservation. Through a series of experiments, we will systematically analyze and assess the performance of our framework in these specific areas. This section is structured as follows: Sect. 5.1 provides an overview of the experimental preparation conducted. Section 5.2 presents a comparative analysis of our method and other state-of-the-art (SOTA) approaches using evaluation metrics. Lastly, Sect. 5.3 delves into an ablation study to further investigate and analyze the individual components of our proposed method.

5.1   Experiment Settings

Datasets. Anonymization process is conducted on two datasets: (i) **CelebA-HQ** [19], which comprises $30,000$ high-resolution (1024×1024) facial images of celebrities sourced from the CelebA dataset. These images exhibit diverse demographic attributes, including age, gender, and race. Additionally, each image is annotated with 40 attribute labels. (ii) **FFHQ** [14], is a curated collection of high-quality images depicting human faces. This dataset comprises $70,000$ PNG images with a resolution of 1024×1024 pixels. Notably, FFHQ exhibits significant diversity in terms of age, ethnicity, and background settings, providing a comprehensive range of facial characteristics.

State-Of-The-Art (SOTA). We conduct an evaluation of our anonymization framework against three SOTA methods, namely FALCO [1], CIAGAN [20], and DeepPrivacy [13]. In particular, we focus on the FALCO method, which employs latent code optimization techniques for dataset anonymization, thereby sharing similar objectives with our task.

Evaluation Metrics. We briefly introduce the evaluation metrics we used:

- *Face detection accuracy (Face Dete)* refers to the precision with which a face detection system or algorithm can identify and locate faces in an image. Accurate face detection is crucial for training machine learning models and is fundamental for various applications such as face recognition, emotion detection, facial expression analysis, and video surveillance. Therefore, facial datasets must provide a solid foundation for face detection tasks while maintaining diversity and robustness. In this study, we utilized state-of-the-art face detection models, specifically MTCNN [45] and dlib [15], to evaluate the performance of different datasets regarding face detection accuracy. An ideal anonymized dataset would demonstrate the same level of accuracy in face detection as the original dataset, indicating its potential to train face detection models comparable to those trained on the original data.

- *Identity anonymity (ID Anon)* refers to the extent to which individuals' personal identities are concealed or protected within a system or framework. Identity anonymity upholds privacy and confidentiality, especially in data sharing or analysis scenarios. This study employs FaceNet [27] to quantify identity anonymity. An ideal anonymized dataset would achieve 100% anonymity in face identification, indicating effective de-identification of each face.

- *Image quality* refers to an image's visual fidelity, clarity, and excellence. High-quality images accurately represent the original scene with minimal distortion or loss of details, facilitating accurate analysis and communication of visual information. This study employs the Fréchet Inception Distance (FID) [10] to quantify image quality, widely used in the field of generative adversarial networks (GANs) to assess the diversity and quality of generated images by analyzing their statistical properties. The Fréchet distance measures the dissimilarity between real and anonymized datasets. A lower FID score indicates higher image quality in the anonymized dataset.

- *Facial attribute preservation* ensures that the system retains and accurately represents distinctive facial characteristics post-anonymization. This balance between privacy and accurate characterization enables responsible facial data usage while maintaining individual privacy and application utility. Evaluation involves qualitative inspection and quantitative metrics like attribute classification accuracy. Our study adheres to [1], utilizing a pre-trained MobileNetV2 model for fair and consistent experiment comparisons.

5.2 Comparison to State-Of-The-Art (SOTA)

In this section, we present a comparative analysis of our proposed method against three SOTA techniques, namely CIAGAN [20], DeepPrivacy [13], and FALCO [1]. By benchmarking our proposed method against the most advanced and widely recognized approaches in the field, we aim to comprehensively evaluate its effectiveness and performance. This comparison enables us to assess the advancements made by our method and highlight its unique contributions in pushing the boundaries of the current SOTA.

Quantitative Evaluation. In this section, a comprehensive quantitative analysis is conducted to compare the effectiveness of our proposed method with other approaches. The specific numerical comparison results are presented in Tables 1 and 2.

Table 1. Anonymized image dataset evaluation results. Face Detection Accuracy (Face Dete), Identity Anonymity (ID Anon), Image Quality, and Attributes Classification.

	Face Dete ↑		ID Anon ↑		Image quality ↓	Attributes classification ↑	
	dlib(%)	MTCNN(%)	CASIA(%)	VGG(%)	FID	Inner face(%)	Outer face(%)
Original	98.55	99.91	0.00	0.00	0.00	85.64	85.21
CIAGAN [20]	−3.14	−0.35	**97.95**	**99.61**	31.11	80.22	80.72
DeepPrivacy [13]	+0.47	−0.13	97.29	98.12	28.32	80.29	79.55
FALCO [1]	+1.45	+0.09	96.71	97.27	28.19	80.16	79.5
Ours	**−0.29**	**−0.08**	97.55	98.21	**23.42**	**82.73**	**82.5**

Face Detection Accuracy (Face Dete). In this paragraph, we present a detailed explanation of the face detection accuracy values. The specific numerical comparison results for face detection accuracy, including our proposed method and other approaches, are provided in the first two columns of Table 1. The first row of Table 1 represents the percentage of accurately detected faces in the original dataset, as detected by the face detectors dlib [15] and MTCNN [45]. The last row illustrates the percentage of faces detected in our anonymized dataset by the same face detectors. The intermediate rows present the detection ratios achieved by employing SOTA anonymization methods.

It is worth noting that the original dataset exhibits a few instances where certain faces were not detected by the employed face detectors, resulting in a detection ratio of 98.55% for dlib [15] and 99.91% for MTCNN [45]. This can be attributed to the diverse range of facial poses in the original dataset, encompassing extreme angles and substantial occlusions. However, these challenging examples contribute to enhancing the robustness of face detectors when training them with such diverse data. We visualize this phenomenon in Fig. 3 and provide a further elaborate in Sect. 5.2.

Therefore, in the first two columns of Table 1, a detection rate closer to that of the original dataset indicates that the employed anonymization method better preserves the diversity of data. Remarkably, our proposed method outperforms other approaches, yielding detection rates of 98.26% (0.29% lower than the original dataset) and 99.83% (0.08% lower than the original dataset), respectively.

Identity Anonymity (ID Anon). To evaluate the efficacy of identity anonymity, we conducted facial identity verification using the FaceNet [27] network pretrained on two datasets, namely VGG [24] and CASIA [42]. The corresponding success rates for concealing identity information are summarized in Table 1.

Our approach achieved the second-highest success rate among the SOTA techniques. Notably, the CIAGAN method, which serves as a face inpainting technique, achieved the highest score. However, this accomplishment was accompanied by compromised image quality and a trade-off in preserving facial attributes.

Image Quality. The image quality results (FID [10]) of our approach in comparison to others are documented in the fifth column of Table 1.

A lower FID distance, also known as the FID score, indicates higher quality and diversity of the anonymized images. This implies that the distribution of the anonymized images closely aligns with the distribution of real images.

The image quality results are presented in Table 1. Our method demonstrated the lowest FID score of 23.42, showcasing a significant improvement of approximately 17% compared to the second lowest FALCO [1] method, which obtained a score of 28.19.

Facial Attribute Preservation. To evaluate the metric of 'facial attribute preservation' in the anonymized datasets, we conducted training on a CNN network using the anonymized datasets as the training set. Specifically, the pre-trained MobileNetV2 model is utilized in this study. The original dataset served as the ground truth for training and evaluation purposes and is split into training and testing sets. All models trained using different datasets were evaluated on the testing set of the original dataset. To ensure fairness, consistent strategies are applied across the different datasets, including data size and training parameter settings.

The facial attribute categories are derived from the attribute labels in the CelebA dataset [19]. To compare the effectiveness of facial attribute preservation with face inpainting methods, such as DeepPrivacy [13] and CIAGAN [20], we categorize the 38 attribute labels (excluding 'Attractive' and 'Blurry') into two groups: inner face labels and outer face labels, based on the corresponding facial regions represented by each attribute category. The specific categorizations are presented in Table 2.

The quantitative results of facial attribute preservation are presented in Tables 1 and 2, where the values indicate the percentage of correct classifications, representing the ratio of correctly predicted labels to the total number of predictions. The original dataset is considered as the ground truth in the tables, serving as the baseline for comparing the anonymized datasets. The higher classification accuracy indicates better performance of the models trained on a particular dataset. This implies that the anonymized dataset possesses a facial attribute distribution that is closer to the ground truth, thereby enabling more effective support for various machine-learning tasks.

In Table 1, the last two columns illustrate the classification accuracies for inner face and outer face attributes, respectively. Under this setting, the original dataset achieves accuracies of 85.64% and 85.21% for inner face and outer face attributes, respectively. Our dataset achieves accuracies of 82.73% and 82.5% under the same settings. Compared with the second-best results, our dataset demonstrates an improvement of 2.44% in accuracy for inner face attributes (compared to DeepPrivacy [13]) and an increase of 1.78% for outer face attributes (compared to CIAGAN [20]).

Table 2 presents the classification accuracies for different datasets across all labels. A total of 38 label categories are utilized for evaluation. Our approach achieves the highest prediction accuracy for 31 label categories, while CIAGAN [20] achieves the highest accuracy for 6 categories, and FALCO [1] attains the highest score for the '5 o'Clock Shadow' label.

Table 2. Facial attribute preservation results.

		Original Ours	FALCO [1]	Deep [13]	CIAGAN [20]	
Inner face region	5 o Clock Shadow	<u>85.45</u>	81.21	**82.05**	79.56	77.60
	Arched Eyebrows	<u>86.43</u>	**84.13**	80.01	81.58	82.03
	Bags Under Eyes	<u>85.69</u>	**82.34**	79.54	79.15	80.55
	Big Lips	<u>85.15</u>	**82.59**	79.70	80.54	80.62
	Big Nose	<u>85.14</u>	**82.12**	79.60	78.72	78.69
	Bushy Eyebrows	<u>84.87</u>	80.98	81.18	78.72	**81.85**
	Eyeglasses	<u>83.74</u>	**81.71**	79.77	79.98	80.10
	Goatee	<u>83.46</u>	79.38	79.32	76.08	**80.22**
	Heavy Makeup	<u>87.51</u>	**85.33**	80.91	83.03	80.32
	High Cheekbones	<u>86.13</u>	83.22	78.75	80.16	**84.03**
	Male	<u>85.64</u>	**82.37**	82.19	80.17	82.02
	Mouth Slightly Open	<u>86.21</u>	**83.03**	79.21	80.72	79.50
	Mustache	<u>83.13</u>	**79.35**	78.47	75.88	77.13
	Narrow Eyes	<u>83.81</u>	**80.82**	77.89	78.41	77.08
	No Beard	<u>87.22</u>	**85.00**	82.01	83.05	80.82
	Pale Skin	<u>86.07</u>	**84.49**	82.72	83.27	77.93
	Pointy Nose	<u>85.81</u>	**83.26**	80.36	81.40	79.93
	Rosy Cheeks	<u>86.59</u>	**83.28**	77.07	79.88	81.75
	Smiling	<u>86.03</u>	**82.76**	78.55	79.76	80.26
	Wearing Lipstick	<u>87.36</u>	**85.29**	81.27	83.12	81.80
	Young	<u>86.96</u>	**84.55**	82.72	82.84	80.35
Outer face region	Bald	<u>83.03</u>	**80.38**	77.20	75.57	79.06
	Bangs	<u>86.63</u>	**84.02**	80.01	81.47	82.59
	Black Hair	<u>85.12</u>	**82.63**	82.37	79.60	80.76
	Blond Hair	<u>88.35</u>	**85.93**	81.47	84.99	82.35
	Brown Hair	<u>86.21</u>	**84.03**	81.35	82.19	80.92
	Chubby	<u>82.61</u>	**79.91**	76.60	75.18	79.32
	Double Chin	<u>83.32</u>	**79.81**	75.25	74.88	79.29
	Gray Hair	<u>85.55</u>	**83.55**	78.45	78.39	78.64
	Oval Face	<u>85.54</u>	**83.12**	80.37	81.04	79.98
	Receding Hairline	<u>84.32</u>	82.08	78.94	78.36	**82.14**
	Sideburns	<u>84.56</u>	79.81	80.27	76.77	**81.42**
	Straight Hair	<u>85.94</u>	**83.14**	81.87	81.01	80.50
	Wavy Hair	<u>86.4</u>	**83.98**	80.27	82.88	81.39
	Wearing Earrings	<u>85.83</u>	**84.20**	78.82	80.60	82.11
	Wearing Hat	<u>85.9</u>	**82.72**	81.08	81.59	79.85
	Wearing Necklace	<u>85.17</u>	**82.86**	78.24	80.47	79.32
	Wearing Necktie	<u>84.03</u>	80.40	78.89	77.42	**82.58**

Through comparison, it is evident that our dataset exhibits superior facial attribute preservation performance.

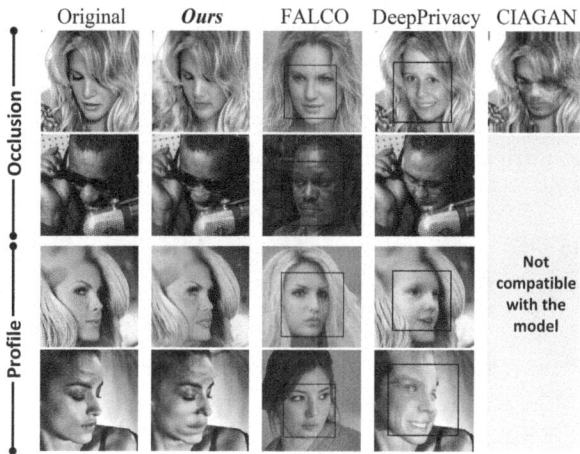

Fig. 3. Comparative analysis of visual effects: Our method vs. FALCO [1], DeepPrivacy [13], and CIAGAN [20], in the context of original images undetectable by the dlib [15] frontal face detector.

Qualitative Evaluation. In this section, we conduct a comprehensive qualitative evaluation to compare the efficacy of our proposed method with other approaches.

In the context of qualitative evaluation, visual quality serves as a crucial criterion for comparing our approach with other SOTA methods and demonstrating the observed effects. The primary visual performances utilized for presentation are displayed in Figs. 3 and 4. Notably, Fig. 3 focuses specifically on showcasing unique instances extracted from the original dataset, and we delve into its analysis in the subsequent discussion.

Figure 3 illustrates the results of face detection using dlib's frontal face detector (dlib [15]). The face detection boxes, delineated by black boxes, indicate successful detections, while images without the black boxes signify instances where the face detection failed. It was noted that certain face images in the real dataset were not detected by the dlib frontal face detector [15]. Consequently, we conducted a comprehensive investigation of this particular subset of data, identifying two broad categories: faces turned at an extreme angle and faces occluded by obstacles. It is pertinent to mention that although these images constitute only a small fraction of the face dataset employed in our experiments, they are frequently encountered in real-world scenarios. Therefore, it becomes imperative for sophisticated machine learning models to effectively handle these unique samples during the training phase by assigning appropriate weights, thereby optimizing the overall performance.

Upon meticulous examination of these samples, an intriguing observation emerged: our anonymized dataset exhibits detections that closely resemble those of the original dataset. In contrast, when processing these images, both the FALCO [1] and the DeepPrivacy models [13] tend to treat them as normal face images, erroneously introducing facial regions that were not present in the original images. Moreover, the data preprocessing of CIAGAN [20], which relies on facial key points information, proved inefficient in handling these images. Thus, we demonstrate that our anonymized dataset outperforms alternative approaches in preserving the intricate nature of facial poses.

Fig. 4. Comparative analysis of visual effects: Our method vs. FALCO [1], DeepPrivacy [13], and CIAGAN [20].

Next, we present additional visual comparison images, as depicted in Fig. 4. Subsequently, we explain the comparison results from three distinct perspectives. Firstly, regarding image quality, our generated and FLACO images demonstrate a superior level of realism compared to the images produced by others. This enhancement can be attributed to the remarkable capabilities of the generation model in retaining intricate details, such as the accurate handling of teeth processing. Secondly, concerning the diversity of generated faces, our approach showcases a higher retention of attributes from the original dataset, encompassing expressions, skin colour, age, and other pertinent features. This advantage can be attributed to the inherent strengths of the Diffusion model in effectively handling complex attributes during the generation process. Lastly, we prioritize privacy protection by effectively safeguarding potentially sensitive information in the image backgrounds. This distinguishes our method from DeepPrivacy [13] and CIAGAN [20], which may overlook such privacy concerns. Concurrently, we aim to maintain the dataset's diversity to the utmost extent possible, setting our approach apart from FALCO [1], which may inadvertently compromise diversity during the generation process.

Original *Ours IRS* *Ours IIS* FALCO DeepPrivacy CIAGAN

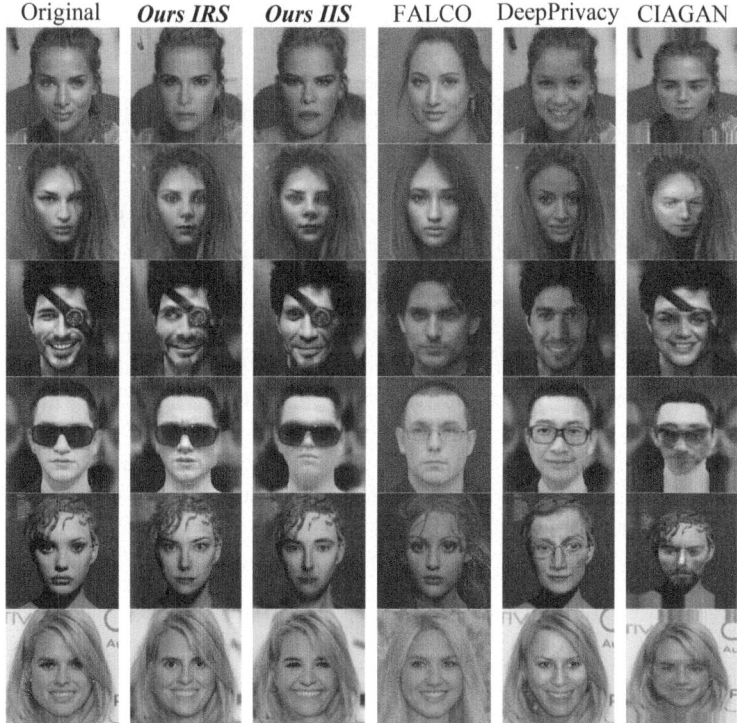

Fig. 5. Visual quality of ablation study. **Ours** compared to FALCO [1], DeepPrivacy [13], CIAGAN [20].

5.3 Ablation Study

This section presents an ablation study conducted to investigate optimization methods and identify the most effective approach for generating anonymized datasets.

In the domain of image privacy protection utilizing optimization techniques, two primary categories of optimization methods have emerged: the class replacement scheme (CRS) and the class indistinguishable scheme (CIS). These methods differ in their fundamental approach to handling the classes to be protected. The CRS adopts a new label as the optimization target, aiming to induce the misclassification of the protected class into the new class by the DNN. Conversely, the CIS seeks to maximize the class loss function to achieve the incorrect classification.

For our specific task of safeguarding facial identity information, we explore the impact of using different optimization methods on the dataset. To facilitate discussions, we refer to these two approaches as the identity replacement scheme (IRS) and the identity indistinguishable scheme (IIS). More specifically, the main difference between the IRS and the IIS lies in their respective approaches to handling identity loss. In the case of the IRS, we select a target face from the fake dataset and aim to minimize the identity distance between the anonymized face and the specified target face. The identity loss of the IRS is defined as follows:

$$\mathcal{L}_{IRS} = 1 - \left| \frac{E_{\mathcal{A}}(\mathbf{x}_A^i) \cdot E_{\mathcal{A}}(\mathbf{x}_F^i)}{\left\| E_{\mathcal{A}}(\mathbf{x}_A^i) \right\|_2 \cdot \left\| E_{\mathcal{A}}(\mathbf{x}_F^i) \right\|_2} \right| = 1 - \left| \cos(E_{\mathcal{A}}(\mathbf{x}_A^i), E_{\mathcal{A}}(\mathbf{x}_F^i)) \right|, \quad (9)$$

where $E_{\mathcal{A}}$ is a pre-trained ArcFace [5] identity encoder. $\mathbf{x}_A^i \in \mathcal{X}_A$ and $\mathbf{x}_F^i \in \mathcal{X}_F$ are the images from the anonymized dataset and the images from the fake dataset respectively. $\cos(\cdot, \cdot)$ denotes the cosine similarity. The loss yields a value ranging from 0 to 1, where 0 indicates that the vectors are identical, while 1 signifies no similarity.

In the IIS, our objective is to minimize the cosine similarity between the images from the real and anonymized datasets. This is achieved by optimizing the identity loss, as depicted in (7).

Table 3. Ablation study evaluation results. Face Detection Accuracy (Face Dete), Identity Anonymity (ID Anon), Image Quality, and Attributes Classification.

| | Face Dete ↑ | ID Anon ↑ | | Image quality ↓ | Attributes ↑ |
	dlib(%)	CASIA(%)	VGG(%)	FID	Average(%)
Ours IRS	**+0.13**	95.33	96.10	**21.17**	**83.12**
Ours IIS	−0.29	**97.55**	**98.21**	23.42	82.63

Based on the results presented in Table 3, each of the two optimization methods, namely the IRS and the IIS, possesses its own advantages and limitations. Specifically, IRS demonstrates certain improvements in image quality, whereas IIS outperforms IRS in preserving facial identity.

The relatively lower privacy protection rate observed in the IRS can be attributed to the possibility of selected facial features in the images being similar to those in the original dataset. We will delve deeper into this aspect in a subsequent section for further analysis. Moreover, Fig. 5 shows visual comparisons with SOTA methods, offering a visual performance achieved by the different approaches.

6   Conclusions

In this paper, we propose a novel framework for image anonymization that targets the optimization of latent space vectors with the help of pre-trained diffusion models. Our approach leverages an identity loss function and an attribute preserving loss function to operate directly in the latent space of the pre-trained diffusion model, thereby eliminating the necessity for training intricate networks. Through extensive experimentation, we demonstrate the efficacy of our method in successfully anonymizing the identity of images while preserving facial attributes compared to existing SOTA techniques.

References

1. Barattin, S., Tzelepis, C., Patras, I., Sebe, N.: Attribute-preserving face dataset anonymization via latent code optimization. In: Proceedings of the IEEE Conference on Computer Vision and Pattern Recognition (CVPR), June 2023 (2023)
2. Cao, J., Liu, B., Wen, Y., Xie, R., Song, L.: Personalized and invertible face de-identification by disentangled identity information manipulation. In: Proceedings of the IEEE/CVF International Conference on Computer Vision (ICCV), October 2021, pp. 269–272 (2021)
3. Cao, J., Liu, B., Wen, Y., Xie, R., Song, L.: Achieving privacy-preserving multi-view consistency with advanced 3D-aware face de-identification. In: Proceedings of ACM Multimedia Asia, pp. 1–6. ACM (2023)
4. Cao, J., Liu, B., Wen, Y., Zhu, Y., Xie, R., Song, L.: Hiding among your neighbors: face image privacy protection with differential private k-anonymity. In: Proceedings of the IEEE International Symposium on Broadband Multimedia Systems and Broadcasting (BMSB), pp. 1–6. IEEE (2022)
5. Deng, J., Guo, J., Xue, N., Zafeiriou, S.: ArcFace: additive angular margin loss for deep face recognition. In: Proceedings of the IEEE/CVF Conference on Computer Vision and Pattern Recognition (CVPR), June 2019 (2019)
6. Du, L., Zhang, W., Fu, H., Ren, W., Zhang, X.: An efficient privacy protection scheme for data security in video surveillance. J. Vis. Commun. Image Represent. **59**, 347–362 (2019). https://doi.org/10.1016/j.jvcir.2019.01.027
7. Eskimez, S.E., Maddox, R.K., Xu, C., Duan, Z.: Generating talking face landmarks from speech. In: Deville, Y., Gannot, S., Mason, R., Plumbley, M.D., Ward, D. (eds.) LVA/ICA 2018. LNCS, vol. 10891, pp. 372–381. Springer, Cham (2018). https://doi.org/10.1007/978-3-319-93764-9_35
8. Gerstner, T., DeCarlo, D., Alexa, M., Finkelstein, A., Gingold, Y., Nealen, A.: Pixelated image abstraction with integrated user constraints. Comput. Graph. **37**(5), 333–347 (2013)
9. Goodfellow, I., et al.: Generative adversarial networks, October 2020
10. Heusel, M., Ramsauer, H., Unterthiner, T., Nessler, B., Hochreiter, S.: GANs trained by a two time-scale update rule converge to a local Nash equilibrium (2018)
11. Ho, J., Jain, A., Abbeel, P.: Denoising diffusion probabilistic models. In: Proceedings of the 34th International Conference on Neural Information Processing Systems, NIPS 2020. Curran Associates Inc., Red Hook, NY, USA (2020)
12. Ho, J., Saharia, C., Chan, W., Fleet, D.J., Norouzi, M., Salimans, T.: Cascaded diffusion models for high fidelity image generation. J. Mach. Learn. Res. **23**(47), 1–33 (2022)
13. Hukkelås, H., Mester, R., Lindseth, F.: DeepPrivacy: a generative adversarial network for face anonymization (2019)
14. Karras, T., Laine, S., Aila, T.: A style-based generator architecture for generative adversarial networks (2019)
15. King, D.E.: Dlib-ml: a machine learning toolkit. J. Mach. Learn. Res. **10**, 1755–1758 (2009)

16. Li, H., et al.: SRDiff: single image super-resolution with diffusion probabilistic models. Neurocomputing **479**, 47–59 (2022)
17. Li, T., Lin, L.: AnonymousNet: natural face de-identification with measurable privacy, June 2019
18. Liu, B., et al.: DP-image: differential privacy for image data in feature space (2021)
19. Liu, Z., Luo, P., Wang, X., Tang, X.: Deep learning face attributes in the wild, December 2015
20. Maximov, M., Elezi, I., Leal-Taixe, L.: CIAGAN: conditional identity anonymization generative adversarial networks. In: 2020 IEEE/CVF Conference on Computer Vision and Pattern Recognition (CVPR), June 2020. IEEE (2020)
21. McPherson, R., Shokri, R., Shmatikov, V.: Defeating image obfuscation with deep learning (2016)
22. Mirza, M., Osindero, S.: Conditional generative adversarial nets (2014)
23. Oh, S.J., Benenson, R., Fritz, M., Schiele, B.: Faceless person recognition: privacy implications in social media. In: Leibe, B., Matas, J., Sebe, N., Welling, M. (eds.) ECCV 2016. LNCS, vol. 9907, pp. 19–35. Springer, Cham (2016). https://doi.org/10.1007/978-3-319-46487-9_2
24. Parkhi, O.M., Vedaldi, A., Zisserman, A.: Deep face recognition. In: British Machine Vision Conference (2015)
25. Preechakul, K., Chatthee, N., Wizadwongsa, S., Suwajanakorn, S.: Diffusion autoencoders: toward a meaningful and decodable representation. In: IEEE Conference on Computer Vision and Pattern Recognition (CVPR) (2022)
26. Richardson, E., et al: Encoding in style: a StyleGAN encoder for image-to-image translation (2021)
27. Schroff, F., Kalenichenko, D., Philbin, J.: FaceNet: a unified embedding for face recognition and clustering. In: Proceedings of the IEEE Conference on Computer Vision and Pattern Recognition (CVPR), June 2015 (2015)
28. EU Personal Data Protection in Policy and Practice. ITLS, vol. 29. T.M.C. Asser Press, The Hague (2019). https://doi.org/10.1007/978-94-6265-282-8_9
29. Shan, S., Wenger, E., Zhang, J., Li, H., Zheng, H., Zhao, B.Y.: Fawkes: protecting personal privacy against unauthorized deep learning models. In: Proceedings of the USENIX Security (2020)
30. Sinha, A., Song, J., Meng, C., Ermon, S.: D2C: diffusion-decoding models for few-shot conditional generation. In: Ranzato, M., Beygelzimer, A., Dauphin, Y., Liang, P., Vaughan, J.W. (eds.) Advances in Neural Information Processing Systems, vol. 34, pp. 12533–12548. Curran Associates, Inc. (2021)
31. Song, J., Meng, C., Ermon, S.: Denoising diffusion implicit models. In: International Conference on Learning Representations (2021)
32. Sun, Q., Ma, L., Oh, S.J., Gool, L.V., Schiele, B., Fritz, M.: Natural and effective obfuscation by head inpainting (2018)
33. Tolosana, R., Vera-Rodriguez, R., Fierrez, J., Morales, A., Ortega-Garcia, J.: Deep-Fakes and beyond: a survey of face manipulation and fake detection. Inf. Fus. **64**, 131–148 (2020)
34. Tov, O., Alaluf, Y., Nitzan, Y., Patashnik, O., Cohen-Or, D.: Designing an encoder for StyleGAN image manipulation. arXiv preprint arXiv:2102.02766 (2021)
35. Wen, Y., Liu, B., Cao, J., Xie, R., Song, L.: Divide and conquer: a two-step method for high quality face de-identification with model explainability. In: Proceedings of the IEEE/CVF International Conference on Computer Vision (ICCV), October 2023, pp. 269–272 (2023)

36. Wen, Y., Liu, B., Cao, J., Xie, R., Song, L., Li, Z.: IdentityMask: deep motion flow guided reversible face video de-identification. IEEE Trans. Circ. Syst. Video Technol. **32**(12), 8353–8367 (2022). https://doi.org/10.1109/TCSVT.2022.3191982
37. Wen, Y., Liu, B., Ding, M., Xie, R., Song, L.: IdentityDP: differential private identification protection for face images. Neurocomputing **501**, 197–211 (2022)
38. Wen, Y., Liu, B., Song, L., Cao, J., Xie, R.: Face De-identification: Safeguarding Identities in the Digital Era. Springer, Heidelberg (2024)
39. Xue, H., Liu, B., Din, M., Song, L., Zhu, T.: Hiding private information in images from AI. In: 2020 IEEE International Conference on Communications (ICC), ICC 2020, pp. 1–6 (2020)
40. Xue, H., Liu, B., Yuan, X., Ding, M., Zhu, T.: Face image de-identification by feature space adversarial perturbation. Concurrency Comput. Pract. Exp. **35**(5), e7554 (2023)
41. Yeh, R.A., Chen, C., Lim, T.Y., Schwing, A.G., Hasegawa-Johnson, M., Do, M.N.: Semantic image inpainting with deep generative models (2017)
42. Yi, D., Lei, Z., Liao, S., Li, S.Z.: Learning face representation from scratch. arXiv preprint arXiv:1411.7923 (2014)
43. Yu, X., Chinomi, K., Koshimizu, T., Nitta, N., Ito, Y., Babaguchi, N.: Privacy protecting visual processing for secure video surveillance. In: 2008 15th IEEE International Conference on Image Processing, pp. 1672–1675 (2008)
44. Zhang, G., Liu, B., Zhu, T., Zhou, A., Zhou, W.: Visual privacy attacks and defenses in deep learning: a survey. Artif. Intell. Rev. (2022)
45. Zhang, K., Zhang, Z., Li, Z., Qiao, Y.: Joint face detection and alignment using multitask cascaded convolutional networks. IEEE Sig. Process. Lett. **23**(10), 1499–1503 (2016)
46. Zhang, Y., Lu, Y., Nagahara, H., Taniguchi, R.i.: Anonymous camera for privacy protection. In: 2014 22nd International Conference on Pattern Recognition, pp. 4170–4175 (2014)
47. Zhao, Y., Chen, J.: A survey on differential privacy for unstructured data content. ACM Comput. Surv. **54**(10s) (2022)
48. Zhao, Y., Yuan, D., Du, J.T., Chen, J.: Geo-ellipse-indistinguishability: community-aware location privacy protection for directional distribution. IEEE Trans. Knowl. Data Eng., 1–11 (2022)
49. Zheng, Y., et al.: General facial representation learning in a visual-linguistic manner. In: Proceedings of the IEEE/CVF Conference on Computer Vision and Pattern Recognition (CVPR), June 2022, pp. 18697–18709 (2022)
50. Zhu, Y., Cao, J., Liu, B., Chen, T., Xie, R., Song, L.: Identity-consistent video de-identification via diffusion autoencoders. In: Proceedings of the IEEE International Symposium on Broadband Multimedia Systems and Broadcasting (BMSB), pp. 1–6. IEEE (2024)

A Modular Sharing Scheme for EMRs Using Consortium Blockchain and Proxy Re-encryption

Jiazheng Quan⑩, Niefeng Wu⑩, Hu Chen⑩, Yamei Wang⑩,
and Yuexin Zhang$^{(\boxtimes)}$⑩

Fujian Provincial Key Laboratory of Network Security and Cryptology, College of
Computer and Cyber Security, Fujian Normal University, Fuzhou, China
yxzhang@fjnu.edu.cn

Abstract. In medical referrals, it is critical to securely and efficiently share Electronic Medical Records (EMRs) for timely and accurate treatments. The decentralization and tamper-proof of blockchain provides new options for secure sharing of diagnosis and treatment data. In this paper, we propose a modular sharing scheme for EMRs using consortium blockchain and proxy re-encryption. Specifically, to meet security and privacy requirements of medical referrals under the healthcare consortium model, cloud servers are employed to store the encrypted EMRs. Additionally, modular encryption is designed to realize fine-grained access control. In our scheme, consortium blockchain is employed to store patients' index key group, and it is responsible for performing proxy re-encryption operations. Such design ensures secure access control of patients' EMRs, and doctors within the medical consortium model can access and/or update EMRs after obtaining authorization from patients. We analyse the security and performance of our scheme, and compare it with a few relevant schemes. According to the comparisons, our scheme has more functionalities, and it achieves fine-grained access control of patients' EMRs. Moreover, resource consumptions of patients and doctors are light. Specifically, time cost of our scheme is reduced at least 25%.

Keywords: Proxy Re-Encryption (PRE) · consortium blockchain · fine-grained access control · Electronic Medical Records (EMRs)

1   Introduction

With rapid developments of cryptography, blockchain, parallel and distributed technologies, digitization of medical information is gradually gaining popularity. Specifically, medical consortium model is vigorously promoted in recent years. Typically, the medical consortium has a tertiary hospital as the lead unit, combined with a number of secondary hospitals, rehabilitation hospitals, nursing hospitals and community health service centers, to build "$1 + X$" medical consortium. The main role of medical consortium model is to support medical referrals

© The Author(s), under exclusive license to Springer Nature Singapore Pte Ltd. 2025
T. Zhu et al. (Eds.): ICA3PP 2024, LNCS 15251, pp. 60–79, 2025.
https://doi.org/10.1007/978-981-96-1525-4_4

for timely and accurate treatment. For instance, when a patient has a referral, the receiving hospital requests information of the patient's past medical history, such as ultrasound examination and blood tests, which is particularly important for the patient's prognosis and treatment. In practice, highly sensitive health-related information exist in patients' EMRs, such as histories of medications. Thus, it is never a trivial task to securely and efficiently share patients' EMRs in medical consortium model.

In this paper, we develop a modular sharing scheme for EMRs using consortium blockchain and proxy re-encryption. Specifically, consortium blockchain is employed to store the indexed key groups of patients, and cloud servers are employed to store the encrypted EMRs. Such design ensures bidirectional and efficient referrals between upper and lower level hospitals in medical consortium model. Moreover, doctors in the same department jointly decide a patient's EMRs needed in their department, and the doctors in other departments cannot access the patient's EMRs, which achieves fine-grained access control while protecting security and privacy of patients' EMRs. Main contributions of this paper are summarized as follows:

1. *Improved flexibility and scalability.* We propose a modular sharing scheme for EMRs using consortium blockchain and proxy re-encryption. Specifically, our scheme utilizes cloud servers to store the encrypted EMRs, and employs consortium blockchain to store the indexed key groups. According to evaluations, the design greatly reduces storage burden of consortium blockchain, and improves flexibility and scalability.

2. *Fine-grained access control.* Our scheme provides mechanism for modular encryption of EMRs based on the tags of doctors' affiliated departments. The mechanism achieves fine-grained access control. Typically, it allows any doctor within medical consortium model to update and access patients' EMRs after obtaining patients' authorization, avoiding information occlusion and improving the utilization of medical resources.

3. *High performances.* In our scheme, patient's ownership of EMRs is strengthened by combining consortium blockchain and proxy re-encryption. Additionally, the symmetric key group is uplinked and stored after connecting index items, and patients do not need to manage keys. We analyse the security and performance of our scheme, and compare it with relevant schemes. According to the comparisons, resource consumptions of patients and doctors are light. Specifically, time cost of our scheme is reduced at least 25%.

The rest of this paper is organized as follows. In Sect. 2, a few related works are reviewed. Section 3 introduces preliminaries employed in this scheme. Section 4 presents the detailed operations of our scheme. Section 5 provides security and performance analysis, and makes comparisons. The conclusions are drawn in Sect. 6.

2   Related Work

The characteristics of blockchain, including decentralization, tamper-proof, and traceability, have made secure sharing of patients' EMRs becomes possible in a semi-trusted environment [1]. Typically, multi-factor authentication technology is widely deployed in mobile devices applications, and it provides the first line of defense for secure sharing of EMRs within medical consortium model [2]. For instance, Liu et al. [3] proposed a blockchain-based multi-keyword searchable encryption scheme, say MKIPSE, to provide adequate privacy protection and efficient ciphertext retrieval of EMRs. In this scheme, local servers of medical organizations were considered as the trusted entities, and they formed a consortium blockchain. The scheme supported keyword retrieval in blockchain while storing ciphertexts in cloud servers. Such a collaborative model effectively reduced computational and storage costs of cloud servers.

Additionally, Azaria et al. [4] proposed MedRec, an access control scheme using Attribute-Based Encryption (ABE) and blockchain. Specifically, [4] utilized attribute-based cryptography to encrypt keys for fine-grained access control. At the same time, two private blockchains were employed for secure sharing of EMRs. In practical applications, the operation and maintenance costs of private blockchains are high. Comparing with private and public blockchains, consortium blockchain is more in line with actual deployment requirements of secure EMRs sharing under medical consortium model. For example, Zhang et al. [5] employed a dual-blockchain structure and designed a secure sharing scheme for medical records, i.e. EMRSBC, which separated storage and sharing of medical data. Specifically, a consortium blockchain was employed by medical organizations to store source data of medical records, and a private blockchain was employed to store digital summaries of records. In practice, however, data compatibility among medical organizations are poor. As a result, it is not easy to realize bidirectional and efficient referrals between upper and lower level hospitals.

Wang et al. [6] designed a secure and lightweight cloud IoT user authentication scheme. It ensured remote control and real-time data access in cloud-assisted IoT systems. Additionally, it improved the execution efficiency by offloading resource-consuming tasks to cloud servers. Moreover, Xu et al. in [7] proposed EPPFM for protecting EMRs in a multi-user environment. Specifically, [7] provided different keys for each patient and used matrix-based proxy re-encryption to achieve flexible authorization and revocation. Yao et al. proposed IBPRE-SHCD-MHCE in [8]. It was a proxy re-encryption scheme that supported both single-hop conditional delegation and multi-hop ciphertext evolution. According to the analysis, it achieved CCA-security and was resistant to conspiracy attacks. However, its computational cost was relatively high. Most of existing schemes utilized a single blockchain to achieve data sharing. However, in referral applications, the amount of patients' medical data are huge. As a result, storing a large amount of diagnosis and treatment information lead to a drastic increase in storage burden of blockchain.

Gautam et al. [9] combined obfuscation algorithm with RSA algorithm, and designed EMRs storage framework based on clouds. However, the computational cost of [9] was proportional to key length [10]. Additionally, Kwame et al. [11] proposed a blockchain-based secure proxy re-encryption scheme for IoT data sharing. However, this scheme did not realize fine-grained access control of data by requesters, and it failed to effectively protect the privacy of data owners. Peng et al. [12] proposed a secure sharing architecture for EMRs based on a dual blockchain system, and developed an identity-based three-party authentication key negotiation scheme TAKA. This scheme used a dual blockchain system constructed using medical blockchain and regulatory blockchain to protect patients' privacy and data security. Moreover, it enabled doctors to urgently access the EMRs of comatose patients. Niu et al. [13] designed EMRs sharing scheme based on CP-ABE algorithm, and it achieved fine-grained access control and secure medical data sharing.

3   Preliminaries

In this section, we review preliminaries including blockchain, proxy re-encryption, Nyberg-Rueppel digital signature, and system model.

3.1   Consortium Blockchain

Blockchain integrates cryptography, distributed data storage, consortium mechanism and other computer application technologies into a transparent public ledger. It has the characteristics of decentralization, non-repudiation, non-falsification and non-tampering. Typically, blockchain networks can be categorized into public, consortium and private blockchains according to the mode of participation. Specifically, in consortium blockchain, multiple institutional members establish a consortium blockchain through the negotiated rules. All member entities jointly participate in maintaining the distributed ledger, and only consortium members can participate in bookkeeping based on internal rules of consortium model. Read and write privileges of the ledger are executed according to the negotiated rules.

Compared with public and private blockchains, consortium blockchain has stricter access control, higher security, and more complex permission design requirements. It is believed that, using consortium blockchain to manage various types of keys achieves a fairly high level of security. Illegal changes are virtually impossible due to the consensus mechanism, which requires an attacker to hold at least 51% of the arithmetic power in consortium blockchain [14].

In scenario of medical referrals, consortium blockchain can be established between hospitals and patients in order to realize secure sharing of patients' EMRs while maximizing the protection of patients' privacy. Only after a doctor's legitimate request is authorized by patients, the relevant encrypted EMRs data can be obtained from cloud servers. Due to the complexity of actual network environment, consortium blockchain is a semi-trusted third party. In order to protect on-chain key as well as various types of information, on-chain data are encrypted. Such design ensures that blockchain honestly accomplishes key management function and data forwarding. It also prevents entities with ulterior motives from obtaining information from consortium blockchain and inferring useful information about the encrypted data.

3.2 Proxy Re-encryption

PRE is an algorithm that securely transforms a ciphertext. The conceptual model of PRE was introduced by Blaze et al. [15]. The core of this technique lies in the fact that, delegator encrypts a message with his own key in order to obtain a ciphertext, which can be transformed to another ciphertext and keep the corresponding plaintext unchanged. Specifically, the transformed ciphertext can be decrypted with the key of the authorized person for the purpose of achieving secure sharing. In practice, the encrypted EMRs data are stored in cloud servers. However, cloud server providers cannot obtain the complete trust of users, i.e., cloud server providers may pose security threats to patients' data. In addition, potential security threats of blockchain cannot be ruled out. For patients, it is wise to keep their health-related EMRs data firmly under their control, i.e., any unauthorized entity cannot access patients' EMRs when executing our scheme.

The features of proxy re-encryption can meet security requirements, and avoid integrity and confidentiality of EMRs from being compromised by semi-trusted third parties. For instance, a semi-trusted party re-encrypts the ciphertext using a re-encryption key, which is generated by an authorized entity. In the process, the semi-trusted party cannot obtain any information of plaintext by making use of its own holdings, while the authorized person can decrypt and obtain the plaintext using its own private key. The use of re-encryption technique does not need authorizer to perform re-encryption operations, but offloading re-encryption operation to semi-trusted party. Thus, the workload of authorizer can be significantly reduced. Our scheme employs cloud servers to store the encrypted EMRs and employs consortium blockchain to conduct re-encryption operations. Such design ensures the security of EMRs while reduces resource consumptions of patients and doctors. Typically, PRE consists of the following algorithms:

1. **Setup**(k) → S_1, P_1, S_2, P_2. Inputting security parameter k, setup algorithm **Setup**$(\cdot)$ outputs two public-private key pairs (S_1, P_1) and (S_2, P_2) for two communicating parties.
2. **Enc**$(S_1, P_1, trait, M)$ → C. Inputting (S_1, P_1), characteristic markers $trait$ and plaintext M, encryption algorithm **Enc**$(\cdot)$ outputs ciphertext C.
3. **ReKeyGen**$(P_1, P_2, trait)$ → $rk_{1\rightarrow 2}$. Inputting $P_1, P_2, trait$, re-encryption key generation algorithm **ReKeyGen**$(\cdot)$ outputs re-encryption key $rk_{1\rightarrow 2}$.
4. **ReEnc**$(rk_{1\rightarrow 2}, C)$ → C'. Inputting $rk_{1\rightarrow 2}$ and C, re-encryption algorithm **ReEnc**$(\cdot)$ outputs the re-encrypted ciphertext C'.
5. **Dec**(S_2, C') → M. Inputting S_2, C', decryption algorithm **Dec**$(\cdot)$ outputs plaintext M.

3.3   Nyberg-Rueppel Digital Signature

Nyberg and Rueppel proposed the Nyberg-Rueppel message recovery digital signature in [16]. Specifically, in the process of verifying a digital signature, verifier can recover the original message that is being signed, and signer does not need to send the signed message to verifier. Nyberg-Rueppel digital signature scheme is designed based on discrete logarithmic hard problem. In addition, Nyberg-Rueppel digital signature utilizes random numbers to encrypt signature process, which can effectively prevent attackers from predicting signature process or forging signatures. Meanwhile, [16] can accomplish the tasks of digital signature and plaintext data acquisition without exposing the encrypted information, thus, it protects users' private information. Nyberg-Rueppel digital signature scheme consists of the following algorithms:

1. **KeyGen**$(p, q, g) \rightarrow x, y$. Inputting two large prime numbers p, q and generating element g, where $q \mid (p - 1)$, $g \in {}_R Z_p^*$ and $g^q = 1 (mod\ p)$, key generation algorithm **KeyGen**$(\cdot)$ outputs private key $x \in {}_R Z_p^*$ and public key $y = g^x (mod\ p)$.

2. **SigGen**$(m, k, g, x, y) \rightarrow (e, s)$. Inputting message m, random number $k(0 < k < q)$, generator g and public-private key pair (x, y), the scheme computes $\tilde{m} = R(m)$, $r = g^{-k} (mod\ p)$, where redundancy function R is a single mapping that is easy to invert. Further computation yields $e = \tilde{m} r (mod\ p), s = xe + k (mod\ p)$. Completing these operations, signature generation algorithm **SigGen**$(\cdot)$ outputs (e, s) as digital signature of message m.

3. **SigVer**$(e, s, y) \rightarrow m$. Inputting digital signature (e, s) and public key y, the scheme verifies if signature satisfies $0 < e < q$ and $0 \leq s < q$. If the verification passes successfully, then the scheme calculates $v = g^s y^{-e} (mod\ p)$ and $m' = ve (mod\ p)$, and verifies if $m' \in R(m)$, where $R(m)$ denotes the value domain of R. If the verification passes successfully, signature verification algorithm **SigVer**$(\cdot)$ recovers and outputs message $m = R^{-1}(m')$. Correctness can be checked via $m' = ve (mod\ p) = g^s y^{-e} e (mod\ p) = g^{xe+k-xe} e (mod\ p) = g^k e (mod\ p) = \tilde{m}$.

3.4   System Model

This paper designs a modular sharing scheme based on consortium blockchain and proxy re-encryption in order to securely and efficiently access patients' EMRs in medical referral scenarios. The main participants of our scheme include patients, doctors, cloud servers, consortium blockchain, and Registration Authority (RA). Among them, cloud servers and consortium blockchain are semi-trusted third parties. Namely, they execute the scheme honestly, but they may save the intermediate computational state of the scheme and try to obtain other information. Additionally, RA is a trusted third party that assists patients and doctors. Specifically, RA generates system parameters and stores public keys for patients and doctors after the indexing connection. During the execution of scheme, patients and doctors need to obtain other's public keys through RA.

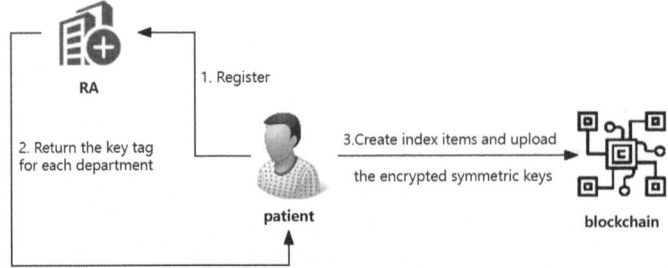

Fig. 1. Flow chart of patients.

Patients are the owners of EMRs data, and they have absolute access control authority over the data. Typically, at the first visit, patients need to register. A patient generates his/her own symmetric key group and encrypts the symmetric keys of different departments. Then, based on tags of each department provided by RA, index items are created for the encrypted symmetric key group, and finally it is uploaded to consortium blockchain. When a patient is referred to a receiving hospital, a doctor may need to request patient's EMRs. Consortium block-chain receives the request and asks the patient for authorization. The patient authorizes and uploads the re-encrypted key to consortium blockchain. Figure 1 shows operation flows of patients.

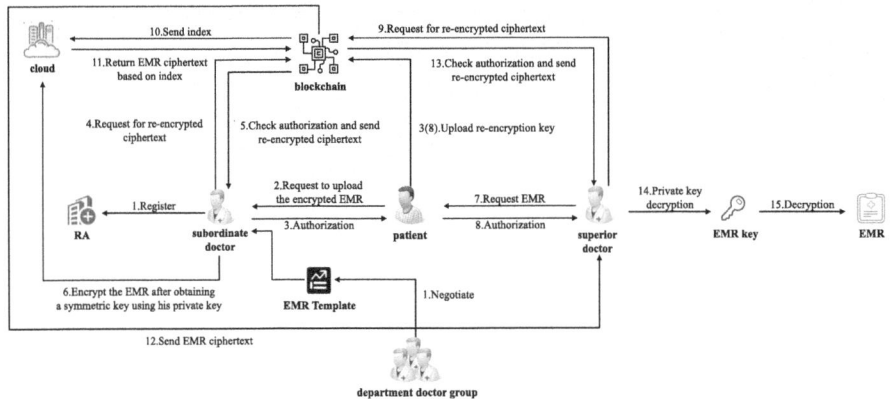

Fig. 2. Flow chart of doctors.

Doctors are data requesters of EMRs, and in scenario of medical referrals, doctors need to obtain patients' past medical history as well as relevant physiological indicator data in order to provide timely and accurate treatment. Firstly, the doctor needs to send a request to the patient. After the patient authorizes the doctor through consortium blockchain, the doctor downloads the re-encrypted

EMRs through client of healthcare organization. Then, the doctor decrypts the re-encrypted EMRs using his own private key and obtains the required diagnosis and treatment data. In our scheme, we provide attribute-based modular access to each department in hospital, and the doctor uses his private key to decrypt the re-encrypted EMRs. As a result, the doctor can only access the diagnosis and treatment data needed by this section, and cannot obtain the patient's other EMRs data. Additionally, when a doctor needs to update the patient's EMRs, the doctor needs to request consortium blockchain to obtain the patient's encrypted symmetric key. Only after the patient's authorization, consortium blockchain returns the re-encrypted ciphertext. Then, the doctor re-encrypts the updated diagnosis and treatment data, and uploads the results to cloud servers. Figure 2 depicts operation flows that the doctor needs to perform.

4   Our Scheme

This section overviews our scheme and introduces the detailed operations.

4.1   Overview

In our scheme, a patient establishes an index link for the key group using hospital's departmental tags. After the link is established, the patient uploads the encrypted key group to consortium blockchain. Specifically, when the patient visits hospital for the first time, the doctor generates EMRs for the patient and requests the symmetric key for their department from consortium blockchain. After obtaining patient's authorization, the doctor modularly encrypts EMRs and uploads the data to cloud servers. Recall that semi-trusted third party cannot decrypt the stored ciphertext correctly. Only when the doctor requests access and the patient confirms authorization, the encrypted EMRs can be successfully decrypted. The architecture of our scheme is shown in Fig. 3, which is mainly divided into sharing interface module, data processing module, and semi-trusted agent module.

Typically, the sharing interface module is mainly used in client side of medical organizations. It provides an operator console for patients and doctors to interact according to the scheme. The data processing module integrates all data processing algorithms and random number generators needed in our scheme, which is easy for patients, doctors and semi-trusted third parties to call through inter-face module. Specifically, it mainly includes the following modules: 1) AES symmetric encryption algorithm module; 2) Nyberg-Rueppel digital signature proxy re-encryption module; 3) Consortium blockchain data upload storage module; 4) Cloud servers ciphertext indexing module. The semi-trusted agent module plays the role of a semi-trusted third party under the assumption of semi-trusted model. Typically, it contains two main parts, including a consortium blockchain and a cloud server. Consortium blockchain is employed to store patients generated symmetric key group. Therefore, patients in our scheme do not need to perform key management. In addition, consortium blockchain stores the storage path of the encrypted EMRs data. In order to reduce storage pressure and

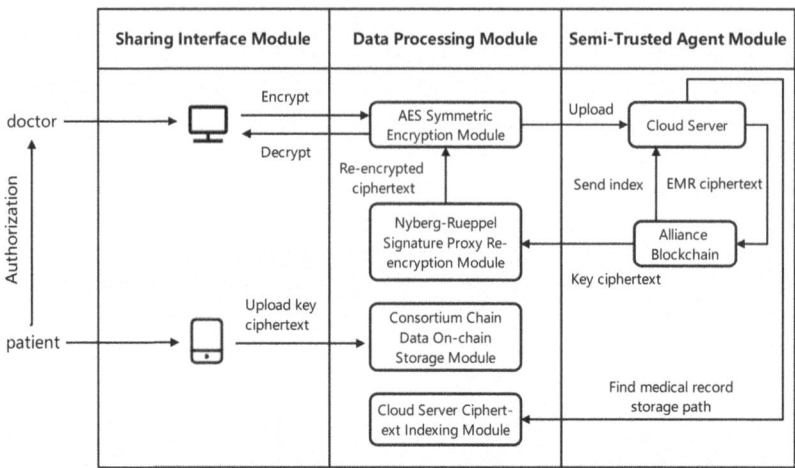

Fig. 3. The system architecture.

operation cost of consortium blockchain, the encrypted EMRs in our scheme are stored in cloud servers. After cloud servers receive the request from consortium blockchain and the storage path of EMRs, it returns the encrypted EMRs to consortium blockchain for proxy re-encryption. When the encrypted EMRs are re-encrypted, consortium blockchain sends the results to doctor. Doctor uses his private key to decrypt and extract the symmetric key of his department. Completing the above operations, the doctor uses the extracted symmetric key to decrypt ciphertext, and obtains the plaintext of modularized EMRs.

4.2 Notations and Descriptions

Table 1 shows notations and descriptions used in our scheme.

4.3 The Detailed Operations

Our scheme consists of three phases, including **Initialization phase, Encryption and Data Uploading phase**, and **Downloading and Decrypting Clinic Data phase**. In this subsection, we provide the detailed operations.

Initialization Phase
In this phase, patients and doctors need to register through RA, thus, following operations need to be completed:

1. *Generate system parameters*
   - Inputting security parameter k, RA chooses a large prime number, generates a large prime number p. Specifically, it should satisfy $q|(p-1)$ such that $g \in_R Z_q^*$ and satisfies $g^q = 1(mod\ p)$;
   - RA chooses a redundancy function φ, where φ is a single mapping and it is an easily invertible function.

Table 1. Scheme notations and descriptions.

Symbol	Description
P	Patient
D	Doctor
RA	Registration Authority
k	System security parameter
q	A large prime number of k bits
Z_q^*	Field of positive integers of module q
UID_P	Patient's ID
UID_D	Doctor's ID
$\widetilde{M}$	Use a redundant function to encrypt M
H	One-way hash function
(P_P, S_P)	Patient's public-private key pair
(P_D, S_D)	Doctor's public-private key pair
$EMRs$	Electronic medical records
$EMRkey$	A symmetric key used to encrypt EMRs
$rk_{P \to D}$	Patient generated re-encryption key

2. *Patient P and Doctor D register*
   - Patient P calls Nyberg-Rueppel signature algorithm **NR.KeyGen**($\cdot$) and obtains public-private key pair (S_P, P_P). Then patient P sends (UID_P, P_P) to RA;
   - Doctor D calls Nyberg-Rueppel signature algorithm **NR.KeyGen**($\cdot$) and obtains public-private key pair (S_D, P_D). Then doctor D sends (UID_D, P_D) to RA;
   - RA distributes $DTag_1, DTag_2, DTag_3, \ldots, DTag_N$, i.e. hospital department tags, to P, where N is the number of hospital departments.

Encryption and Data Uploading Phase

1. *Establish the index*
   In order to reduce the burden of operation and maintenance of patients, patients generate the key group and establish the index. Specifically, patient P executes the following operations:
   - Generates N symmetric keys $EMRkey_n$ with 128-bit, where N is the number of departments in hospital, and $n \in [1, N]$;
   - Indexes symmetric key groups using hospital department tags $DTag_1, DTag_2, DTag_3, \ldots, DTag_N$.
2. *Upload the indexed key group*
   Given $EMRkey_n$ and timestamp T_0, Patient P utilizes its own public key P_P to perform encryption of $EMRkey_n$, and uploads the indexed key group. Specifically, patient P executes the following operations:

- Generates characteristic markers $trait = (UID_P||T_0)$;
- Inputs the message $M_n = EMRkey_n$;
- Chooses a random number i $(0 < i < q)$, and calculates $R = g^{-i} mod \ p$;
- Calculates partial ciphertext $C_P = M_n \oplus H(trait||P_P^i)$;
- Calls Nyberg-Rueppel signature algorithm **NR.SigGen**($\cdot$), and obtains $h_P = \widetilde{M_n}R(mod \ p)$ and $z_P = i + h_P S_P(mod \ q)$. If $z_P = 0$, returns to step (3);
- Calculates $C_{EMRkey_n} = (C_P, trait, h_P, z_P)$;
- Obtains the indexed key group $(C_{EMRkey_1}||DTag_1), (C_{EMRkey_2}||DTag_2),$ $..., (C_{EMRkey_n}||DTag_n)$;
- Uploads the indexed key group to consortium blockchain.

3. *Encrypt EMR_n and upload C_{EMR_n}*

   In order to realize fine-grained access control and secure sharing of EMRs, doctors need to make a request to consortium blockchain. Then, doctors can obtain a symmetric key after getting authorization from patient. Completing these operations, doctors use the symmetric key to modularly encrypt EMRs and upload the encrypted patient's EMRs. Thus, following operations need to be executed:

   - According to EMRs template jointly negotiated by doctors in the same department, EMRs are modularized to obtain $EMR_n, n \in [1, N]$. Then, a doctor D requests for authorization from patient P;
   - Receiving the request, patient P returns authorization UID_{EMR_n} if doctor D can be authorized to upload EMRs, where $UID_{EMR_n} = DTag_n||UID_P$;
   - The doctor D sends a request with authorization UID_{EMR_n} to consortium blockchain for the purpose of obtaining symmetric key $EMRkey_n$;
   - Receiving the request, consortium blockchain verifies the authorization UID_{EMR_n}. If authorization passes verification successfully, consortium blockchain sends the symmetric key $EMRkey_n$, generated based on the proxy re-encrypted algorithm, to doctor D. Then, doctor D encrypts EMRs module and obtains $C_{EMR_n} = E_{EMRkey_n}(EMR_n)$, where $n \in [1, N]$;
   - Doctor D uses UID_{EMR_n} as index items and connects to C_{EMR_n}, then uploads the results to cloud servers.

Downloading and Decrypting Clinic Data Phase

When patient P is referred to other hospital, doctor $\mathcal{D}$ may need to access patient P's historical medical information in order to provide timely and accurate treatment. To ensure the security and privacy of patient's EMRs, only after doctor $\mathcal{D}$'s legitimate request is authorized by patient P, the relevant EMRs data can be obtained. The specific processes are as follows:

1. *Authorization*

   - Doctor $\mathcal{D}$ requests access from patient P. Authorization is successfully completed when patient P uploads UID_P to consortium blockchain;
   - Consortium blockchain forwards UID_P to cloud servers;

- Cloud servers search UID_{EMR_n} using UID_P, and return the resulting modularized EMRs ciphertext C_{EMR_n} to consortium blockchain (which connected by UID_{EMR_n}).

2. *Re-encrypt the modular EMRs ciphertext*

To re-encrypt the modular EMRs ciphertext, following operations need to be executed:

- For the original ciphertext $C_{EMRkey_n} = (C_P, trait, h_P, z_P)$, patient P runs re-encryption key generation algorithm **PRE.ReKeyGen(·)** and generates a re-encryption key $rk_{P \to D}$ for doctor $\mathcal{D}$. Specifically, patient P recovers the random number $i = z_P - h_P S_P$, acquires public key P_D of doctor $\mathcal{D}$ from RA, calculates re-encryption key $rk_{P \to D} = H(trait||P_P^i) \oplus H(trait||P_D^{-i})$;
- Patient P sends re-encryption key to consortium blockchain;
- Consortium blockchain runs re-encryption algorithm **PRE.ReEnc(·)** and computes the partial re-encrypted ciphertext $C_D = rk_{P \to D} \oplus C_P$, and outputs re-encrypted ciphertext $C'_{EMRkey_n} = (C_D, trait, h_P, UID_D, z_P)$.

In summary, Fig. 4 provides a flowchart of encryption and decryption process of the scheme. This completes the description of our modular sharing scheme. In the next section, we will analyse security and performance of our scheme.

3. *Signature verification and EMRs data recovery*

- Consortium blockchain sends C_{EMR_n}, C'_{EMRkey_n} to doctor $\mathcal{D}$;
- Doctor $\mathcal{D}$ acquires the public key P_P from RA;
- Obtaining public key P_P, doctor $\mathcal{D}$ calls Nyberg-Rueppel signature algorithm **NR.SigVer(·)**. Specifically, the algorithm calculates $R' = g^{z_P} P_P^{-h_P} (mod\ p)$ and $M'_n = C_D \oplus H(trait||R'^{-S_D})$, and obtains the partial ciphertext $h'_P = \widetilde{M'_n} R'^{-1} (mod\ p)$. If $h'_P = h_P$, the algorithm accepts M'_n. Let $EMRkey_n = M'_n$. Using symmetric key $EMRkey_n$,

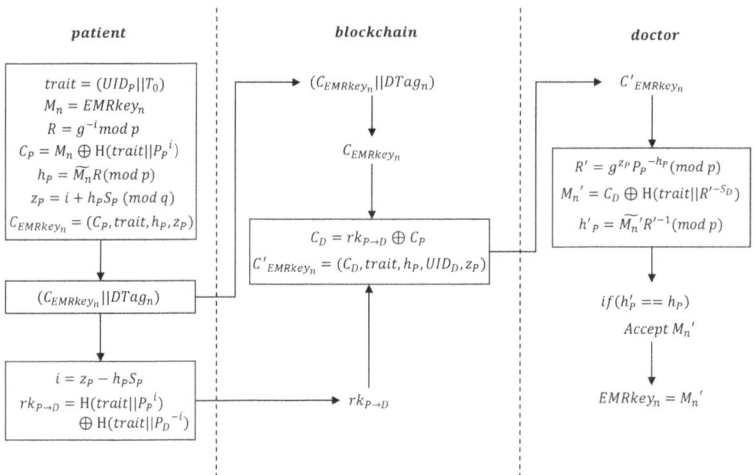

Fig. 4. The process diagram of encryption and decryption.

the algorithm decrypts ciphertext C_{EMR_n} and obtains the corresponding plaintext module of department's EMR_n. If $h'_P \neq h_P$, the algorithm terminates executions immediately and outputs the undefined symbol "$\perp$".

5   Security and Performance Analyses

In this section, we analyse the security of our scheme in terms of correctness, confidentiality, and integrity. Then, we evaluate the performance of our scheme and compare it with a few relevant schemes.

5.1   Security Analyses

Correctness. By verifying h_P and h'_P, we can verify the decryption step of our scheme is correct, which is analyzed as follows.

Recall that $h_P = \widetilde{M}_n R(mod\ p), h'_P = \widetilde{M}'_n R'^{-1}(mod\ p)$. In order to verify $R = R'^{-1}$, we can obtain $M = M'$ by verifying $h_P = h'_P$.

According to $P_P = g^{S_P}, z_P = i + h_P S_P(mod\ q), R = g^{-i} mod\ p$, we have

$$\begin{aligned}
R'^{-1} &= (g^{Z_P} P_P^{-h_P})^{-1} \\
&= g^{-(i+h_P S_P)}(g^{S_P})^{h_P} \\
&= g^{-i-h_P S_P} g^{h_P S_P} \\
&= g^{-i} \\
&= R.
\end{aligned} \tag{1}$$

Therefore, we obtain $M = M'$ when $h_P = h'_P$.

We further verify that with correct calculation procedure, we obtain $M = M'$. Recall

$$\begin{aligned}
M' &= C_D \oplus H(trait||R'^{-S_D}) \\
&= rk_{P \to D} \oplus C_P \oplus H(trait||R'^{-S_D}) \\
&= H(trait||P_P^i) \oplus H(trait||P_D^{-i}) \oplus M \\
&\quad \oplus H(trait||P_P^i) \oplus H(trait||R'^{-S_D}) \\
&= H(trait||P_D^{-i}) \oplus M \oplus H(trait||R'^{-S_D})
\end{aligned}$$

According to Eq. (1), we have

$$R'^{-S_D} = R^{S_D} = (g^{-i})^{S_D} = (g^{S_D})^{-i} = P_D^{-i}.$$

Therefore, we obtain $H(trait||P_D^{-i}) = H(trait||R'^{-S_D})$, and

$$\begin{aligned}
M' &= H(trait||P_D^{-i}) \oplus M \oplus H(trait||R'^{-S_D}) \\
&= H(trait||P_D^{-i}) \oplus M \oplus H(trait||P_D^{-i}) \\
&= M.
\end{aligned}$$

This verifies that with the correct computational procedure, there are $M = M'$, i.e., this scheme is correct.

Confidentiality. In our scheme, ciphertext in consortium blockchain network cannot be decrypted correctly without the symmetric key of each department. The patient stores the encrypted symmetric keys to semi-trusted entity, i.e. consortium blockchain. Consortium blockchain performs key search and ciphertext storage path search through the index item. Specifically, the adversary cannot determine the patient to whom the ciphertext belongs through index item. At the same time, the adversary cannot decrypt the ciphertext of EMRs, i.e., our scheme ensures EMRs data's confidentiality. Now, we prove that adversary $\mathcal{A}$ cannot obtain plaintext corresponding to the ciphertext in consortium blockchain network, and the proof details are shown as follows.

Assuming the existence of a PPT adversary $\mathcal{A}$ that can obtain plaintext corresponding to the ciphertext in consortium blockchain. Then, we construct a simulator S to flip a coin $c \in \{0, 1\}$ to determine the reply message.

1. Setup: the simulator S selects a random number k_0 as a key to the CPA secure encryption algorithm E.
2. Challenge: adversary $\mathcal{A}$ Selects the plaintext that want to be encrypted $m_0, m_1 \in G$. Challenger C randomly generates b (b is 0 or 1), sends g^b and the challenge ciphertext $Enc_{k_0}(m_b)$ to adversary $\mathcal{A}$.
3. Guess: $\mathcal{A}$ determines which plaintext is encrypted. If $b' \neq b$ then the algorithm outputs $v = 1$. Thus, we have

$$Pr[b' \neq b | v = 1] = \frac{1}{2}.$$

Otherwise, the algorithm outputs 0. Thus, we have

$$Pr[b' = b | v = 0] = \frac{1}{2} + Adv\mathcal{A}.$$

Since this algorithm is CPA secure, thus

$$|Pr[b' \neq b | v = 1] - Pr[b' = b | v = 0]| \leq negl(\lambda).$$

In other words, $Adv\mathcal{A} \leq negl(\lambda)$.

From the above analysis, we prove that adversary $\mathcal{A}$ cannot determine the patient to whom the ciphertext belongs by making use of the index items. At the same time, the adversary $\mathcal{A}$ cannot recover the ciphertext of EMRs, i.e., our scheme guarantees the confidentiality of EMRs.

Integrity. Our scheme defends against EMRs forgery and deletion attacks. Specifically, any changes made to EMRs need to be authorized by the patient. After obtaining authorization from patient, the doctor obtains the corresponding symmetric key from consortium blockchain, and these operations are recorded and logged into the transaction of consortium blockchain along with the hash value of EMRs ciphertext. Any unauthorized access or changes to patients' EMRs will result in a change in the hash value.

Assuming that the AES encryption algorithm is secure, i.e., the probability of adversary $\mathcal{A}$ breaks the security of AES algorithm is $\epsilon(k)$. Assuming that the length of index item is j, the probability that adversary $\mathcal{A}$ can pass the verification is $\frac{Q_1(n)}{|2^j|} \cdot \frac{Q_2(n)}{2^k} + \epsilon(k)$ using the constructed plaintext and index item, where $Q_1(n)$ and $Q_2(n)$ are the number of queries executed by adversary $\mathcal{A}$ in polynomial time.

Moreover, assuming that adversary $\mathcal{A}$ obtains index item in consortium blockchain network. In order to obtain the legitimate data encrypted by AES algorithm and thus tamper with EMRs stored in cloud servers, adversary $\mathcal{A}$ can only utilize the plaintext generated by itself to forge legitimate ciphertext. Therefore, adversary $\mathcal{A}$ has the probability of $\frac{Q(n)}{2^k} + \epsilon(k)$ to break the integrity of our scheme, i.e., the probability that the PPT adversary $\mathcal{A}$ successfully breaks the secure of AES algorithm and forges EMRs is negligible. Thus, our scheme ensures the integrity of EMRs.

5.2   Performance Analyses

With Python extension module gmpy2 (gmpy2-2.1.5), we performed a few experiments on Dell Laptop G15-5510 equipped with Intel(R) Core(TM) i5-102-00H CPU @ 2.40 GHz 2.40 GHz, 16.00G RAM and Windows 11 operating system. The final results were plotted in MATLAB R2021a.

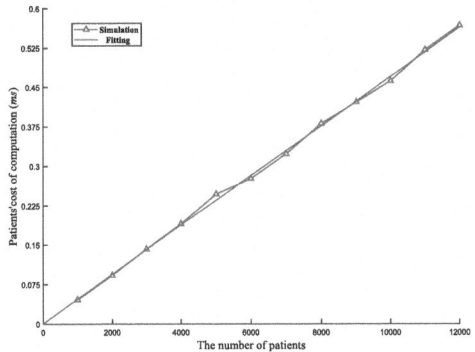

Fig. 5. Performance analyses and comparisons in terms of patients' time cost.

In order to evaluate the computational cost of our scheme, we choose four phases that have the greatest impact on patient referral process for simulation, which are the EMRs encryption phase, the re-encryption key generation phase, proxy re-encryption phase, and EMRs decryption phase. In practical applications, the computational cost of patients should be as light as possible. We first evaluate the computational cost when 12,000 patients visit hospital, and take the average value through several simulations in order to minimize the effect of error on experimental results. Then, the experimental results are shown in Fig. 5.

According to the figure we can see that, the computational cost is linearly related to the number of patients. For example, when the number of referrals is 5,000, the computational cost of the patient group is 0.247ms. When the number of referrals is 12,000, the computational cost of patient group is 0.567 ms. It indicates that in the face of actual referral scenarios with a large number of EMRs, our scheme can still keep the computational cost of patients in a low range.

In practice, each patient may have many EMRs. Additionally, the number of times EMRs need to be shared may be much larger than the number of times the system is initialized. Therefore, the computational cost of system initialization and user registration is neglected. We approximate the total time cost as: total time cost = data owner encryption time cost + $x \times$ (re-encryption key generation time cost + third party encryption time cost + data user decryption time cost), where x is the number of times EMRs are shared. The estimation of total time cost is shown in Fig. 6. According to Fig. 6 we can see that, compared with other related schemes, total time cost of our scheme is the smallest. Specifically, our scheme can save at least 25% time cost.

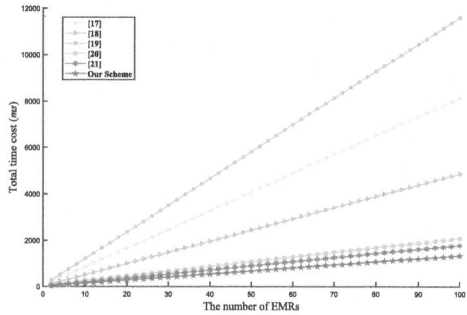

Fig. 6. Performance analyses and comparisons in terms of total time cost.

To facilitate comparison of computational cost, notations are shown in Table 2.

Table 2. Notations and abbreviations in computation.

Symbol	Description
$|A_O|$	The number of attributes owned by data owner
$|A_U|$	The number of attributes owned by data user
G_0	Exponentiation and XOR operation in group G_0
G_1	Exponentiation and XOR operation in group G_1
G_e	Pairing operation in group G_0
G_m	Multiplication operation in group G_0
AES	AES encryption/decryption operation

Recall that system initialization phase and EMRs update phase do not necessarily occur every time, thus, we ignore computational cost introduced by these two phases. Additionally, semi-trusted third parties (including cloud servers and consortium blockchain) are considered to have sufficient resources. Thus, we focus our analysis on evaluating the computational cost to be borne by patients and doctors in referral process, and the results are shown in Table 3.

According to Table 3 we can see that, computational cost of the encrypted EMRs in [17,18] is borne by patients, and computational cost of doctor in [19] is proportional to the number of attributes possessed by patients. Similar to [20], in our scheme, computational cost of doctor is constant and independent of the number of attributes owned by patients. [21] mitigates computational cost of data owner and the applicant by outsourcing cryptographic processing overhead to fog nodes. Similarly, our scheme employs consortium blockchain to bear cryptographic costs of the proxy re-encryption. As a result, both patients and doctors' computational costs are significantly reduced.

Table 3. Comparisons of computational cost.

Scheme	Patient	Doctor
[17]	$2G_0 + G_e + 2G_m$	$G_e + G_m$
[18]	$6G_0 + G_m$	$4G_0 + G_m$
[19]	$3G_0 + G_e + 2G_m$	$|A_O|(G_e + G_m)$
[20]	$4G_0$	$11G_0 + 5G_e$
[21]	$AES + XOR$	$AES + XOR$
Ours	$2G_0 + XOR$	$3G_0 + 2G_m + XOR$

Furthermore, we compare the functionality of our scheme with the related schemes, the results are shown in Table 4.

In classical blockchain based EMRs sharing schemes, data such as EMRs, transactions, and contracts are stored by blockchain, it results in poor scalability and reduces throughput. In [17], a certificate-less proxy re-encryption algorithm is employed to realize secure sharing of patients' EMRs and protect their privacy. The scheme of [18] is designed without the participation of a fully trusted third party. It realizes secure sharing of patients' diagnosis and treatment data while protecting patients' privacy. Additionally, [19] introduces a dual-chain structure to store data transaction records and raw data separately. Such design improves blockchain throughput while ensuring data privacy. However, all these schemes utilize public blockchain to realize access control of data. Recall that the transaction data of public blockchain is visible to all participants, which cannot guarantee the identity of each node is credible. Thus, it introduces security risk of data leakage. Compared with [17–19], our scheme adopts consortium blockchain for proxy re-encryption and employs semi-trusted cloud servers to store patients' EMRs. The design reduces the storage pressure of consortium blockchain, and achieves high flexibility and scalability.

The authors in [20] propose a lightweight medical data sharing scheme based on private blockchain. It provides a secure matching mechanism for patients to communicate their conditions. Recall that [17–20] only support EMRs owners to update the consultation data during data storage process, however, none of them can realize fine-grained access for data requesters. In medical consortium model, frequent EMRs updating will lead to significant time cost and operational complexity for patients. To alleviate this issue, our scheme achieves fine-grained access control by enabling doctors to access and/or updates EMRs after obtaining authorization from patients. In addition, our scheme strengthens patients' control over EMRs (recall that doctors need to be authorized by patients). As a contrast, patients' control over EMRs is weak in [17–20]. Though patients' control, in terms of data uploading phase, is strengthen in [21], doctors can access EMRs without patients' authorization.

Table 4. Functionality comparisons.

Properties	[17]	[18]	[19]	[20]	[21]	Our scheme
Fine-grained access control	No	No	No	No	Yes	Yes
Patient's access control permissions	Weak	Weak	Weak	Weak	Average	Strong
Support for data requester to update	No	No	No	No	Yes	Yes
Privacy preservation	Yes	Yes	Yes	Yes	Yes	Yes
Release main blockchain pressure	Yes	Yes	Yes	Yes	Yes	Yes
The type of blockchain	Public	Public	Public	Private	Public	Consortium

Similar to [20], our scheme supports EMRs sharing during patients' referrals process. In this scenario, data requesters are not limited to attending physicians, they can be doctors from other departments within that healthcare system. In [21], fog nodes are utilized to execute CP-ABE encryption, data owner only needs to complete symmetric encryption. Scheme of [21], however, uses a public blockchain for access control and authentication. As a contrast, in our scheme, proxy re-encryption operations are delegated to consortium blockchain. In addition, patients in our scheme upload the key group after connecting index items to consortium blockchain. Namely, patients have no need to manage keys. Thus, our design minimizes resource consumptions of patients in referrals process.

From the above analysis we can see that, comparing with other related schemes [17–21], our scheme has more functionalities (as shown in Table 4). At the same time, in our scheme, resource consumptions of patients and doctors are light. Specifically, the time cost of our scheme is reduced at least 25%. It is achieved by delegating resource consuming operations to be completed by semi-trusted entities. Namely, cloud servers are employed to store the encrypted EMRs, and consortium blockchain is employed to manage keys and execute re-encryption operations.

6   Conclusions

In recent years, medical consortium becomes an important development direction in medical field. Specifically, in medical consortium model, a number of medical institutions work together through a negotiated mechanism, with the aim of providing secure and efficient medical referral services. As a core resource that needs to be securely shared in medical referrals, EMRs have a crucial impact on patient's diagnosis and treatment process. This paper proposes a modular EMRs sharing scheme based on consortium blockchain and proxy re-encryption. Cloud servers are employed to store the encrypted patients' EMRs. Additionally, proxy re-encryption is utilized to strengthen patients' access control over EMRs. Our scheme modularly encrypts EMRs based on departmental tags to which the doctor belongs. Thus, it realizes fine-grained access control. Namely, only after doctor's legitimate requests are authorized by patient, then doctor can access and/or update EMRs. Such mechanism protects security and privacy of patients' EMRs. We analysis the security and performance of our scheme, and compare it with a few related schemes. According to the comparisons, our scheme has more functionalities, and resource consumptions of patients and doctors are light. For instance, time cost of our scheme is reduced at least 25%. In our future work, we will introduce smart contracts with tamper-proof and auto-executing properties, such that it can reduce computational burden of consortium blockchain.

Acknowledgment. This work is supported by the National Natural Science Foundation of China (No. 61902289), and the Natural Science Foundation of Fujian Province (No. 2023J01534).

References

1. Arbabi, M.S., Lal, C., Veeraragavan, N.R., Marijan, D., Nygård, J.F., Vitenberg, R.: A survey on blockchain for healthcare: challenges, benefits, and future directions. IEEE Commun. Surv. Tutor. **25**(1), 386–424 (2023)
2. Wang, Q., Wang, D.: Understanding failures in security proofs of multi-factor authentication for mobile devices. IEEE Trans. Inf. Forensics Secur. **18**, 597–612 (2023)
3. Liu, J., Fan, Y., Sun, R., Liu, L., Celimuge, W., Mumtaz, S.: Block-chain-aided privacy-preserving medical data sharing scheme for e-healthcare system. IEEE Internet Things J. **10**(24), 21377–21388 (2023)
4. Azaria, A., Ekblaw, A., Vieira, T., Lippman, A.: MedRec: using blockchain for medical data access and permission management. In: 2016 2nd International Conference on Open and Big Data (OBD), Vienna, Austria, pp. 25–30 (2016)
5. Zhang, L., Lan, F., Jiang, P., Jiang, T.: A secure medical record storage and sharing scheme based on dual-blockchain. Comput. Eng. Sci. **41**(9), 1581–1587 (2019)
6. Wang, C., Wang, D., Duan, Y., Tao, X.: Secure and lightweight user authentication scheme for cloud-assisted internet of things. IEEE Trans. Inf. Forensics Secur. **18**, 2961–2976 (2023)
7. Chang, X., Chan, Z., Zhu, L., Zhang, C., Rongxing, L., Guan, Y.: EPPFM: efficient and privacy-preserving querying of electronic medical records with forward privacy in multiuser setting. IEEE Trans. Sustain. Comput. **8**(3), 492–503 (2023)

8. Yao, S., Dayot, R.V.J., Ra, I.H., Xu, L., Mei, Z., Shi, J.: An identity-based proxy re-encryption scheme with single-hop conditional delegation and multi-hop ciphertext evolution for secure cloud data sharing. IEEE Trans. Inf. Forensics Secur. **18**, 3833–3848 (2023)

9. Gautam, P., Ansari, M.D., Sharma, S.K.: Enhanced security for electronic health care information using obfuscation and RSA algorithm in cloud computing. Int. J. Inf. Secur. Priv. **13**(1), 59–69 (2019)

10. Wang, Q., Wang, D., Cheng, C., He, D.: Quantum2FA: efficient quantum-resistant two-factor authentication scheme for mobile devices. IEEE Trans. Dependable Secure Comput. **20**(1), 193–208 (2023)

11. Agyekum, K.O.B.O., Xia, Q., Sifah, E.B., Cobblah, C.N.A., Xia, H., Gao, J.: A proxy re-encryption approach to secure data sharing in the internet of things based on blockchain. IEEE Syst. J. **16**(1), 1685–1696 (2022)

12. Peng, G., Zhang, A., Lin, X.: Patient-centric fine-grained access control for electronic medical record sharing with security via dual-blockchain. IEEE Trans. Netw. Sci. Eng. **10**(6), 3908–3921 (2023)

13. Niu, S., Fei, Yu., Chen, L., Wang, C.: A data sharing scheme for encrypted electronic health record. Comput. Eng. Sci. **44**(9), 1610–1619 (2022)

14. Aponte-Novoa, F.A., Orozco, A.L.S., Villanueva-Polanco, R., Wightman, P.: The 51% attack on blockchains: a mining behavior study. IEEE Access **9**, 140549–140564 (2021)

15. Blaze, M., Bleumer, G., Strauss, M.: Divertible protocols and atomic proxy cryptography. In: Nyberg, K. (ed.) EUROCRYPT 1998. LNCS, vol. 1403, pp. 127–144. Springer, Heidelberg (1998). https://doi.org/10.1007/BFb0054122

16. Nyberg, K., Rueppel, R.A.: Message recovery for signature schemes based on the discrete logarithm problem. In: De Santis, A. (ed.) EUROCRYPT 1994. LNCS, vol. 950, pp. 182–193. Springer, Heidelberg (1995). https://doi.org/10.1007/BFb0053434

17. Eltayieb, N., Sun, L., Wang, K., Li, F.: A certificateless proxy re-encryption scheme for cloud-based blockchain. In: Shen, B., Wang, B., Han, J., Yu, Y. (eds.) FCS 2019. CCIS, vol. 1105, pp. 293–307. Springer, Singapore (2019). https://doi.org/10.1007/978-981-15-0818-9_19

18. Noh, S.-W., Park, Y., Sur, C., Shin, S.-U., Rhee, K.-H.: Block-chain based user-centric records management system. Int. J. Control Autom. **10**(11), 133–144 (2017)

19. Wang, Z., Tian, Y., Zhu, J.: Data sharing and tracing scheme based on blockchain. In: Proceedings of the 8th International Conference on Logistics, Informatics and Service Sciences (LISS), pp. 1–6 (2018)

20. Thwin, T.T., Vasupongayya, S.: Blockchain-based access control model to preserve privacy for personal health record systems. Secur. Commun. Netw. **2019**, Article ID 8315614 (2019)

21. Fugkeaw, S., Wirz, L., Hak, L.: Secure and lightweight blockchain-enabled access control for fog-assisted IoT cloud based electronic medical records sharing. IEEE Access **11**, 62998–63012 (2023)

PFDF: Privacy Preserving Federated Decision Forest for Classification

Tongyaqi Li[1,2] (ID), Qingqiang Qi[1] (ID), Chengyu Hu[1,2,3](✉) (ID), Xuelei Li[4] (ID),
Peng Tang[1,2,3] (ID), and Shanqing Guo[1,2,3] (ID)

[1] School of Cyber Science and Technology, Shandong University,
Qingdao 266237, Shandong, China
hcy@sdu.edu.cn
[2] Quan Cheng Laboratory, Jinan 250103, Shandong, China
[3] The Key Laboratory of Cryptologic Technology and Information Security,
Ministry of Education, Shandong University, Jinan 250100, Shandong, China
[4] Inspur (Beijing) Electronic Information Industry Co., Ltd., Beijing, China

Abstract. The rapid development of data analysis technology and easily accessible datasets enable the construction of comprehensive analysis models, promoting decision-making processes in various fields. Decision trees, known for their high interpretability and cost-effectiveness, has become a common choice for decision-making in areas like housing price prediction and medical forecasting. However, in federated learning, there exists a risk of individual privacy leakage. To address this challenge, we propose a new privacy-preserving decision tree boosting model (PFDF). To protect data holders' privacy, we adopt differential privacy technology to perturb sensitive data that might lead to privacy leakage. Under the premise of privacy preservation, this model includes a novel approach for designing the global attribute selection and leaf node judgment scheme, considering the data imbalance among data holders. Additionally, continuous splitting values are generated using a clustering method. To enhance classification accuracy, our "multi-perspective" decision tree boosting scheme considers the optimal attributions of previously constructed trees. Accuracy tests on several benchmark datasets demonstrate that our scheme outperforms classical and the state-of-the-art approaches currently known to us in both centralized and federated learning.

Keywords: Decision tree · Federated learning · Differential privacy

1 Introduction

Decision tree constitutes a non-parametric supervised learning method [17] well-suited for classification and regression tasks. It relies on labeled datasets to construct a decision model capable of classifying sample data with unknown labels,

---

T. Li and Q. Qi—Contribute equally to this work.

© The Author(s), under exclusive license to Springer Nature Singapore Pte Ltd. 2025
T. Zhu et al. (Eds.): ICA3PP 2024, LNCS 15251, pp. 80–99, 2025.
https://doi.org/10.1007/978-981-96-1525-4_5

without making any assumptions regarding the distribution of the underlying data. In comparison to alternative supervised learning methods, decision tree offers distinctive advantages, such as enhanced human interpretability [19], relatively low computational cost [17], and remarkable applicability in handling heterogeneous data [15]. These merits make decision tree a favorable choice across a wide range of data mining applications, including house price prediction, medical forecasting and search queries.

However, in practical decision tree applications, the data required for model construction often pertain to data holders' private information. Directly utilizing such sensitive information to build decision trees introduces the potential risk of privacy leakage [3]. Furthermore, in federated learning, the data may be held by different data holders who lack mutual trust, which makes it more difficult to build decision trees without revealing privacy.

Illustrated in Fig. 1, the scenario presents a health organization seeking a predictive and analytical model for a specific disease. Multiple hospitals situated in diverse regions possess characteristic information about their patients and the corresponding labels denoting disease presence or absence. Given that these information is private to the patients, direct sharing during model training must be avoided to prevent privacy leakage. Additionally, the data held by each hospital varies considerably based on their geographical locations. Under these conditions, the health organization aims to construct a related decision tree model to predict the disease, facilitating a comprehensive exploration of the association between characteristics and disease manifestation.

In the aforementioned scenario, which necessitates the protection of the privacy of data holders, the heterogeneity in the number and distribution of data held by each data holder introduces new challenges in selection of optimal attribute, handling of continuous attributes, and obtaining lables of leaf nodes during decision tree construction. Furthermore, the classification accuracy achieved by a single decision tree typically falls short of the accuracy attained through an ensemble of multiple decision trees. As a result, we design a decision forest model. We consider the optimal attributes of previously constructed trees, with the aim of enabling each decision tree to contribute to decision-making from diverse perspectives. Moreover, differential privacy emerges as an effective solution for mitigating privacy leakage issues in our work. Differential privacy enforces stringent privacy guarantees against individual privacy leaks, ensuring that results derived from the dataset remain insensitive to changes in individual records [9]. To address the privacy leakage challenge in decision tree, researchers have combined differential privacy technology with decision tree construction in several works [13,29]. However, these solutions have been largely limited to the centralized scenario, with only a few works extending applicability to the horizontal federated learning [24,26]. Even the increasingly popular Gradient Boosted Decision Trees (GBDTs) [5,21] rarely involves the scenario of horizontal federated learning [16], because the training needs to consider the order of samples. However, it is worth mentioning that our scheme is not only used for federated learning, but also has good performance in the centralized scenario.

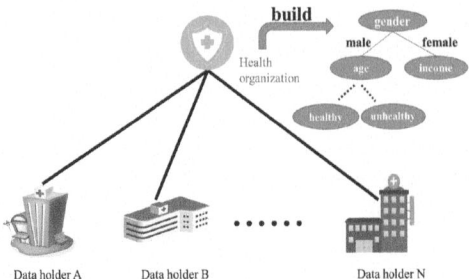

Fig. 1. An example of building a decision tree in federated learning

Based on the aforementioned problems and scenario assumptions, this paper proposes a privacy-preserving decision tree boosting model. The proposed approach takes into account the difference in data held by data holders and the weight problem in the aggregation process. We design a global optimal attribute selection and leaf node determination scheme. Moreover, we leverage multiple clustering method to effectively address the challenge of handling discrete continuous attributes in the multi-participant scenario. Furthermore, the frequency of using optimal attributes is considered to achieve diverse perspectives in constructing each decision tree. Multiple distinct decision trees are then collectively employed to determine the final classification. The model is applicable to horizontal federated learning and demonstrates favorable predictive performance. Furthermore, the implementation of this model in a centralized scenario also exhibits superior classification accuracy compared to classical approaches such as traditional decision tree and random forest.

In this paper, our main contributions are summarized as follows:

- The implementation of a privacy-protected "multi-angle" decision forest in federated learning, with heterogeneous data distribution. This mainly involves the design of three specific schemes, selection of global optimal attribute, handling of continuous attributes, and obtaining global labels for leaf nodes.
- Ensuring that the constructed decision forest satisfies the differential privacy, allocating the privacy budget of different components according to the pre-specified privacy budget. This ensures the privacy protection of data holders during decision forest construction.
- Conducting a series of comparative experiments and ablation experiments on benchmark datasets. The experimental results demonstrate its superior performance for classification and the rationality of our proposed approach. Furthermore, our approach exhibits strong performance not only in federated learning with heterogeneous data but also in the centralized scenario.

2   Related Work

For different application scenarios, many variants of decision tree have been proposed, such as ID3 [28], C4.5 [30] and CART [25] and the aggregation of multiple trees into a forest or a single tree can be realized by random forest [4] algorithm , GBDTs algorithm and other methods. In this paper, we mainly focus on classification task of decision trees, and the method which aggregate multiple decision trees into forest.

However, the training of the decision tree mostly relies on private dataset, and most data holders do not trust each other and are unwilling to upload their private data to the central server for centralized training. Therefore, people consider combining federated learning and the construction of decision tree model. While federated learning ensures local training and secure information transfer, only relying on federated learning doesn't fully guarantee data privacy. To strengthen privacy preservation, researchers seek other privacy protection methods to strengthen privacy-preserving of the federated learning.

Recent research has focused on the implementation of federated decision trees, such as SecureBoost under security training in federated learning [7]. These methods often rely on cryptographic techniques such as secret sharing, homomorphic encryption [6], or secure aggregation. While this allows for secure joint training of decision tree models without any participants directly publishing their data, the model may not necessarily be private and does not guarantee formal differential privacy [10]. For example, the splitting decision in a tree can directly display sensitive information about the data holders' data. Moreover, the reliance on these heavy-weight cryptographic technologies often makes the method computationally intensive or requires a large number of communication rounds [33], making it impossible for them to scale beyond multiple data holders [1].

For the combination of differential privacy and decision tree, Blum et al. [2] proposed a simple ID3-based algorithm that decomposes the split function into two queries per feature to satisfy differential privacy. However, in the work, too much redundant noise is added in the tree, affecting its prediction accuracy. Friedman et al. [14] improved on previous work by optimizing queries based on parallelism, comparing different splitting criteria, and trying to split continuous features. Zhu et al. [34] proposed a privacy-preserving data release algorithm for decision tree. Rana et al. [29] propose differentially private random forest to aggregate trees, which can improve generalization performance. However, all these works are carried out in the centralized scenario, and so far, few works have considered the construction of decision forest and decision tree combined with differential privacy in federated learning involving multiple data holders and using a central server for aggregation.

Some previous related researches mainly focus on constructing GBDTs satisfying differential privacy in federated learning. However, for GBDTs, the order of samples needs to be known during the training process, and most of the work focuses on the construction of GBDTs in vertical federated learning [23,32]. In this paper, we propose a "multi-angle" forest construction scheme, which can build a privacy-preserving decision forest incorporating both continuous and discrete attributes based on differential privacy in horizontal federated learning. This approach requires only a few training rounds and demonstrates better classification performance.

3   Preliminaries

3.1   Differential Privacy

According the definition of differential privacy, the modification of a single record in the dataset has little effect on the result of computation. Therefore, it is almost impossible for an attacker to obtain an accurate single record by observing the result. The general definitions of differential privacy and local differential privacy are given below:

Definition 1. (Differential Privacy (DP) [11]) A randomized mechanism $\mathcal{M}$: $\mathbb{X}^n \to \mathbb{Y}$ satisfies (ϵ, δ)-DP if for any two neighboring datasets $X \simeq_r X' \in \mathbb{X}$ and any subset $S \subseteq \mathbb{Y}$

$$\Pr[\mathcal{M}(x) \in S] \le e^\epsilon \cdot \Pr[\mathcal{M}(x') \in S] + \delta \tag{1}$$

Definition 2. (Local Differential Privacy (LDP) [8]) A randomized mechanism $\mathcal{M}$ satisfies (ϵ, δ)-LDP if for any pair input x, $x' \in \mathbb{X}$ and any output $y \in \mathbb{Y}$ of $\mathcal{M}$,

$$\Pr[\mathcal{M}(x) = y] \le e^\epsilon \cdot \Pr[\mathcal{M}(x') = y] + \delta \tag{2}$$

For the above two definitions, when $\delta = 0$, they individually satisfy ϵ-DP and ϵ-LDP.

In the following, we present the "automatic" composition theorem of differential privacy.

Theorem 1. [12] Let $\mathcal{M}_i \colon \mathbb{N}^{|\mathcal{X}|} \to \mathcal{R}_i$ be an $(\varepsilon_i, \delta_i)$-differentially private algorithm for $i \in [k]$. Then if $\mathcal{M}_{[k]} : \mathbb{N}^{|\mathcal{X}|} \to \prod_{i=1}^k \mathcal{R}_i$ is defined to be $\mathcal{M}_{[k]}(x) = (\mathcal{M}_1(x), \dots, \mathcal{M}_k(x))$, then $\mathcal{M}_{[k]}$ is $\left(\sum_{i=1}^k \varepsilon_i, \sum_{i=1}^k \delta_i\right)$-differentially private.

In this paper, we mainly use two mechanisms of differential privacy: the Laplace mechanism and the Exponential Mechanism. Among them, the Laplace mechanism is mainly applied to scenario where the output result is a real number; while the Exponential mechanism is applied to scenarios where the output result is an element in the set.

The following are the Laplace mechanism and the exponential mechanism:

Theorem 2. [11] A query $\mathcal{M}$ satisfies ϵ-differential privacy if it outputs

$$\mathcal{M}(x) = f(x) + L \tag{3}$$

, where $f : f(x) \to \mathbb{R}$, and for any adjacent datasets D and D', define sensitivity as:

$$\triangle(f) = max_{D,D'} \|f(D) - f(D')\| \tag{4}$$

$L \sim Lap(\Delta(f)/\epsilon)$ *is an i.i.d. random variable drawn from the Laplace distribution with mean 0 and scale* $\Delta(f)/\epsilon$. *We shorten* $L \sim Lap(\Delta(f)/\epsilon)$ *to* $Lap(\Delta(f)/\epsilon)$ *when our meaning is clear from context.*

Theorem 3. *[27] Using a scoring function* $u : u(z, x) \rightarrow \mathbb{R}$ *where u has a higher value for more preferable outputs* $z \in Z$, *a query* $\mathcal{M}$ *satisfies* ϵ-*differential privacy if it outputs z with probability proportional to* $\exp\left(\frac{\epsilon u(z,x)}{2\Delta(u)}\right)$. *That is,*

$$Pr(M(x) = z) \propto \exp\left(\frac{\epsilon \times u(z, x)}{2\Delta(u)}\right) \tag{5}$$

3.2 Information Gain

Information gain is the difference between information entropy and conditional entropy, which represents the degree of reduction of information uncertainty after the selected split feature. Suppose the dataset is D, the selected feature is A, use $H(D)$ to represent the information entropy of dataset D, use $H(D|A)$ to represent the conditional entropy after the selected feature A divides the dataset D, Then the information gain can be expressed by the following formula:

$$InfoGain(D, A) = H(D) - H(D|A) \tag{6}$$

When selecting the split attribute of the decision tree, the information gain is often used as the meature index, and the attribute with the largest information gain is selected as the selected split attribute.

3.3 Threat Model

In this study, similar to other research in the federated learning, we adopt an honest-but-curious model, wherein the data holders do not trust others. Specifically, assuming the existence of n data holders and a server, they strictly adhere to the protocol to compute and transmit information, yet they possess the ability to analyze all accessible data to infer as much information as possible. Moreover, we also presuppose that the server will not engage in collusion with any data holders. This security framework aligns with the standard setup employed in various federated learning works.

4 Methodology

This work proposes the Privacy-preserving Federated Decision Forest (PFDF) model, which addresses the challenge of constructing decision forest in the federated learning while preserving data privacy. The PFDF model consists of four main steps: data holder uploads samples' size, the selection of the global optimal attribute, handling of continuous attributes, and obtaining the global label of leaf nodes. Each step is carefully designed to ensure privacy preservation and achieve high classification accuracy. Moreover, this model can also be employed in the centralized scenario and has superior performance.

4.1 Overview

The construction of the PFDF model consists of four main steps, each of which identifies and addresses the challenges in building decision trees within the context of federated learning:

In the first step (Sect. 4.1), to alleviate the challenge of data imbalance problem and protect data holders' privacy, data holders upload sample size and introduce differential privacy noise to perturb it. Weighted aggregation will be conducted based on the data size of data holders. Next, in Sect. 4.2, catering to the scenario of horizontal federated learning, we design a globally optimal attribute selection scheme satisfying differential privacy. Furthermore, we consider the frequency of optimal attribute usage to enhance diversity within the decision trees. In Sect. 4.3, to tackle the challenge of discretizing continuous attributes in federated learning, we employ multiple clustering methods for discretization. We also introduce differential privacy noise to the locally obtained discrete values to ensure privacy preservation. Finally, in Sect. 4.4, accounting for data holders' weights and privacy protection, we determine global labels of leaf nodes by aggregating accuracies through weighted aggregation.

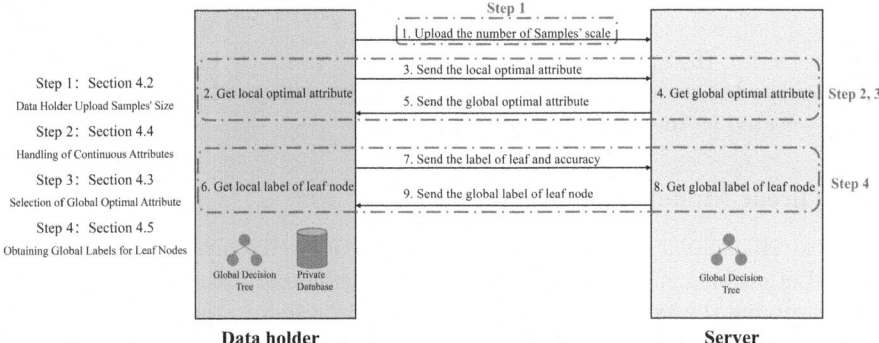

Fig. 2. Privacy Preserving Federated Decision Forest for Classification

Figure 2 illustrates the primary steps involved in the interaction between a data holder and the server in the process of constructing decision trees. Please note that we have omitted the step for handling of continuous attributes in the figure.

In addition, the previously mentioned steps outline the main procedures for constructing a global decision tree. To enhance decision-making performance, we employ a boosting method to build multiple decision trees, collectively forming a decision forest to make the final decision. Let A represent the attribute set in the dataset. The complete construction process of this approach is presented in Algorithm 1, and the detailed steps for building the decision forest are explained in the following four sections. Table 1 displays commonly used symbols for convenience.

Table 1. Symbol Description

Symbol	Description
$|A|$	Number of attributes
ϵ	Privacy budget
K	Number of trees
r_i	Data size of the i-th data holder
F^j	Global information gain for attribute j
m	Index of the m-th global decision tree
G	Information gain of a certain attribute
H^l	Aggregated value for the global label l of a leaf node
x	Number of times an attribute is selected as the global optimal attribute
G_i^j	Perturbed information gain for attribute j at data holder i
$S(r,i)$	Scoring function for data holder i and data size r

4.2   Data Holder Upload Samples' Size

In the federated learning, variations in the size and distribution of data held by each data holder can be substantial, leading to diverse contributions to the construction of the global decision tree. To address this issue, we have designed a strategy to upload information regarding the samples' size held by each data holder before initiating the decision tree construction. During the subsequent parameter aggregation, this information will be utilized for weighted aggregation to mitigate the phenomenon of imbalanced data distribution. Furthermore, as the samples size of data holders is considered privacy information, we will perturb it using differential privacy noise before sending it to the server.

Specifically, we divide the data holder's data size into R intervals, and each data holder calculates the distance between their own data size and the median of each interval. For data size held by each data holder, if the size exceeds the maximum value, it is treated as the maximum value, and if it is less than the minimum value, it is treated as the minimum value. Subsequently, the reciprocal of the distance is utilized as the scoring function S:

$$S(r,i) = \frac{1}{||D_i| - M_r|} \tag{7}$$

where $r \in \{1, ..., R\}$ and $i \in \{1, .., n\}$, M_r denotes the median of r-th interval.

Afterward, the data holder applies the Exponential mechanism for perturbation and sends the disturbed data size level to the server. For data holder i, the

Algorithm 1. Privacy Preserving Federated Decision Forest for Classification (training)

  Input: Data holders' samples ;
  Output: Decision forest
1: # **4.2:** Data Holder Upload Samples' Size
2: **Data holders**: calculate r_i and sends (publishes) it to Server
3: **for** $k = 1$ to K **do**
4:   **while** node is not leaf **do**
5:    # **4.3:** Selection of Global Optimal Attribute
6:    **Data holders:**
7:    **if** A_j is continuous attribute **then**
8:     # **Algorithm 2**
9:    **else**
10:     **for** $j \leftarrow$ discrete attributes **do**
11:      Calculate $\widetilde{G}_i^j \leftarrow G_i^j + Lap(1/\epsilon_2)$
12:     **end for**
13:     $best \leftarrow argmax\widetilde{G}_i^j$
14:     Send $\widetilde{G}_i^{best}$ and A_i^{best} to Server
15:    **end if**
16:    **Server:**
17:    Get $A_{best} \leftarrow argmax\frac{\sum_j r_i \tilde{G}_i^j}{2x^j}$
18:    Send A_{best} to Data holders
19:   **end while**
20:   # **Algorithm 3**
21: **end for**

---

probability $Pr(r_i)$is given by:

$$Pr(r_i) \propto \exp\left(\frac{\epsilon_1 \times S(r,i)}{2}\right), r_i \in \{1, ..., R\} \tag{8}$$

where $r_i \in \{1, ..., R\}$ and ϵ_1 is the privacy budget.

4.3 Selection of Global Optimal Attribute

Considering the scenario of federated learning, the optimal attribute selected by each data hoder based on their individual dataset may differ, making it impractical to directly employ the greedy decision tree's optimal attribute selection algorithm. Hence, we design a globally optimal attribute selection scheme tailored to this situation.

  The process of global optimal attribute selection consists of two stages: local selection of optimal attributes for each data holder and global selection of the optimal attribute. The specifics are as follows:

1) **Local selection for optimal attribute.** For each data holder, we employ the fundamental optimal attribute selection approach: calculating the information gain for each attribute and choosing the attribute with the highest

information gain as the local optimal attribute. Subsequently, we transmit the selected attribute along with its corresponding information gain to the server. Furthermore, to ensure privacy preservation, we add noise to the information gain using the Laplace mechanism before choosing the local optimal attribute, as indicated in line 10–12 of Algorithm 1.

2) **Global selection for optimal attribute.** The server receives the optimal attribute selected by each data holder and the corresponding perturbed information gain $G_i^j (i \in \{1, 2, ..., n\}, j \in \{1, 2, ..., u\})$, calculates $F_i^j = G_i^j r_i$, where r_j represents the weight of the data holder's contribution. The server then aggregates F_i^j for the same optimal attribute to obtain the F^j for the global information gain of each attribute. Finally, the attribute with the highest F^j is selected as the globally optimal attribute.

However, using the above approach to construct each tree will result in identical global decision tree, rendering the constructed forest meaningless. To address this issue, we consider the number of times each attribute is selected as the global optimal attribute, and decrease the likelihood of reselecting an already used attributes. This ensures that each decision tree is constructed from different perspectives. The specifics are as follows:

If an attribute is selected as the global optimal attribute in the previous m global decision trees and has been used x times, when constructing the $(m+1)$th decision tree, we divide it by $2x$ to obtain the final global information gain for this attribute, as depicted in line 14 of Algorithm 1. By implementing this setting, the probability of an attribute being selected repeatedly diminishes, thereby achieving greater diversity in the constructed decision trees.

Additionally, we impose a constraint on the maximum number of attributes used by the branches of each tree, i.e., the maximum tree height. In Sect. 5, we investigate the impact of tree height on accuracy.

4.4   Handling of Continuous Attributes

In the context of decision tree construction, handling discrete attributes is relatively straightforward since the values of such attributes is generally considered public, and values can be directly used as splitting nodes. However, dealing with continuous attributes requires finding suitable splitting values. In a centralized scenario, various methods for discretizing continuous attributes exist. However, in the federated learning, the samples possessed by different data holders may exhibit varying ranges for continuous attributes, posing a new challenge. Specifically, the challenges lie in discretizing the continuous attributes for each data holder and subsequently aggregating multiple locally discretized values into globally applicable values. To address these challenges, we employ multiple clustering methods for discretizing continuous attributes.

Initially, we need to specify the maximum and minimum values for each continuous attribute. For sample held by each data holder, if the attribute value exceeds the maximum value, it is treated as the maximum value, and if it is less than the minimum value, it is treated as the minimum value. We then proceed

---

Algorithm 2. Handling of Continuous Attributes

    Input: Data holder's samples ;
    Input: The global splitting values for a continuous attribute
1: **Data holders:**Clustering to get centroids C_i^j and using it to get G_i^j
2: **if** A_j is local optimal attribute **then**
3:     $\tilde{C}_i^j \leftarrow \{c + Lap(1/\epsilon_3)|c \in C_i^j\}$
4:     $\widetilde{G_i^j} \leftarrow G_i^j + Lap(1/\epsilon_2)$
5:     Send A_j, $\tilde{C}_i^j$ and $\widetilde{G_i^j}$ to Server
6: **end if**
7: **Server:**
8: **if** A_j is global optimal attribute **then**
9:     Clustering $\left\{\tilde{C}_i^j, i \in 1, ..., n\right\}$ to get global splitting values
10:    Send global splitting values to Data holders
11: **end if**

---

with three steps: local selection of the optimal attribute, determination of global splitting values, and local dataset partition. The specific steps are outlined below:

1) **Local Selection of the Optimal Attribute.** Each data holder utilizes a clustering algorithm to divide the data of continuous attributes into different clusters, and obtains the cluster center. This center represents the division values for each local continuous attribute, effectively discretizing the continuous attribute. Subsequently, akin to discrete attributes, we calculate the information gain of the attribute for optimal attribute selection.

2) **Determination of Global Splitting Values.** Once a continuous attribute is globally selected as the optimal attribute, each data holder introduces differential privacy noise to the local splitting values of the attribute and sends it to the server. The server then clusters all the splitting values and obtains the cluster center as the global splitting values of the attribute. This global partition value is communicated back to all data holders.

For clustering, we adopt the elbow method [31] in our experiments, which automatically selects a suitable number of clusters and obtains the cluster centers. However, users may also manually set the number of clusters or opt for other clustering methods based on specific requirements.

By now, we have addressed the handling of continuous attributes and presented the complete process in Algorithm 2. It should be noted that before adding noise, we will normalize all continuous attribute values based on the settings of the maximum and minimum values, so the sensitivity in the Laplace mechanism in the third row of Algorithm 2 is 1. And it is also important to note that dealing with continuous attributes incurs additional privacy budget consumption compared to discrete attributes. Nevertheless, our experiments show that this expenditure is justifiable, especially when compared to randomly selected splitting values.

4.5   Obtaining Global Labels for Leaf Nodes

In the centralized scenario, the training of decision tree requires access to all samples associated with each leaf node. However, in the federated learning, the samples are dispersed among data holders, who neither trust each other nor wish to disclose their private samples, posing a new challenge. To address this challenge, we divide the judgment of leaf nodes into two steps: the data holder uploads the local judgment of the leaf node, and the server determines the global label of the leaf node. Below, we will provide a detailed description of this process.

In federated learning, each data holder preserves the global model, enabling them to judge whether a node is a leaf node based on their own dataset and determine the label of the leaf node. Specifically, this scheme entails three conditions for judging a node as a leaf node:

1) when this node has no samples;
2) when a branch has exhausted the specified maximum number of attributes;
3) when all labels of samples in a node are the same.

Once all data holders judge the node as a leaf node, each data holder uploads the label selection of the leaf node (if the process ends prematurely, the category of the leaf node should also be uploaded) and the accuracy of the corresponding category, denoted as Acc_i^l, where l represents the selected label. Moreover, this information is perturbed using the Laplace mechanism. Upon receiving this information, the server multiplies the data size level of each data holder by the accuracy after perturbation and aggregates them when they have the same classification. Mathematically, this is expressed as $H^l = \sum r_i Acc_i^l$, where $argmax H^l$ denotes the label corresponding to the global leaf node. Algorithm 3 outlines the tasks that data processors and server should undertake.

Algorithm 3. Obtaining Global Labels for Leaf Nodes

    Input: Data holders' samples ;
    Output: Global label of leaf node
1: **Data holders**:
2: **if** $|D_i| = 0$ **then**
3:     Send majority labels of its parent node to Server
4: **end if**
5: **if** $|A_{rest}| = |A| - Maximum$ **then**
6:     Send it's majority label of the sample to Server
7: **end if**
8: **if** All samples' labels are same **then**
9:     Send label of the samples to Server
10: **end if**
11: Send $\widetilde{Acc_i^l} \leftarrow Acc_i^l + Lap(1/\epsilon_4)$ to Server
12: **Server**:
13: **if** All Data holders judge that the node is a leaf **then**
14:     Calculate $\sum \widetilde{Acc_i^l} r_i$ for the same label
15:     Send $argmax H^l$ to Data holders
16: **end if**

---

5   Privacy Analysis

In this section, we demonstrate that our model satisfies ϵ-DP.

Let t denote the total number of trees constructed in the PFDF model. As each tree is built on the same dataset of all data holders, according to the sequential composition theorem, each tree can be allocated a privacy budget of ϵ/t, where ϵ represents the total privacy budget.

Suppose d represents the maximum height of the tree, and c is the number of continuous attributes in the dataset. When dividing nodes, the dataset is divided into disjoint parts, and it is evident that each node on one branch conforms to the sequential composition theorem, while the different branches comply with the parallel composition theorem. One internal node on a branch consume ϵ_2 privacy budget when selecting a global optimal attribute and ϵ_3 privacy budget to perturb the splitting values if the global optimal attribute is a continuous attribute.

Consequently, we deduce that the PFDF model satisfies $\epsilon = \epsilon_1 + t((d-1)\epsilon_2 + \min\{(d-1),c\}\epsilon_3 + \epsilon_4)$-differential privacy.

6   Experiments

6.1   Experimental Settings

Datasets. We have carried out tests on various UCI datasets with a different number of samples and attributes for classification tasks. More specifically, there are three datasets commonly used in decision tree performance testing: **Adult** [22], **Heart Disease** [20] and **Credit** [18]. Dataset description is provided in Table 2. For each dataset, we randomly select 70% of the samples as the training set of data holders, and 30% of the samples as the test set.

Measure Index. We measure the performance of the proposed method using accuracy as the evaluation metric.

Our Experiments

– **Centralized Decision Forest (CDF).** Implementation of our model in the centralized scenario.
– **Privacy Preserving Federated Decision Forest (PFDF).** The main algorithm proposed in this paper, designed to achieve privacy-preserving decision forest in the horizontal federated learning.

Baselines. To validate the effectiveness of our proposed model, we conduct comparative experiments involving the following four algorithms:

– **Decision Tree.** Classical Decision Tree Algorithm in the Centralized Scenario.

- **Random Forest (RF).** Random Forest Construction Algorithm in the Centralized Scenario.
- **LDP-Hist-based xgboost [26].** Have each data holder add Gaussian noise before publishing their gradient information.
- **DP-RF [26].** DP-RF corresponds to using Totally Random(TR) split method, the averaging weight update, and uniform split candidates. Use the Gaussian mechanism under RDP accounting to perturb leaf weight.

Table 2. Dataset Description

Dataset	Number of Samples	Number of Attributes	Number of Categories
Adult	48,882	14	2
Heart Disease	297	14	2
Credit	30,000	24	2

Ablation Experiment. We conduct two ablation experiments on the mentioned datasets.

First, we aim to verify the effectiveness of our approach in handling continuous attributes using multiple clustering method. We compare two methods: one directly selects random values as splitting values, and the other adopts our proposed clustering scheme.

In the second experiment, we consider a federated learning scenario with three data holders, and set the proportion of data held by the three data holders as: $\{1:1:1, 1:1:2, 1:3:6, 1:1:8\}$. We compare our solution with the average aggregation scheme that does not take into account the data holder's data size. The purpose is to showcase the performance of our approach under imbalanced data distribution.

Through these ablation experiments, we delve into the impact of various components of our proposed model on the results, highlighting the advantages of our solution in handling continuous attributes and data heterogeneity in the federated learning. These experiments provide empirical analysis for our model and offer more comprehensive and reliable support for our work.

Hyperparameter Settings. In this paper, we vary the maximum height of the tree, considering three different tree height values: $\{|A|/2, |A|/3, 2|A|/3\}$, where $|A|$ denotes the size of the attribute set. Additionally, we experiment with different numbers of trees, namely $\{1, 3, 5, 7, 9\}$, to investigate their respective impacts on the scheme's accuracy.

By adjusting hyperparameters and privacy budget $\epsilon \in \{1, 1.5, 2, 2.5, 3, 3.5\}$, we aim to comprehensively explore the trade-offs between privacy protection strength and predictive performance. The results of these experiments provide valuable insights into the optimal hyperparameter settings for our model, offering guidance for practical deployment in real-world scenarios.

6.2 Experimental Results

Comparative Experiments in the Centralized Scenario. In the centralized scenario, we successfully implement our proposed scheme and conduct comparative experiments to evaluate its performance. In contrast to conventional decision tree construction, we employ a boosting method that utilizes multiple trees to achieve better decision-making. Furthermore, in comparison to the random forest, our optimal attribute selection method takes into account the attributes selection for constructed decision trees when constructing each decision tree, thereby enhancing the diversity of the decision trees.

To address the challenges posed by continuous attributes, we employ clustering algorithms to obtain more suitable splitting values, leading to improved accuracy. The experimental results conducted on three datasets (Adult, Heart Disease, and Credit) show that our scheme outperforms traditional decision tree and random forest approaches, as shown in Fig. 3.

Figure 3 presents a comparison of accuracy in the centralized scenario between our proposed approach and traditional decision tree and random forest methods. Notably, our approach outperforms the other two methods in terms of

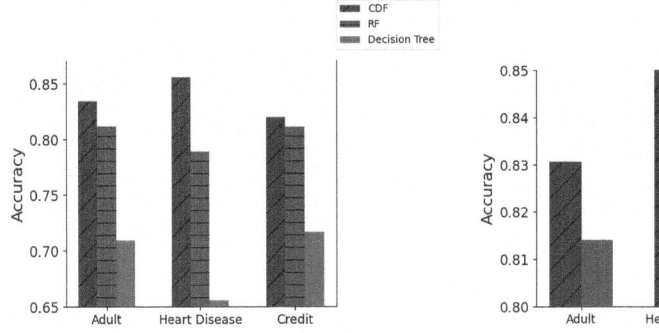

Fig. 3. In the centralized scenario, the accuracy of the three models is compared on the three datasets

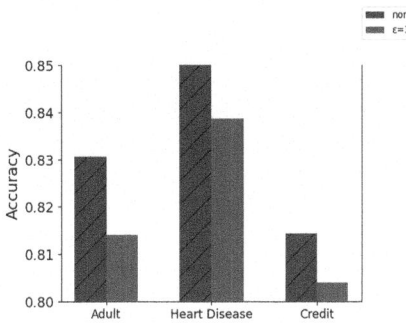

Fig. 4. The accuracy of the PFDF at $\epsilon = 3$ and the accuracy without differential privacy noise on the three datasets.

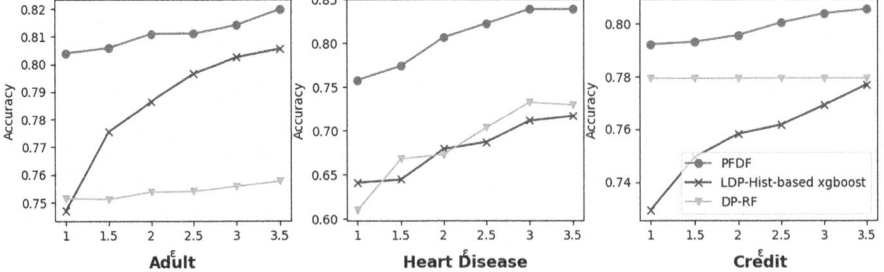

Fig. 5. On three datasets, the accuracy comparison of the three schemes

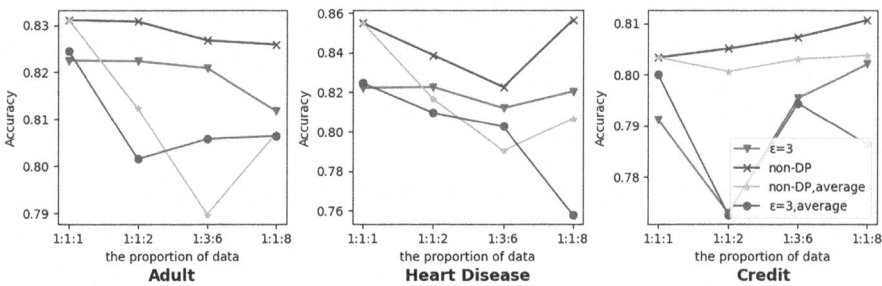

Fig. 6. On three datasets, when the data size distribution is unbalanced, the accuracy of weighted aggregation and average aggregation in the two cases of no differential privacy noise and $\epsilon = 3$

predictive results across all three datasets. Our approach consistently achieving at least a 1% accuracy enhancement over the random forest approach. These findings underscore the superior predictive performance of our method, highlighting the efficacy of our foundational model. In the subsequent sections, we delve further into verifying the advantages of our approach within the context of federated learning.

Accuracy Evaluation of PFDF in Federated Learning. We conduct experiments on the same three datasets as mentioned above. Figure 4 illustrates the comparative accuracy results of our PFDF algorithm with and without differ-

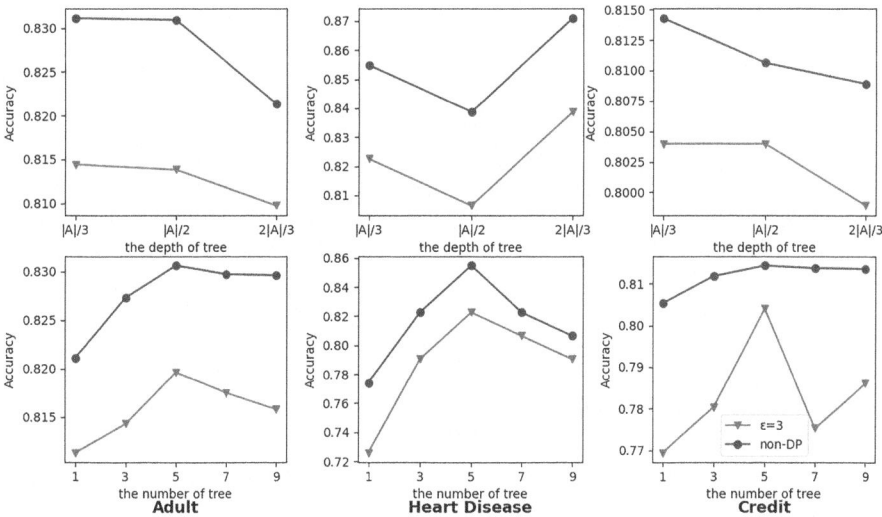

Fig. 7. The effect of the depth and number of trees in the forest on the accuracy on three datasets

ential privacy noise perturbation in the federated learning. Compared to the scenario without noise, our approach incurs approximately 1% accuracy reduction when using a privacy budget of $\epsilon = 3$. For other privacy budget settings, the accuracy results are presented in Fig. 5.

Comparison of Privacy-Preserving Methods in the Federated Learning. In this experiment, we conduct a comparative analysis of privacy-preserving methods in the federated learning. We set the maximum depth of decision trees to $|A|/3$, the number of trees to 5, and varied the privacy budget within the range of $\{1, 1.5, 2, 2.5, 3, 3.5\}$. The three models are our proposed Privacy Preserving Federated Decision Forest (PFDF) scheme, LDP-Hist-based xgboost, and DP-RF schemes, and all comparative schemes are executed using the provided source codes. Figure 5 illustrates the results of the comparison. Notably, our PFDF scheme consistently outperform the other two methods in terms of accuracy across all three datasets under the same privacy budget settings.

Ablation Experiment

– **Comparison of Accuracy for Handling Continuous Attributes** We perform tests on three datasets and compare the accuracy results of our multiple clustering scheme and the random value approach with and without differential privacy noise, as presented in Table 3. Despite the privacy budget is consumed by our clustering method, our approach achieves at least 1% higher accuracy than the random value approach on these three benchmark datasets under the same total privacy budget, demonstrating the effectiveness of our method.

Table 3. Accuracy Comparison of Different Approaches for Handling Continuous Attributes

Method	Adult	Heart Disease	Credit
Non-DP + Cluster	0.8296	0.8548	0.8143
Non-DP + Random	0.8173	0.8387	0.8030
$\epsilon = 3$ + Cluster	0.8158	0.8226	0.8040
$\epsilon = 3$ + Random	0.7949	0.7903	0.7819

– **Weighted Aggregation** At the beginning of training, each data holder is required to upload the perturbed data, and weighted aggregation is considered when choosing the global optimal attribute and obtaining leaf node labels. If the individual contributions of each data holder are not considered, the average aggregation scheme consumes less privacy budget but ignores the impact of data imbalance. To validate the effectiveness of our weighted aggregation scheme, we compare the two approaches under both noisy and non-noisy settings, as shown in Fig. 6. The results confirm the efficacy of our weighted aggregation scheme. It is worth noting that in the average aggregation scheme, when the data size ratio is 1:3:6, the accuracy after adding Laplace noise is sometimes even better than the result without the noise, indicating that average aggregation significantly reduces accuracy, while the perturbed results sometimes outperform the unperturbed average aggregation, underscoring the necessity of our weighted aggregation scheme.

Effect of Decision Tree Depth on Accuracy. We also test the impact of different decision tree depths on accuracy on the three datasets, as shown in Fig. 7. In the experiments, we set the number of decision trees to 5 and observe that the accuracy does not always increase with the depth. In fact, there are cases where the accuracy shows a significant decrease. For the case without adding differential privacy noise, we attribute this result to overfitting phenomenon; for the case with differential privacy noise, in addition to overfitting, the decrease in accuracy may also be attributed to the fact that, under the same number of decision trees, smaller depths result in larger privacy budgets allocated to each tree, leading to higher accuracy.

Effect of Decision Trees' Number on Accuracy. We also test the impact of different number of decision trees on accuracy on the same datasets. By varying the number of decision trees in the experiments, we obtain accuracy, as shown in Fig. 7. Interestingly, we observe variations in accuracy with different numbers of decision trees, and all three datasets achieve the highest accuracy when the number of decision trees is set to 5. We present the results in Fig. 7. Based on our approach to selecting the optimal attribute, after constructing a certain number of decision trees, the contribution of the chosen optimal attribute to the classification task may diminish to a certain extent, and its use in classification could even lead to overfitting. This experiment contributes to a better understanding of the influence of decision tree quantity on accuracy in our proposed scheme and provides guidance for selecting an appropriate number of decision trees.

7   Conclusion

In this paper, we propose a novel privacy-preserving decision forest construction scheme that addresses privacy protection concerns related to data holders' privacy data and tackles the issue of data imbalance, while effectively handling both discrete and continuous attributes. Our decision forest are constructed from diverse perspectives, and the classification results of all trees are combined to accomplish the classification task of decision trees, exhibiting high accuracy in both centralized and horizontal federation learning. By implementing specific experiments, we conduct accuracy tests on various datasets and perform multiple comparative experiments, demonstrating superior performance compared to existing methods. Moreover, our ablation experiments provide further support for the effectiveness of the designed scheme.

Acknowledgment. This project is supported in part by National Natural Science Foundation of China (No. 62372268, No. 62002203), Shandong Provincial Natural Science Foundation (No. ZR2022LZH013, No. ZR2021LZH007, No. ZR2020QF045), Department of Science&Technology of Shandong Province grant (No. SYS202201), Quan Cheng Laboratory grant (No. QCLZD202302), Jinan City "20 New Universities" Funding Project (2021GXRC084), Young Scholars Program of Shandong University.

References

1. Abspoel, M., Escudero, D., Volgushev, N.: Secure training of decision trees with continuous attributes. Cryptology ePrint Archive (2020)
2. Blum, A., Dwork, C., McSherry, F., Nissim, K.: Practical privacy: the sulq framework. In: Proceedings of the Twenty-Fourth ACM SIGMOD-SIGACT-SIGART Symposium on Principles of Database Systems, pp. 128–138 (2005)
3. Brankovic, L., Estivill-Castro, V.: Privacy issues in knowledge discovery and data mining. In: Australian Institute of Computer Ethics Conference, pp. 89–99. Citeseer (1999)
4. Breiman, L.: Random forests. Mach. Learn. **45**, 5–32 (2001)
5. Chen, T., Guestrin, C.: Xgboost: a scalable tree boosting system. In: Proceedings of the 22nd ACM SIGKDD International Conference on Knowledge Discovery and Data Mining, pp. 785–794 (2016)
6. Chen, X., et al.: Fed-eini: an efficient and interpretable inference framework for decision tree ensembles in vertical federated learning. In: Chen, Y., et al. (eds.) 2021 IEEE International Conference on Big Data (Big Data), Orlando, FL, USA, 15–18 December 2021, pp. 1242–1248. IEEE (2021). https://doi.org/10.1109/BIGDATA52589.2021.9671749
7. Cheng, K., et al.: Secureboost: a lossless federated learning framework. IEEE Intell. Syst. **36**(6), 87–98 (2021)
8. Duchi, J., Wainwright, M.J., Jordan, M.I.: Local privacy and minimax bounds: sharp rates for probability estimation. In: Advances in Neural Information Processing Systems, vol. 26 (2013)
9. Dwork, C.: Differential privacy. In: International Colloquium on Automata, Languages, and Programming, pp. 1–12. Springer (2006)
10. Dwork, C.: Differential privacy: a survey of results. In: International Conference on Theory and Applications of Models of Computation, pp. 1–19. Springer (2008)
11. Dwork, C., McSherry, F., Nissim, K., Smith, A.: Calibrating noise to sensitivity in private data analysis. In: Halevi, S., Rabin, T. (eds.) TCC 2006. LNCS, vol. 3876, pp. 265–284. Springer, Heidelberg (2006). https://doi.org/10.1007/11681878_14
12. Dwork, C., Roth, A., et al.: The algorithmic foundations of differential privacy. Found. Trends® Theor. Comput. Sci. **9**(3–4), 211–407 (2014)
13. Fletcher, S., Islam, M.Z.: A differentially private decision forest. AusDM **15**, 99–108 (2015)
14. Friedman, A., Schuster, A.: Data mining with differential privacy. In: Proceedings of the 16th ACM SIGKDD International Conference on Knowledge Discovery and Data Mining, pp. 493–502 (2010)
15. Gorishniy, Y., Rubachev, I., Khrulkov, V., Babenko, A.: Revisiting deep learning models for tabular data. Adv. Neural. Inf. Process. Syst. **34**, 18932–18943 (2021)
16. Grislain, N., Gonzalvez, J.: Dp-xgboost: private machine learning at scale. arXiv preprint arXiv:2110.12770 (2021)

17. Han, J., Kamber, M., Mining, D.: Concepts and Techniques, vol. 340, pp. 94104–3205. Morgan Kaufmann (2006)
18. Hofmann, D.H.: German credit dataset: UCI machine learning repository (1994)
19. Huysmans, J., Dejaeger, K., Mues, C., Vanthienen, J., Baesens, B.: An empirical evaluation of the comprehensibility of decision table, tree and rule based predictive models. Decis. Support Syst. **51**(1), 141–154 (2011)
20. Janosi, A., Steinbrunn, W., Pfisterer, M., Detrano, R.: Heart disease data set, UCI machine learning repository (1988). https://archive.ics.uci.edu/ml/datasets/heart+Disease
21. Ke, G., et al.: Lightgbm: a highly efficient gradient boosting decision tree. In: Advances in Neural Information Processing Systems, vol. 30 (2017)
22. Kohavi, R., Becker, B.: Adult dataset. UCI machine learning repository (1996)
23. Le, N.K., et al.: Fedxgboost: privacy-preserving xgboost for federated learning. arXiv preprint arXiv:2106.10662 (2021)
24. Liu, Y., Chen, M., Zhang, W., Zhang, J., Zheng, Y.: Federated extra-trees with privacy preserving. arXiv preprint arXiv:2002.07323 (2020)
25. Loh, W.Y.: Classification and regression trees. Wiley Interdisc. Rev. Data Mining Knowl. Discov. **1**(1), 14–23 (2011)
26. Maddock, S., Cormode, G., Wang, T., Maple, C., Jha, S.: Federated boosted decision trees with differential privacy. In: Proceedings of the 2022 ACM SIGSAC Conference on Computer and Communications Security, pp. 2249–2263 (2022)
27. McSherry, F., Talwar, K.: Mechanism design via differential privacy. In: 48th Annual IEEE Symposium on Foundations of Computer Science (FOCS 2007), pp. 94–103. IEEE (2007)
28. Quinlan, J.R.: Induction of decision trees. Mach. Learn. **1**, 81–106 (1986)
29. Rana, S., Gupta, S.K., Venkatesh, S.: Differentially private random forest with high utility. In: 2015 IEEE International Conference on Data Mining, pp. 955–960. IEEE (2015)
30. Salzberg, S.L.: C4. 5: Programs for Machine Learning by J. Ross Quinlan. Morgan Kaufmann Publishers, Inc., 1993 (1994)
31. Thorndike, R.L.: Who belongs in the family? Psychometrika **18**(4), 267–276 (1953)
32. Wang, R., Ersoy, O., Zhu, H., Jin, Y., Liang, K.: Feverless: fast and secure vertical federated learning based on xgboost for decentralized labels. IEEE Trans. Big Data (2022)
33. Wu, Y., Cai, S., Xiao, X., Chen, G., Ooi, B.C.: Privacy preserving vertical federated learning for tree-based models. arXiv preprint arXiv:2008.06170 (2020)
34. Zhu, T., Li, G., Zhou, W., Philip, S.Y.: Differentially private data publishing and analysis: a survey. IEEE Trans. Knowl. Data Eng. **29**(8), 1619–1638 (2017)

Data-Free Encoder Stealing Attack in Self-supervised Learning

Chuan Zhang[1], Xuhao Ren[1], Haotian Liang[1], Qing Fan[1], Xiangyun Tang[2],

Chunhai Li[3], Liehuang Zhu[1], and Yajie Wang[1(✉)]

[1] Beijing Institute of Technology, Beijing 100081, China
wangyajie19@bit.edu.cn
[2] Minzu University of China, Beijing 100081, China
[3] Guilin University of Electronic Technology, Guilin 541004, Guangxi, China

Abstract. Self-supervised learning technology has rapidly developed in making full use of unlabeled images, using large amounts of unlabeled data to pre-train encoders, which has led to the rise of Encoder as a Service (EaaS). The demands of large amounts of data and computing resources put pre-trained encoders at risk of stealing attacks, which is an easy way to acquire encoder functionality cheaply. Conventional attacks against encoders assume the adversary can possess a surrogate dataset with a distribution similar to that of the proprietary training data employed to train the target encoder. In practical terms, this assumption is impractical, as obtaining such a surrogate dataset is expensive and difficult. In this paper, we propose a novel data-free encoder stealing attack called DaES. Specifically, we introduce a generator training scheme to craft synthetic inputs used for minimizing the distance between the embeddings of the target encoder and surrogate encoder. This approach enables the surrogate encoder to mimic the behavior of the target encoder. Furthermore, we employ gradient estimation methods to overcome the challenge posed by limited black-box access to the target encoder, thereby improving the attack's efficiency. Our experiments conducted across various encoders and datasets illustrate that our attack enhances state-of-the-art accuracy by up to 6.20%.

Keywords: Encoder Stealing Attacks · Encoder as a Service · Self-supervised learning · Data-free

1 Introduction

Self-supervised learning has gained attention in recent years as an effective approach to training encoders [1,29]. These encoders can acquire embeddings, also termed feature vectors, from inputs even with minimal or absent labeled data [15,50]. These embeddings find broad application in extracting information representations [16]. Combining these pre-trained encoders with corresponding downstream classifiers can be leveraged for various downstream tasks [4,49]. For example, the pre-trained SimCLR [4], trained on unlabeled data from the ImageNet, when used in conjunction with a downstream classifier, outperforms the supervised learning approach of AlexNet [44,48].

© The Author(s), under exclusive license to Springer Nature Singapore Pte Ltd. 2025
T. Zhu et al. (Eds.): ICA3PP 2024, LNCS 15251, pp. 100–120, 2025.
https://doi.org/10.1007/978-981-96-1525-4_6

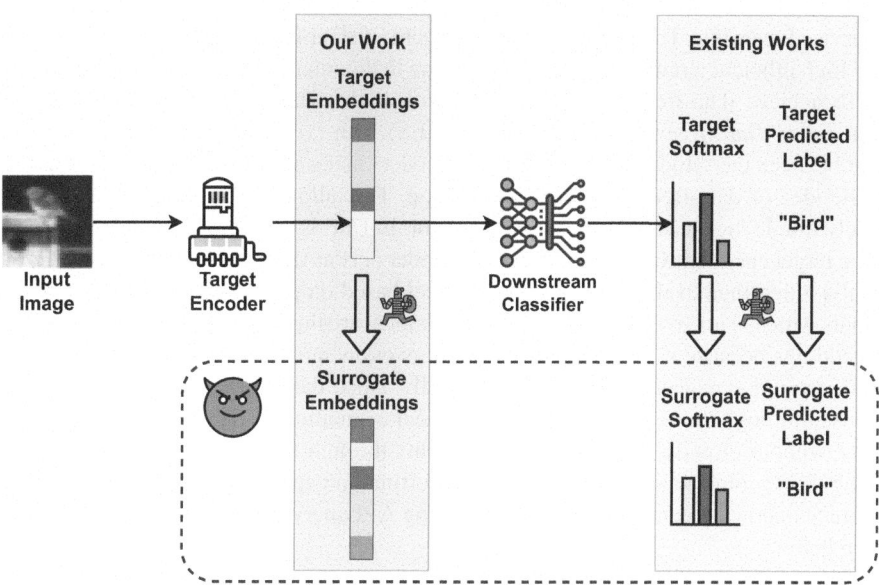

Fig. 1. Comparison between our work and existing works. Existing works primarily focus on stealing classifiers using softmax or predicted labels, while our work centers around stealing target encoders through their embeddings.

Although pre-trained encoders have demonstrated remarkable performance in various downstream tasks, the training process itself is resource-intensive, requiring abundant datasets and considerable computational resources [26,31]. Consequently, many AI companies consider their pre-trained encoders as intellectual properties and solely offer services through cloud platform APIs for financial gain. Therefore, the concept of Encoder-as-a-Service (EaaS) has gained popularity in recent years [32,45]. For example, OpenAI provides API access to their GPT-3 [25], a potent encoder utilized in various NLP [9,35] tasks.

Researchers have discovered that simply providing APIs is insufficient to defend against potential stealing attacks [7,14,25]. These attacks not only violate the model's intellectual copyright but also enable additional adversarial activities, such as backdoor injection, membership inference, and adversarial samples [3,21,24]. While previous works [22,33] primarily concentrated on supervised classifiers, with the adversary utilizing prediction results or softmax outputs (as shown in Fig. 1) to carry out the attack, Gao et al. [25] were the pioneers in introducing the concept and method of stealing the functionality of pre-trained encoders. Additionally, authors of [10] also use the data augmentation technique to enhance the performance and efficiency of stealing attacks. Unfortunately, these methods require the adversary to have surrogate datasets for a successful attack. However, acquiring a suitable surrogate dataset for stealing attacks is highly costly in terms of time and money [39]. Furthermore, these methods may also require about 10^5 queries to successfully execute an attack. To tackle this problem, many studies have concentrated on data-free model-stealing attacks with high efficiency [22,31]. Nevertheless, these methods all rely on the contexts that are associated with supervised learning. Therefore, the development of data-free encoder stealing attacks with high efficiency remains an important issue that demands attention and resolution.

To achieve data-free encoder stealing with high efficiency, we propose a novel work called *Data-free Encoder Stealing* (DaES). Our work maximizes the disagreement between the "student" and "teacher" encoders using a generator, which is inspired by the idea of data-free knowledge distillation. This allows the "student" encoder to closely mimic the "teacher" encoder's behavior. In DaES, the "teacher" encoder serves as the target encoder, while the "student" encoder acts as the surrogate encoder. Significantly, rather than evaluating the disagreement based on prediction results or softmax outputs which is not available for encoders, we measure the distance between the output embeddings because it offers a more precise assessment of the disagreement between the target and surrogate encoders. Additionally, since the target encoder can only be accessed through the API, we employ a gradient estimation technique to train the generator without directly computing the gradients through the target encoder. By combining this training scheme with the gradient estimation approach, DaES achieves high accuracy in downstream tasks while minimizing API query costs. Our contributions are given below:

- We explore data-free scenarios encoder stealing attacks in self-supervised learning and introduce a new attack called DaES. Through EaaS APIs, the adversary can steal the functions of a pre-trained encoder without a private pre-training dataset.
- We propose a generator training scheme to optimize the surrogate encoder, allowing it to emulate the behavior of the target encoder. Additionally, we employ gradient estimation techniques to address the challenge posed by the limited black-box access to the target encoder.
- We evaluate the performance of DaES across different image encoders. The experimental results show that DaES outperforms the accuracy of state-of-the-art methods by up to 6.20%, while reducing the API query budgets by up to 74.42%.

2   Related Works

In this section, we mainly introduce stealing attacks and defensive strategies against stealing attacks.

2.1   Stealing Attacks

Most stealing attacks solely concentrate on the downstream classifiers. For example, Tramer et al. [38] investigated model stealing attacks in supervised learning, in which they demonstrated the feasibility of stealing the functionality of high-performing machine learning models deployed online through APIs. Since then, extensive research has been conducted on various aspects of these attacks [36,46]. For instance, if the adversary manages to obtain similar or in-distribution data that resemble the surrogate dataset used for attacks, they can leverage data augmentation or active learning techniques by combining the datasets to query the target model [27,30]. However, this approach may prove insufficient in scenarios where the tasks involve significant commercial value and the associated training dataset is considered highly confidential.

Existing works mainly focus on *data-free* model stealing methods to tackle this challenge. In this scenario, the adversary has no information or knowledge about the dataset used to train the target models. Two recently developed techniques, DFME [39] and MAZE [18], have been specifically designed to extract the functionality of target models in this constraint. However, these studies impose significant query budgets on the adversary, making them impractical in real-world scenarios. To address this problem, Lin et al. [22] incorporated Generative Adversarial Networks (GAN) to leverage weak image priors and employed deep reinforcement learning techniques to improve query efficiency. In the case of the "hard label" setting, where the adversary can only access the top-1 prediction, Sanyal et al. [34] and Peng et al. [31] respectively introduced their model stealing attacks constrained by this limitation. Additionally, a method called "AugSteal" [10] is introduced, which utilizes data augmentation to improve the performance and efficiency of the attack.

In the realm of emerging pre-trained encoders, Liu et al. [25] developed a pioneering attack called "StolenEncoder" to steal the functionality of pre-trained encoders. Their work formulates the attack as an optimization problem and resolves it using standard stochastic gradient descent. The attack incorporates data augmentations to optimize the query budget to enhance its effectiveness. Furthermore, similar work is also discussed by Cong et al. [7]. However, existing works on encoder stealing attacks rely on strong assumptions, particularly the knowledge of the pre-training datasets for the target encoders. In contrast, our work focuses on encoder stealing attacks in the data-free setting, which are highly relevant in real-world scenarios.

2.2 Defensive Strategies

There are two major categories of methods for defending against stealing attacks. The first category focuses on detecting anomalous or malicious queries [17], while the second category perturbs the results obtained from the models or pre-trained encoders [19,28]. The first type of defensive method is based on the observation that attacks from adversaries often exhibit different mathematical statistics compared to normal queries, such as high query frequency or abnormal inputs [37]. Meanwhile, the second type of method operates on the intuition that the performance of the surrogate model can be significantly degraded if the adversary can only use the perturbed outputs for training [5]. However, in scenarios where pre-trained encoders are deployed to provide services for various downstream classification tasks [23,42], it becomes challenging to accurately define the concept of an "anomalous query" in practice. Additionally, analyzing these queries and identifying the adversary becomes challenging as the adversary may distribute the query frequency among multiple registered accounts. Furthermore, researchers in [25] have also identified the difficulty of balancing security and utility when implementing perturbation-based defensive mechanisms.

3 Preliminaries

In this work, we mainly focus on image encoders. Contrastive learning represents a prevalent category of self-supervised learning methods designed for pre-trained image encoders. Consequently, we will delve into the specifics of contrastive learning, which uses a large number of unlabeled images or image-text pairs to pre-train the image encoder, referred to as the pre-trained dataset. Next, we explore the process of pre-training the image encoder, followed by a discussion on applying this methodology to train downstream classifiers.

3.1 Pre-training an Encoder

In this section, we explore the two most advanced contrast learning algorithms SimCLR [4], MoCo [12], and BYOL [11]. These algorithms utilize unlabeled images for pre-training the image encoder.

SimCLR [4]. SimCLR is a simple comparative learning framework. The structure comprises four components: *data augmentation, the base encoder $f(\cdot)$, the projection head $g(\cdot)$, and the Contrastive loss function.*

The data augmentation module randomly converts a data sample into two enhanced views. Specifically, the augmentations consist of random cropping, random color distortion, and random Gaussian blur. When two augmented views stem from the same data sample, we classify them as positive pairs; otherwise, they are labeled as negative pairs. Moreover, when provided with an input image x, the image encoder f generates the corresponding feature vector $f(x)$. This feature vector then undergoes mapping by the projection head to compute the contrastive loss. The projection head, composed of a hidden layer, functions as a multi-layer perceptron. In SimCLR, a minibatch of N images $\{\mathbf{x}_i | i \in [1, N]\}$ results in two augmented views, generating a total of $2N$ augmented images: $\{\widetilde{x}_k | k \in [1, 2N]\}$. For a positive pair $\{\widetilde{x}_i, \widetilde{x}_j\}$, $\widetilde{x}_i$ is compared against the remaining $2(N-1)$ images to create negative pairs. The contrastive loss definition for this positive pair is as shown below:

$$\mathcal{L}_{i,j} = -\log \frac{exp\left(sim\left(h(f(\widetilde{x}_i)), h(f(\widetilde{x}_j))\right)/\tau\right)}{\sum_{k=1, k \neq i}^{2N} exp\left(sim\left(h(f(\widetilde{x}_i)), h(f(\widetilde{x}_K))\right)/\tau\right)}. \tag{1}$$

Here, sim represents the cosine similarity, while τ stands for a temperature hyperparameter. The total contrastive loss is the summation of $\mathcal{L}_{i,j}$ across all positive pairs. SimCLR conducts pre-training of an encoder f and a projection head h by minimizing the aggregate contrastive loss.

MoCo [12]. MoCo emphasizes that contrastive learning can be perceived as a dictionary lookup task, with the encoder's embeddings representing the "keys". A "query" is considered a match with a key if it originates from the same image. The goal of MoCo is to train an encoder that generates similar embeddings for a query-key pair while producing dissimilar embeddings for other instances. An effective dictionary should be extensive and coherent, encompassing diverse negative images to facilitate the learning of high-quality embeddings. MoCo endeavors to construct such a dictionary using a queue and a momentum encoder.

Given a query x^q, MoCo obtains an encoded query $q = f_q(x^q)$. For other queries represented as x^k, MoCo creates a dynamic dictionary with keys $\{k_i | i \in [0, K]\}$. This dictionary functions as a queue that stores the current mini-batch encoded embeddings while discarding the oldest mini-batch entries. The advantage of utilizing a queue lies in the detachment of the dictionary size from the mini-batch size, enabling flexibility in setting the dictionary size as a hyper-parameter. Assuming k_+ is the key to which q corresponds, the loss function can then be formulated as follows:

$$\mathcal{L} = -\log \frac{exp(q \cdot k_+/\tau)}{\sum_{i=0}^{K} exp(q \cdot k_i/\tau)}. \tag{2}$$

Here τ represents temperature, impacting the model's performance. The training process of the function f_q involves minimizing the contrastive loss and adjusting the parameter θ_q through gradient descent. Due to the presence of a queue, updating the parameter θ_k through back-propagation poses challenges, f_k is updated as:

$$f_k \leftarrow m\theta_k + (1-m)\theta_q,$$

where m represents a momentum coefficient. Ultimately, we retain f_q as the finalized pre-trained encoder.

BYOL [11]. The BYOL architecture comprises two neural networks: the online networks and the target networks. The online networks, defined by θ, comprise an encoder f_θ, a projector g_θ, and a predictor q_θ. In contrast, the target networks include an encoder f_η and a projector q_η. These networks collaborate to bootstrap the embeddings and support mutual learning.

Upon receiving an input sample x, BYOL creates two augmented views, $v \leftarrow a(x)$ and $v' \leftarrow a'(x)$, through image augmentations a and a'. The online networks generate a projection $z_\theta \leftarrow g_\theta(f_\theta(v))$, whereas the target networks produce a target projection $z'_\eta \leftarrow g_\eta(f'_\eta(v'))$. The primary objective of the online networks is to align the prediction $q_\theta(z_\theta)$ with z'_η. This alignment is formally defined as follows:

$$\mathcal{L}_{\theta,\eta} = 2 - 2 \cdot \frac{< q_\theta(z_\theta), z'_\eta >}{||q_\theta(z_\theta)||_2, ||z'_\eta||_2}. \tag{3}$$

In contrast, BYOL returns v' to the online networks and v to the target networks individually, leading to the computation of $\widetilde{L}_{\theta,\eta}$. The ultimate loss function can be expressed as:

$$\mathcal{L}_{BYOL} = \mathcal{L}_{\theta,\eta} + \widetilde{L}_{\theta,\eta}.$$

BYOL adjusts the weights of the online and target networks by:

$$\theta \leftarrow optimizer(\theta, \nabla_\theta L_{\theta,\eta}, \epsilon), \eta \leftarrow \tau\eta + (1-\tau)\theta,$$

where ϵ denotes the learning rate. The weight η of the target networks is updated using a weighted average approach, where $\tau \in [0, 1]$ represents the decay rate of the target encoder. After training, the encoder $f(\theta)$ of the online networks is considered the pre-trained encoder.

C. Zhang et al.

3.2 Model Stealing Attack

The purpose of a model stealing attack is to steal the parameters or functionality of the target model. To achieve this, when presented with a victim model $f(x; \alpha)$, an attacker can submit queries to the victim model and get corresponding responses. The queries and responses then serve as inputs and "labels" for training the surrogate model, denoted as $f(x; \alpha')$. Formally, adversary can train $f(x; \alpha')$ using a query dataset $\mathcal{D}$ by

$$\mathcal{L}_{steal} = \mathbb{E}_{x-\mathcal{D}}[sim(f(x; \alpha), f(x; \alpha'))], \tag{4}$$

where $sim(\cdot, \cdot)$ is a similarity function.

3.3 Training/Testing Downstream Classifiers

Image encoders can generate feature vectors for diverse downstream tasks, focusing on image classification in this work. Utilizing the pre-trained image encoder, we extract the feature vector of the trained image and subsequently train the downstream classifier based on the feature vector and corresponding labels following the supervised learning approach. Given a test image, we output its feature vector using an image encoder and then predict its label using a downstream classifier. The dataset used for both training and testing the downstream classifier is referred to as the *downstream dataset*.

4 Threat Model

In this study, we focus on two main entities: the *service provider* and the *adversary*. The service provider, typically a well-resourced organization such as OpenAI, Google, and Meta, possesses many resources for training encoders. They utilize contrastive learning algorithms to pre-train highly performing encoders, which are then deployed online to offer paid EaaS for commercial purposes. In contrast, the adversary's objective is to illicitly extract the target's functionality. The adversary interacts with the EaaS API by uploading images as input to the target encoder.

4.1 Adversary's Objective

The adversary is concentrated on two objectives:

- **Objective 1: Replicating the functionality.** In this objective, the adversary's goal is to steal the target encoder, ensuring that its surrogate retains the same functionality. The surrogate encoder should be able to perform numerous downstream tasks with comparable or even superior performance compared to the target encoder.
- **Objective 2: Minimizing the cost of queries.** The adversary's incurred monetary cost from accessing the EaaS API is directly influenced by the total number of queries made. Thus, the adversary's goal is to steal the target encoder by minimizing the queries required.

4.2 Adversary's Ability

In our work, the adversary only can make queries to the EaaS API by providing inputs and obtaining the corresponding embedding from the target encoder. Specifically, it is not allowed to modify the architecture or parameters of the encoder of EaaS, eavesdropping, or manipulating the queries of other users, even if doing so would advance the attack strategy.

4.3 Adversary's Background Knowledge

In this part, we analyze two elements of the adversary's background knowledge: the surrogate dataset and the encoder architecture. It is essential to emphasize that DaES can not utilize contrastive learning for training the surrogate encoder. As a result, there is no requirement for the adversary to gather information regarding the particular contrastive algorithm employed in training the target encoder.

Surrogate Dataset. Previous research [25,51] has often assumed that the adversary has access to either the same dataset or a dataset with a similar distribution as the one used to train the target encoder. However, recent studies by Kariyappa et al. [18] and Sanyal et al. [34] have shown that this assumption is unrealistic due to the confidentiality and protection of training datasets by the encoder's owner. Therefore, our work focuses on a practical scenario where the adversary lacks information about the pre-training dataset.

Encoder Architecture. In terms of the encoder architecture, the adversary may encounter two situations. Firstly, it may gain access to the architectural information of the target encoder. For instance, certain EaaS providers may openly share the specifics of their encoder architecture to promote transparency and attract prospective clients. Alternatively, if the architectural details are not publicly available, the adversary might opt for a more complex and deeper network architecture. This choice is motivated by the belief that a more complex architecture will better emulate the target's functionality.

5 Data-Free Encoder Stealing Attack

5.1 Overview

In the DaES framework, a generator training scheme is implemented to concurrently train the generator $\mathcal{G}$ and the surrogate encoder $\mathcal{S}$, which aims to minimize the dissimilarity between the embeddings generated by $\mathcal{S}$ and the target encoder $\mathcal{T}$. This is accomplished through an optimization problem that aims to minimize the divergence between these embeddings. To address the constraint of the adversary lacking access to the pre-trained dataset, DaES employs the generator $\mathcal{G}$ to generate synthetic images as the surrogate dataset $\mathcal{D}_S$. Meanwhile, the surrogate encoder $\mathcal{S}$ acts as the discriminator during training, replicating the behavior of the target encoder $\mathcal{T}$ in terms of the output embeddings. Specifically, for each element $x \in \mathcal{D}_S$, the minimization optimization problem can be formulated as follows:

$$\arg\min_{\theta_S} \mathbb{E}_{x \sim \mathcal{D}_S}[\mathcal{L}(\mathcal{T}(x), \mathcal{S}(x))], \tag{5}$$

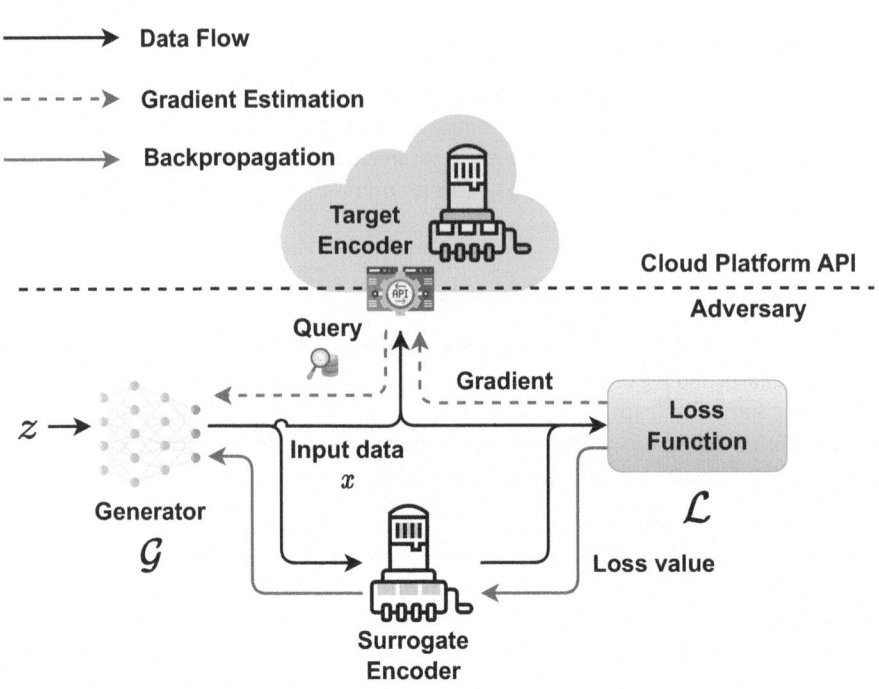

Fig. 2. The structure of the proposed Data-Free Encoder Stealing (DaES) attack.

where $\mathcal{L}$ represents the loss function. As mentioned before, it is used to measure the distance between the embeddings. Previous studies have also utilized this paradigm [39]. However, their approach to evaluating disagreement involves using prediction labels or softmax outputs as input, which are unavailable in the scenario of pre-trained encoders. In contrast, DaES only relies on the distance between output embeddings to measure the disagreement between the target encoder and the surrogate encoder.

As shown in Fig. 2, the process begins by generating a random noise vector z from a standard normal distribution. This vector is then used as input for the generator $\mathcal{G}$, which generates the corresponding input data x. Subsequently, both the surrogate and target encoders compute the respective embeddings based on x. The obtained embeddings are then used to calculate the loss value using the loss function $\mathcal{L}$. The generator and surrogate encoder are optimized during the backpropagation step. The adversary repeats these steps until it reaches its desired threshold.

Surrogate Encoder. Previous studies have demonstrated that encoder stealing attacks can effectively preserve the target's functionality $\mathcal{T}$ and achieve satisfactory performance across diverse downstream tasks, even when the architectures of the encoders differ [25]. Hence, knowledge of the target encoder's architecture is not required for successfully executing the encoder stealing attack. The adversary only needs to choose a network architecture, such as the widely used ResNet-50 [13], that has sufficient capac-

Algorithm 1. Data-Free Encoder Stealing Attack

Require: Surrogate encoder $\mathcal{S}$, total epochs e_T, generator iterations $n_{\mathcal{G}}$, learning rate μ, surrogate iterations $n_{\mathcal{S}}$, step size ε, random directions m

Ensure: Trained surrogate encoder

1: **for** $i \leftarrow 1, 2, \cdots, e_T$ **do**
2:     // Train the generator $\mathcal{G}$.
3:     **for** $j \leftarrow 1, 2, \cdots, n_{\mathcal{G}}$ **do**
4:         $z \sim \mathcal{N}(0, 1)$
5:         $x = \mathcal{G}(z; \theta_{\mathcal{G}})$
6:         Approximate the gradient $\nabla_{\theta_{\mathcal{G}}} \mathcal{L}$
7:         $\theta_{\mathcal{G}} = \theta_{\mathcal{G}} - \mu \nabla_{\theta_{\mathcal{G}}} \mathcal{L}$
8:     **end for**
9:     // Optimize the surrogate encoder $\mathcal{S}$.
10:    **for** $j \leftarrow 1, 2, \cdots, n_{\mathcal{S}}$ **do**
11:       $z \sim \mathcal{N}(0, 1)$
12:       $x = \mathcal{G}(z; \theta_{\mathcal{G}})$
13:       Calculate $\mathcal{T}(x), \mathcal{S}(x), \mathcal{L}, \nabla_{\theta_{\mathcal{S}}} \mathcal{L}$
14:       $\theta_{\mathcal{S}} = \theta_{\mathcal{S}} - \mu \nabla_{\theta_{\mathcal{S}}} \mathcal{L}$
15:    **end for**
16: **end for**

---

ity to store the target encoder's functionality as the surrogate encoder $\mathcal{S}$ [25]. Therefore, in the context of the proposed DaES framework, we select ResNet-50 as the surrogate encoder specifically for conducting encoder stealing attacks.

The loss function $\mathcal{L}$ is utilized to evaluate the dissimilarity between the output embeddings of the surrogate and target encoder. In this work, we primarily adopt the cosine similarity as the loss function, given its proven effectiveness in prior studies [7,25]. Further details regarding the selection of $\mathcal{L}$ are provided in Subsect. 5.2. Importantly, the gradient of the loss with respect to the surrogate encoder's gradients, $\theta_{\mathcal{S}}$, can be computed independently of the target encoder's gradients since the weights of the surrogate encoder do not impact the embeddings produced by the target encoder.

Generator. In DaES, the generator $\mathcal{G}$ is used to produce synthesized images, which are then used as input data for maximizing the disagreement between the surrogate encoder $\mathcal{S}$ and the target encoder $\mathcal{T}$. The loss function $\mathcal{L}$ that is utilized to optimize $\mathcal{G}$ is the same as that of $\mathcal{S}$, with the objective being to maximize it. Therefore, the minimization optimization problem presented in Eq. (5) can be considered as an adversarial game, where $\mathcal{G}$ and $\mathcal{S}$ compete to maximize and minimize the same function, respectively. Essentially, the generator is optimized to generate challenging images for the surrogate encoder, while the surrogate encoder is optimized to preserve the target's functionality with respect to the embeddings. The game can be formulated as follows:

$$\min_{\mathcal{S}} \max_{\mathcal{G}} \mathbb{E}_{z \sim \mathcal{N}(0,1)} [\mathcal{L}(\mathcal{T}(\mathcal{G}(z)), \mathcal{S}(\mathcal{G}(z)))]. \tag{6}$$

As illustrated in Fig. 2, the calculation of the gradient, denoted as $\mathcal{L}$, concerning the parameters $\theta_{\mathcal{G}}$ relies on obtaining the gradient of the target encoder $\mathcal{T}$. However, as the target encoder is only accessible through the provided API (black box), it becomes necessary to employ gradient estimation methods. We will discuss these methods in the Subsect. 5.3.

Algorithm. During each training iteration of the DaES attack, the generator $\mathcal{G}$ and the surrogate encoder $\mathcal{S}$ undergo alternate training. To achieve a more balanced training process between $\mathcal{G}$ and $\mathcal{S}$ for each epoch, the training steps are repeated $n_{\mathcal{G}}$ times for $\mathcal{G}$ and $n_{\mathcal{S}}$ times for $\mathcal{S}$. Once the iteration is complete, the attack proceeds to the next epoch. While increasing $n_{\mathcal{G}}$ can speed up the training of $\mathcal{G}$ and enable it to generate more challenging samples for $\mathcal{S}$, it may be wasteful if $\mathcal{S}$ does not have enough samples for further optimization. Hence, achieving a balance between $n_{\mathcal{G}}$ and $n_{\mathcal{S}}$ requires careful fine-tuning. The hyperparameters ε and m are also involved in the gradient estimation process. The detailed process of DaES is illustrated in Algorithm 1.

5.2 Loss Function

In this section, we explore loss functions for evaluating the disagreement between $\mathcal{T}$ and $\mathcal{S}$. These losses are frequently utilized in contrastive learning because they share similarities with the encoder stealing attacks [6,43]. The loss function plays a critical role in the attack, as it can hinder optimizer convergence by causing gradients computed through $\mathcal{S}$ and $\mathcal{T}$ to vanish when an incorrect loss function is employed.

Cosine Similarity for Disagreement Evaluation. The majority of previous studies in contrastive learning focused on leveraging the cosine similarity to assess the similarity between embeddings for positive and negative pairings [4,6,43]. Therefore, the cosine similarity between the outputs of $\mathcal{S}$ and $\mathcal{T}$ is a suitable choice as the loss function for training the surrogate encoder. For the surrogate encoder $\mathcal{S}$ and the target encoder $\mathcal{T}$, the cosine similarity of the corresponding embeddings based on a given input x is computed as follows:

$$\mathcal{L}_{Cosine}(\mathcal{S}(x), \mathcal{T}(x)) = -\frac{\mathcal{S}(x) \cdot \mathcal{T}(x)}{||\mathcal{S}(x)|| \times ||\mathcal{T}(x)||}. \tag{7}$$

However, the cosine distance solely considers the angle between two vectors and disregards their magnitudes. Consequently, as argued by Truong et al. [39], the corresponding vectors produced by two different models may be considerably dissimilar even if the cosine distance is -1.

$\mathcal{L}_1$ **and** $\mathcal{L}_2$ **Norm Loss for Disagreement Evaluation.** To address this issue that may hinder the convergence of the loss function during the optimization, we propose the use of the $\mathcal{L}_1$ norm loss as a means of measuring the distance between the embedding obtained from the surrogate encoder and the one obtained from the target encoder. However, according to Liu et al. [25], the $\mathcal{L}_1$ norm loss aims to promote sparsity in the differences between the two embeddings, meaning that some dimensions may have no difference while others may have significant differences. To gain a better understanding of the impact of the loss function in DaES, we also incorporate the $\mathcal{L}_1$ and $\mathcal{L}_2$ norm loss to evaluate the similarity between the two embeddings.

5.3   Gradient Estimation

The adversary has limited access to the target encoder $\mathcal{T}$ through a black-box API. However, in order to effectively train the generator $\mathcal{G}$, it is crucial to compute the gradients of the loss $\mathcal{L}$ with respect to the generator's parameters $\theta_{\mathcal{G}}$ for backpropagation. Therefore, we estimate these gradients by interacting with the API and using zeroth-order optimization techniques with lower query budgets.

Images as Proxy. The generator $\mathcal{G}$ typically has a large number of parameters, often in the millions or more. Consequently, directly applying the zeroth-order optimization method to obtain precise gradient approximations in such a high-dimensional space would be computationally intensive and impractical for the adversary. To tackle this problem, the adversary can estimate the gradients with respect to the input data x and utilize these gradients for backpropagation through the generator $\mathcal{G}$ [18]. This technique can reduce the dimensionality of the gradients for estimation, resulting in more accurate zeroth-order approximations.

Forward Difference Method for Estimation. The Forward Difference method [40] estimates gradients by calculating directional derivatives of a function f at a point x in m random directions v_i, $i = 1 \cdots m$. The directional derivatives are obtained by measuring the variation of f along each direction v_i with a small step size ε. These derivatives are then averaged to approximate the gradient $\nabla_{FD} f(x)$. Each directional derivative provides valuable information about the true gradient, and the accuracy of the approximation improves with a greater number of accumulated derivatives. The detail of this method is represented as Eq. (8).

$$\nabla_{FD} f(x) = \frac{1}{m} \sum_{i=1}^{m} d \frac{f(x + \varepsilon v_i) - f(x)}{\varepsilon} v_i, \tag{8}$$

where d represents the step size used in the gradient approximation according to the forward difference method.

One notable benefit of this work is that the quantity of random directions, represented as m, can be selected regardless of the input space dimensionality. This characteristic allows for a better trade-off between gradient accuracy and query utilization. In contrast, another gradient estimation method called Finite Difference, which is also employed for generating adversarial samples [2], requires a substantial number of queries to estimate each gradient.

6   Experiments

6.1   Experimental Settings

Our experiments employ three distinct contrastive learning paradigms: MoCo [12], SimCLR [4], and BYOL [11] for training the target encoders on the CIFAR-10 dataset. The target encoder architecture is based on the widely adopted ResNet-34 network. We construct the surrogate encoder using the more complex ResNet-50 network for the stealing step. Subsequently, we apply the target and surrogate encoders to perform

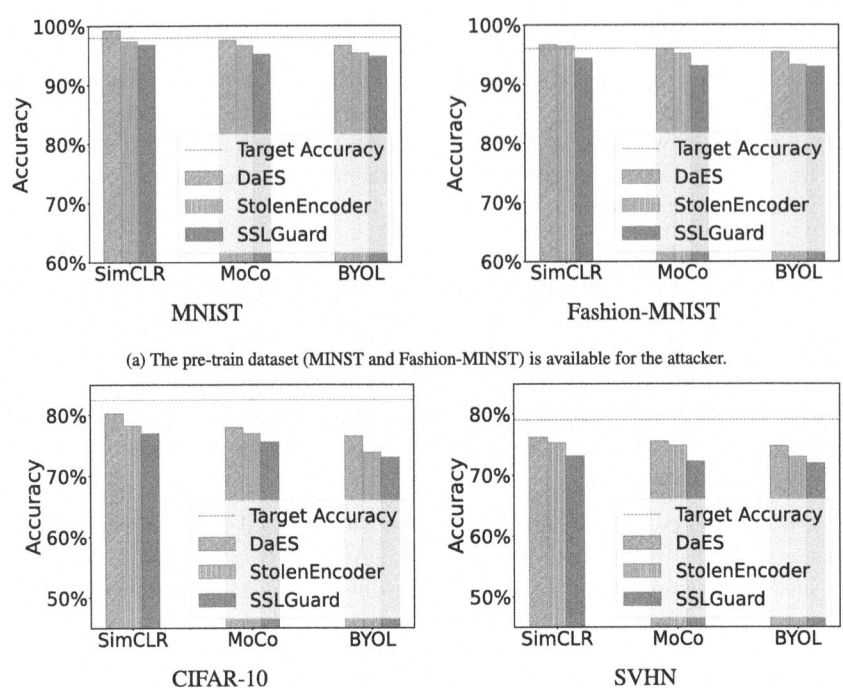

(a) The pre-train dataset (MINST and Fashion-MINST) is available for the attacker.

(b) The pre-train dataset (CIFAR-10 and SVHN) is available for the attacker.

Fig. 3. The performance of DaES is compared to conventional encoder stealing attacks across various downstream classification tasks. The x-axis denotes different contrastive learning algorithms, while the y-axis represents the accuracy of the downstream tasks.

image classification tasks on four real-world datasets: MNIST [8], Fashion-MNIST [41], CIFAR-10 [20], and SVHN [47]. For the experimental platform, we utilized one NVIDIA 4090 GPU equipped with 24 GB of memory, as our computational resource in terms of the hardware.

Regarding hyperparameters, we initialize the learning rate as 0.01, set the step size ε to 10^{-7}, and fix the number of random directions m to 10. Additionally, we specify the total number of epochs e_T as 100, and both the generator and surrogate iterations $n_{\mathcal{G}}$ and $n_{\mathcal{S}}$ are set to 100 as well.

6.2 Performance Comparison

We evaluate the effectiveness of the DaES attack in various downstream image classification tasks. We use the conventional encoder stealing attacks presented in prior works [7,25] as baselines for better comparison. Two distinct scenarios are considered: firstly, when the adversary knows the pre-training dataset, and secondly when the adversary has no prior knowledge of the private pre-training dataset. The corresponding experimental results can be seen in Figs. 3 and 4.

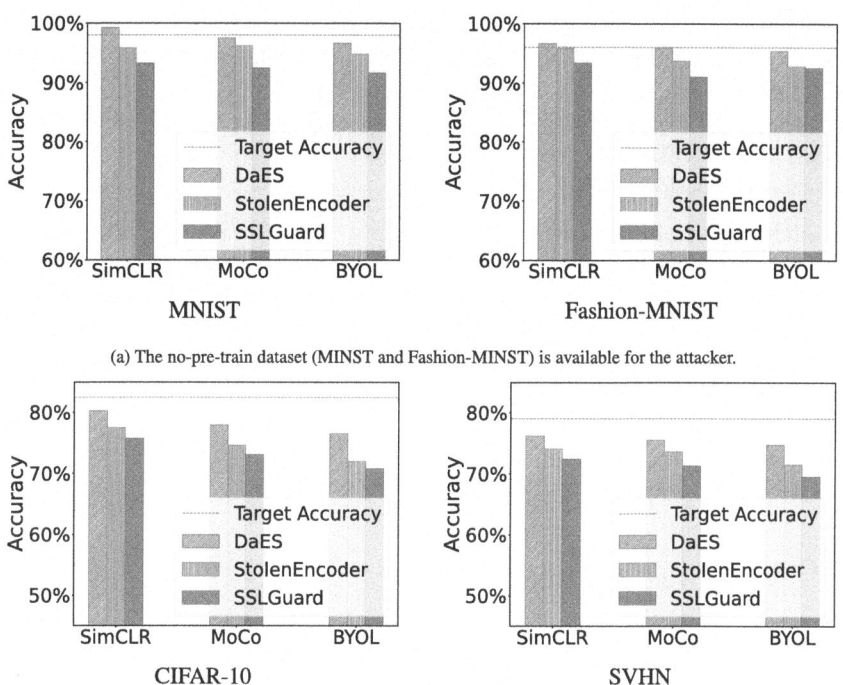

(a) The no-pre-train dataset (MINST and Fashion-MINST) is available for the attacker.

(b) The no-pre-train dataset (CIFAR-10 and SVHN) is available for the attacker.

Fig. 4. DaES performance is rigorously assessed against conventional encoder stealing attacks across a range of downstream classification tasks. The x-axis specifies the different contrastive learning algorithms utilized for the target encoder, while the y-axis indicates the corresponding accuracy of the downstream tasks.

We have compared the performance of DaES with existing encoder stealing attacks like SSLGuard and StolenEncoder as baselines to demonstrate the performance of DaES on downstream tasks. The reason the drops in baselines without the pre-train dataset are not significant is the downstream tasks may based on simple datasets (e.g., MNIST or FashionMNIST). Even if the surrogate encoder is obtained via different datasets with totally different distributions, the distribution of output feature vectors is complex enough to cover the inner characteristics of such datasets for completing downstream classification tasks. Moreover, we use *Target Accuracy* to represent the accuracy of the target encoder on downstream tasks.

Available Pre-training Dataset. Figure 3 illustrates the attack performance on downstream classification tasks using four different datasets when the adversary knows the pre-training dataset. DaES demonstrates strong performance even with variations in the contrastive learning algorithm of the target encoder and the downstream classification tasks. Notably, when the downstream tasks are based on MNIST or Fashion-MNIST, DaES outperforms the target encoder by at least 1% in terms of accuracy. It is noteworthy that even under the strong assumption that the adversary can obtain the same dataset for pre-train the target encoder as the surrogate encoder for the encoder steal-ing attack, DaES consistently performs better than conventional attacks [7,25] on all downstream classification tasks across three contrastive learning algorithms. While the performance may slightly degrade for more complex downstream classification datasets (e.g., CIFAR-10, SVHN), the surrogate encoder is still able to preserve over 90% of the target encoder's performance.

Unavailable Pre-train Dataset. As shown in Fig. 4, when considering a more practical scenario in which the adversary is unaware of the pre-training dataset, the correspond-ing performance of DaES can be seen in Fig. 4a. It is evident that the performance of conventional attacks [7,25] degrades when the pre-training dataset is not available for the adversary. This is reasonable since the pre-training dataset provides valuable infor-mation to the adversary in these attacks [7,25]. However, the performance of DaES is unaffected, as it does not rely on such datasets, and it still outperforms conventional attacks. This indicates that simply keeping the pre-training dataset confidential is insuf-ficient to defend against potential encoder stealing attacks.

6.3 Cost Analysis

The cost of the DaES is also an important aspect worth considering. As previously demonstrated, training pre-trained encoders require a significant amount of time, valu-able datasets, and substantial computational resources. Therefore, adversaries must minimize the overall cost when stealing the encoder for more illegal profits. This part examines the time and monetary costs associated with training an encoder from scratch or using DaES to steal the target's functionality. The monetary cost of DaES includes expenses related to encoder queries and surrogate encoder training. For the analysis, the query cost is assumed to be $1 per 1,000 queries, based on the pricing policy of Amazon Web Services (AWS). Subsequently, we consider a training platform equipped with a single NVIDIA A100 GPU from the Google Cloud, priced at $3.877 per hour.

Table 1. The monetary cost and time cost associated with normal training from scratch and DaES attacks differ. The complete monetary cost of DaES includes both the cost of queries and the cost of training. Considering that the time needed for queries is considerably less than that for training, we can ignore it when calculating the overall time cost of DaES.

Algorithm	Monetary Cost		Time Cost	
	Training ($)	DaES ($)	Training (h)	DaES (h)
SimCLR	86.61	**29.03 (20 + 9.03)**	22.34	**2.33**
MoCo	80.87	**30.15 (20 + 10.15)**	20.86	**2.62**
BYOL	84.40	**29.57 (20 + 9.57)**	21.77	**2.47**

We report the comparison of the monetary and time costs associated with DaES and training an encoder from scratch in Table 1. The results show that DaES can effec-tively extract the target's functionality while significantly reducing both the monetary and time costs compared to training an encoder from scratch. For example, training a ResNet-34 encoder using the SimCLR algorithm on the CIFAR-10 dataset requires an expenditure of \$86.61 and a duration of 22.34 h. In contrast, DaES achieves stealing a pre-trained encoder with comparable performance on downstream tasks using only \$29.03 and 2.33 h. These findings demonstrate that DaES enables the extraction of the target encoder's functionality and the construction of surrogate encoders with comparable performance, all while requiring fewer financial and temporal resources.

Furthermore, in addition to the monetary cost, we also examine the computational resource expenditures associated with DaES and the state-of-the-art attacks. To be specific, we establish the input batch size for both attacks as 64 and assess the costs based on GPU memory usage. The comparison outcomes are presented in Table 2. As we can see, existing state-of-the-art attacks require a minimum of 23 GB of memory to execute, which nearly exhausts the GPU's memory mentioned earlier. In contrast, DaES utilizes a maximum of only 3 GB. Furthermore, existing state-of-the-art attacks require a larger number of queries compared to DaES (78,200 vs. 20,000), resulting in a longer duration and higher monetary costs for completing the encoder stealing process, as indicated in Table 2. Therefore, DaES is a more viable option for adversaries with limited computational resources.

Table 2. The computational resource and time cost for DaES and the conventional attacks.

Algorithm	GPU Memory Usage		Time Cost	
	Conventional (MB)	DaES (MB)	Conventional (h)	DaES (h)
SimCLR	23576	**2343**	8.35	**2.33**
MoCo	23612	**2678**	8.63	**2.62**
BYOL	23784	**2981**	9.02	**2.47**

6.4 Impact of Loss Functions

The central component of DaES involves minimizing the distance between the embeddings obtained from the surrogate encoder and the target encoder. Hence, selecting an appropriate loss function can significantly improve DaES performance on downstream tasks. In this part, we assess the effects of three distinct loss functions: cosine similarity, $\mathcal{L}_1$ norm, and $\mathcal{L}_2$ norm. The results corresponding to each loss function are illustrated in Fig. 5, where the downstream tasks are CIFAR-10 and SVHN, respectively.

As we can see, the performance of DaES consistently improves as the number of epochs increases for all three loss functions. Among them, the cosine similarity demonstrates superior performance compared to the other two loss functions. One possible explanation is that the cosine similarity better captures the similarity between different embeddings compared to the other functions [25].

6.5 Defense Methods

In this part, we examine the influence of perturbation-based methods, which are widely utilized, on DaES. These methods primarily involve perturbing the embeddings derived

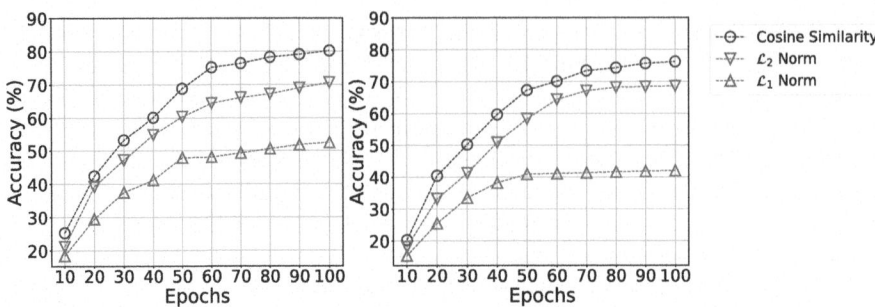

Fig. 5. The performance of the DaES with different loss functions.

from the target encoder for a specific query before presenting it to the user. The underlying idea is that by perturbing the embeddings, the adversary's capacity to train a surrogate encoder is restricted, leading to a less precise model.

Here, we employ three widely used techniques to defend against DaES, namely Top-
K, rounding, and noise poisoning. The adversary exploits the perturbed embeddings to
optimize its surrogate encoder. The corresponding results are illustrated in Table 3. For
simplicity, we use the TA to represent target accuracy, and SA for surrogate accuracy.

Table 3. Performance of DaES on defenses.

Defenses	Pre-training Dataset					
	MNIST		CIFAR-10		SVHN	
	TA (%)	SA (%)	TA (%)	SA (%)	TA (%)	SA (%)
Top-100	**93.45**	**94.21**	**74.23**	**72.19**	**69.09**	**67.52**
Top-50	85.43	83.12	65.34	66.11	59.34	56.71
Top-10	70.15	71.67	53.87	51.53	47.34	48.05
Top-1	41.23	39.90	30.98	25.38	26.45	21.34
m = 3	**95.94**	**96.01**	**79.84**	**80.06**	**74.22**	**73.87**
m = 2	92.48	93.77	78.56	79.91	72.45	73.61
m = 1	83.45	87.10	71.34	76.98	69.83	72.90
$\varepsilon = 0.01$	**92.13**	**92.46**	**78.24**	**77.16**	**73.45**	**72.56**
$\varepsilon = 0.05$	78.23	77.54	57.62	56.87	50.23	48.37
$\varepsilon = 0.1$	56.09	53.42	43.25	44.10	41.98	42.09

If attackers are unaware of the architecture of the target encoder, they tend to opt
for a deeper and more complex surrogate encoder. This is because, as demonstrated
in existing works on deep networks like ResNet, networks with more intricate architectures tend to perform better due to the more complex distributions of their output feature
vectors, potentially achieving higher accuracy compared to the target encoder. However,

the challenge lies in balancing the security and utility of the target encoder. The top-k defense method's results highlight its limitations: smaller values of k lead to inadequate defense, while larger values degrade utility. In the rounding method, where each content in the embedding is rounded to m decimal places, the performance of DaES remains unaffected as m varies from 1 to 3. Additionally, the noise poisoning method's results show that a small magnitude of ϵ is insufficient for protecting the target encoder, while a large magnitude may also impair its performance for normal queries. Consequently, finding the right balance between security and utility for the target encoder is a significant issue that requires further exploration.

7 Conclusion

In this paper, we present DaES, which is a new data-free stealing attack for encoders in self-supervised learning. DaES uses generator training schemes and gradient estimation techniques to efficiently replicate the functions of the target encoder without accessing its private pre-training dataset. The experimental results show that DaES outperforms the current state-of-the-art encoder stealing attacks across various downstream tasks while significantly reducing API query costs. This finding highlights the need for stronger security measures when it comes to protecting the intellectual property of pre-trained encoders. Future work should focus on developing new defense strategies to effectively defend against such data-free stealing attacks.

Acknowledgments. This work was financially supported by the National Natural Science Foundation of China (Grant Nos. 62472032, 62202051, and 62232002); the Open Project Funding of Key Laboratory of Mobile Application Innovation and Governance Technology, Ministry of Industry and Information Technology, (Grant No. 2023IFS080601-K); the Beijing Institute of Technology Research Fund Program for Young Scholars, and the Young Elite Scientists Sponsorship Program by CAST (Grant No. 2023QNRC001). The Young Scientists Fund of the National Natural Science Foundation of China (Grant No. 62402040), the Postdoctoral Fellowship Program of CPSF (Grant No. GZB20230938), the China Postdoctoral Science Foundation (Grant No. 2024T171132 and No. 2023M740246).

References

1. Baevski, A., Hsu, W., Xu, Q., Babu, A., Gu, J., Auli, M.: data2vec: a general framework for self-supervised learning in speech, vision and language. In: Proceedings of Machine Learning Research (ICML), vol. 162, pp. 1298–1312. PMLR (2022)
2. Bhagoji, A.N., He, W., Li, B., Song, D.: Practical black-box attacks on deep neural networks using efficient query mechanisms. In: Ferrari, V., Hebert, M., Sminchisescu, C., Weiss, Y. (eds.) ECCV 2018. LNCS, vol. 11216, pp. 158–174. Springer, Cham (2018). https://doi.org/10.1007/978-3-030-01258-8_10
3. Chen, L., Zhang, Y., Song, Y., Liu, L., Wang, J.: Self-supervised learning of adversarial example: towards good generalizations for deepfake detection. In: CVPR, pp. 18689–18698. Computer Vision Foundation/IEEE (2022)
4. Chen, T., Kornblith, S., Norouzi, M., Hinton, G.E.: A simple framework for contrastive learning of visual representations. In: Proceedings of Machine Learning Research (ICML), vol. 119, pp. 1597–1607. PMLR (2020)

5. Chen, Y., Guan, R., Gong, X., Dong, J., Xue, M.: D-DAE: defense-penetrating model extraction attacks. In: SP, pp. 382–399. IEEE (2023)
6. Chuang, C., et al.: Robust contrastive learning against noisy views. In: CVPR, pp. 16649–16660. Computer Vision Foundation/IEEE (2022)
7. Cong, T., He, X., Zhang, Y.: SSLGuard: a watermarking scheme for self-supervised learning pre-trained encoders. In: CCS, pp. 579–593. ACM (2022)
8. Deng, L.: The MNIST database of handwritten digit images for machine learning research [best of the web]. IEEE Sig. Process. Mag. **29**(6), 141–142 (2012)
9. Feng, S., et al.: Detecting backdoors in pre-trained encoders. In: CVPR, pp. 16352–16362. Computer Vision Foundation/IEEE (2023)
10. Gao, L., Liu, W., Liu, K., Wu, J.: AugSteal: advancing model steal with data augmentation in active learning frameworks. IEEE Trans. Inf. Forensics Secur. (2024)
11. Grill, J., et al.: Bootstrap your own latent - a new approach to self-supervised learning. In: NeurIPS (2020)
12. He, K., Fan, H., Wu, Y., Xie, S., Girshick, R.B.: Momentum contrast for unsupervised visual representation learning. In: CVPR, pp. 9726–9735. Computer Vision Foundation/IEEE (2020)
13. He, K., Zhang, X., Ren, S., Sun, J.: Deep residual learning for image recognition. In: CVPR, pp. 770–778. Computer Vision Foundation/IEEE (2016)
14. Hu, C., Zhang, C., Lei, D., Wu, T., Liu, X., Zhu, L.: Achieving privacy-preserving and verifiable support vector machine training in the cloud. IEEE Trans. Inf. Forensics Secur. **18**, 3476–4291 (2023)
15. Huang, L., You, S., Zheng, M., Wang, F., Qian, C., Yamasaki, T.: Learning where to learn in cross-view self-supervised learning. In: CVPR, pp. 14431–14440. Computer Vision Foundation/IEEE (2022)
16. Jia, J., Liu, Y., Gong, N.Z.: BadEncoder: backdoor attacks to pre-trained encoders in self-supervised learning. In: SP, pp. 2043–2059. IEEE (2022)
17. Juuti, M., Szyller, S., Marchal, S., Asokan, N.: PRADA: protecting against DNN model stealing attacks. In: EuroS&P, pp. 512–527. IEEE (2019)
18. Kariyappa, S., Prakash, A., Qureshi, M.K.: MAZE: data-free model stealing attack using zeroth-order gradient estimation. In: CVPR, pp. 13814–13823. Computer Vision Foundation/IEEE (2021)
19. Kariyappa, S., Qureshi, M.K.: Defending against model stealing attacks with adaptive misinformation. In: CVPR, pp. 767–775. Computer Vision Foundation/IEEE (2020)
20. Krizhevsky, A., Hinton, G., et al.: Learning multiple layers of features from tiny images. Toronto, ON, Canada (2009)
21. Li, C., et al.: An embarrassingly simple backdoor attack on self-supervised learning. In: CVPR, pp. 4367–4378. Computer Vision Foundation/IEEE (2023)
22. Lin, Z., Xu, K., Fang, C., Zheng, H., Jaheezuddin, A.A., Shi, J.: QUDA: query-limited data-free model extraction. In: AsiaCCS, pp. 913–924. ACM (2023)
23. Liu, F., et al.: Integrally migrating pre-trained transformer encoder-decoders for visual object detection. In: CVPR, pp. 6825–6834. Computer Vision Foundation/IEEE (2023)
24. Liu, H., Jia, J., Qu, W., Gong, N.Z.: EncoderMI: membership inference against pre-trained encoders in contrastive learning. In: CCS, pp. 2081–2095. ACM (2021)
25. Liu, Y., Jia, J., Liu, H., Gong, N.Z.: StolenEncoder: stealing pre-trained encoders in self-supervised learning. In: CCS, pp. 2115–2128. ACM (2022)
26. Oliynyk, D., Mayer, R., Rauber, A.: I know what you trained last summer: a survey on stealing machine learning models and defences. CoRR abs/2206.08451 (2022)
27. Orekondy, T., Schiele, B., Fritz, M.: Knockoff nets: stealing functionality of black-box models. In: CVPR, pp. 4954–4963. Computer Vision Foundation/IEEE (2019)

28. Orekondy, T., Schiele, B., Fritz, M.: Prediction poisoning: towards defenses against DNN model stealing attacks. In: ICLR. OpenReview.net (2020)
29. Pang, Y., Wang, W., Tay, F.E.H., Liu, W., Tian, Y., Yuan, L.: Masked autoencoders for point cloud self-supervised learning. In: Avidan, S., Brostow, G., Cissé, M., Farinella, G.M., Hassner, T. (eds.) Computer Vision – ECCV 2022: 17th European Conference, Tel Aviv, Israel, October 23–27, 2022, Proceedings, Part II, pp. 604–621. Springer, Cham (2022). https://doi.org/10.1007/978-3-031-20086-1_35
30. Papernot, N., McDaniel, P.D., Goodfellow, I.J., Jha, S., Celik, Z.B., Swami, A.: Practical black-box attacks against machine learning. In: AsiaCCS, pp. 506–519. ACM (2017)
31. Peng, W., et al.: Are you copying my model? Protecting the copyright of large language models for EaaS via backdoor watermark. In: ACL, pp. 7653–7668. Association for Computational Linguistics (2023)
32. Qu, W., Jia, J., Gong, N.Z.: REaaS: enabling adversarially robust downstream classifiers via robust encoder as a service. In: NDSS. The Internet Society (2023)
33. Rosenthal, J., Enouen, E., Pham, H.V., Tan, L.: Disguide: disagreement-guided data-free model extraction. In: AAAI, pp. 9614–9622. AAAI Press (2023)
34. Sanyal, S., Addepalli, S., Babu, R.V.: Towards data-free model stealing in a hard label setting. In: CVPR, pp. 15263–15272. Computer Vision Foundation/IEEE (2022)
35. Shi, J., Liu, Y., Zhou, P., Sun, L.: BadGPT: exploring security vulnerabilities of ChatGPT via backdoor attacks to InstructGPT. CoRR abs/2304.12298 (2023)
36. Szyller, S., Duddu, V., Gröndahl, T., Asokan, N.: Good artists copy, great artists steal: model extraction attacks against image translation generative adversarial networks. CoRR abs/2104.12623 (2021)
37. Tang, M., et al.: ModelGuard: information-theoretic defense against model extraction attacks. In: USENIX Security Symposium. USENIX Association (2024)
38. Tramèr, F., Zhang, F., Juels, A., Reiter, M.K., Ristenpart, T.: Stealing machine learning models via prediction APIs. In: USENIX Security Symposium, pp. 601–618. USENIX Association (2016)
39. Truong, J., Maini, P., Walls, R.J., Papernot, N.: Data-free model extraction. In: CVPR, pp. 4771–4780. Computer Vision Foundation/IEEE (2021)
40. Villatoro, F.R., Ramos, J.I.: On the method of modified equations. I: asymptotic analysis of the Euler forward difference method. Appl. Math. Comput. **103**(2-3), 111–139 (1999). https://doi.org/10.1016/S0096-3003(98)10031-0
41. Xiao, H., Rasul, K., Vollgraf, R.: Fashion-MNIST: a novel image dataset for benchmarking machine learning algorithms. CoRR abs/1708.07747 (2017)
42. Xu, X., et al.: BridgeTower: building bridges between encoders in vision-language representation learning. In: AAAI, pp. 10637–10647. AAAI Press (2023)
43. Yin, J., Wu, H., Sun, S.: Effective sample pairs based contrastive learning for clustering. Inf. Fus. **99**, 101899 (2023)
44. Yu, W., Yang, K., Bai, Y., Xiao, T., Yao, H., Rui, Y.: Visualizing and comparing AlexNet and VGG using deconvolutional layers. In: Proceedings of the 33rd International Conference on Machine Learning (2016)
45. Yuan, Z., et al.: Pre-trained image encoder for generalizable visual reinforcement learning. Adv. Neural. Inf. Process. Syst. **35**, 13022–13037 (2022)
46. Yue, Z., He, Z., Zeng, H., McAuley, J.J.: Black-box attacks on sequential recommenders via data-free model extraction. In: RecSys, pp. 44–54. ACM (2021)
47. Yuval, N.: Reading digits in natural images with unsupervised feature learning. In: NeurIPS (2011)
48. Zhang, C., Hu, C., Wu, T., Zhu, L., Liu, X.: Achieving efficient and privacy-preserving neural network training and prediction in cloud environments. IEEE Trans. Dependable Secure Comput. **20**(5), 4245–4257 (2023)

49. Zhang, P., et al.: Multi-scale vision Longformer: a new vision transformer for high-resolution image encoding. In: ICCV, pp. 2978–2988. IEEE (2021)
50. Zhang, S., et al.: Align representations with base: a new approach to self-supervised learning. In: CVPR, pp. 16579–16588. Computer Vision Foundation/IEEE (2022)
51. Zhao, S., et al.: Extracting cloud-based model with prior knowledge. CoRR abs/2306.04192 (2023)

Data Poisoning Attack Against Reinforcement Learning from Human Feedback in Robot Control Tasks

Zihui Zhou[1]($\boxtimes$) (iD), Yutong Gao[2,3], and Minfeng Qi[4]

[1] China University of Geosciences(Wuhan), Wuhan 430074, China
zzzzzh@cug.edu.cn
[2] Key Laboratory of Ethnic Language Intelligent Analysis and Security Governance,
Ministry of Education, Minzu University of China, Beijing 100081, China
ytgao92@muc.edu.cn
[3] Hainan International College of Minzu University of China, Li'an International
Education Innovation pilot Zone, Hainan 572499, China
[4] City University of Macau, Macau 999078, China
mfqi@cityu.edu.mo

Abstract. Reinforcement Learning from Human Feedback (RLHF) technology provides a method for agents to learn human preferences and perform actions that satisfy human desires. This technology was originally used to complete robot control tasks in situations where it was difficult to design a reward function. However, in the process of collecting human feedback, malicious human annotators may launch attacks on RLHF, posing a significant challenge to the technology. Previous research has mostly focused on studying the harm caused by different attacks on RLHF in fine-tuning large language models (LLMs). However, our study is focused on the field of robot control. We designed two data poisoning attacks against human feedback datasets and implemented our attacks in three different offline reinforcement learning (RL) environments. The experimental results show that RLHF is vulnerable to data poisoning attacks in robot control tasks.

Keywords: data poisoning attack · Reinforcement Learning from Human Feedback · robot control task

1 Introduction

In recent years, owing to the widespread adoption of large language models (LLMs) [12,34], the key technology underpinning their success, Reinforcement Learning from Human Feedback (RLHF), has also come into the forefront. The predecessor of RLHF, Preference-based Reinforcement Learning (PbRL), was initially proposed by [13] in 2017 and was employed to address situations where designing a reward function for control tasks is complex and challenging. In Christiano's introduction, RLHF involved pairing different trajectories and presenting them to human annotators for labeling, determining their preferences.

© The Author(s), under exclusive license to Springer Nature Singapore Pte Ltd. 2025
T. Zhu et al. (Eds.): ICA3PP 2024, LNCS 15251, pp. 121–140, 2025.
https://doi.org/10.1007/978-981-96-1525-4_7

Subsequently, the acquired preference dataset was utilized to learn a reward function guiding the robot in completing the specified task.

Although RLHF has significant advantages in many fields, it also faces some security and privacy issues. As [8] shown in their paper, RLHF faces many concrete challenges with the **reward model**, the **policy**, and the **human feedback**. For human feedback, challenges can emerge from misaligned evaluators, the difficulty of supervision, the quality of data, and the form of the feedback used as [8] introduced. The attacks mentioned in this paper are mainly implemented based on the problem that it is difficult to supervise the human annotators, which is easy for attackers to infiltrate and poison the data. Our attack is designed to target the RLHF technology used in robot control tasks. In this case, following previous work [13, 20], the trajectory of the robot is first segmented into trajectory segments and presented in the form of a video. Subsequently, these trajectory segments are paired to create trajectory pairs, which are then sent to human experts for labeling based on their preferences. After the labeling process is completed, the trajectory pairs, along with their preference labels, are returned to the trainer. The trainer utilizes this preference dataset to train the reward model, enabling the reward function to predict rewards aligned with human preferences. This, in turn, guides the training of RL effectively.

In our hypothetical scenario, trajectory pairs are distributed among multiple human annotators for labeling preferences within the entire dataset. During this process, we assume that attackers possess the capability to manipulate the preference labels for certain datasets. One prevalent scenario involves the presence of malicious attackers among the human annotators. Compared to label poisoning attacks in large language models, the quality of each label is more critical in robot control tasks due to the relatively smaller number of labels. Therefore, the harm of label poisoning is also greater.

To assess the risks associated with human feedback influenced by poisoned preference datasets for RLHF in robot control tasks, we conducted two types of attacks: **Label Flipping Attack** and **Targeted Label Poisoning Attack**. These attacks aim to investigate the impact of data poisoning during the feedback process. When the preference dataset includes trajectories with poisoned labels, the reward model becomes misinformed and assigns inflated scores. Implementing the Label Flipping Attack results in the robot performing more incorrect actions while opting for the Targeted Label Poisoning Attack leads to the robot executing actions desired by the attacker.

To showcase the effectiveness of our attack, we conducted simulations in three offline Reinforcement Learning (RL) environments: AntMaze, Kitchen, and Walker2d [18]. In the case of the Label Flipping Attack, our goal is to diminish the success rate of the original task. Conversely, for the Targeted Label Poisoning Attack, our objective is not only to decrease the success rate of the original task but, more importantly, to ensure the completion of tasks specified by the attacker. From the experimental outcomes, we observe that the simple Label Flipping Attack can impact the performance of RLHF, but it requires the attacker to possess a larger amount of preference dataset. On the other hand, the Targeted Label Poisoning Attack yields better results, enabling the attacker to achieve the attack with relatively less data while significantly reducing the success rate of the original task. Moreover, we executed a Targeted Label Poisoning Attack according to the configurations outlined in [20], where the reward function generates a non-Markovian reward based on the latest representation, to study the robustness of different strategies. The experimental results indicate that models generating rewards more in line with human preferences exhibit stronger defense against our attacks.

In summary, we have investigated and showcased the susceptibility of RLHF to data poisoning attacks on the preference dataset in the domain of robot control. We aim for our findings to contribute towards enhancing the security of RLHF and developing more robust methods for modeling preferences in the future.

2   Preliminaries

Reinforcement Learning (RL) represents a machine learning paradigm wherein an agent learns decision-making through interactions with its environment. The agent receives feedback in the form of rewards or penalties, depending on its actions. The objective of RL is for the agent to acquire a policy—a strategy or set of rules—that optimizes its cumulative reward over time.

The process of RL is defined as a Markov Decision Process (MDP), represented by the quintuple denoted as the tuple (S, A, T, r, γ). Here, S is the state space, A is the action space, T is the transition matrix that maps state-action pairs to states, r is the reward function, and γ is the discount factor. The goal of reinforcement learning can be expressed as learning a policy π that maximizes cumulative rewards:

$$\mathcal{R}_t = \sum_{i=0}^{\infty} \gamma^i r(\mathbf{s}_{t+i}, \mathbf{a}_{t+i})$$

To assist the agent in acquiring improved strategies, designing an effective reward function is crucial. However, in certain scenarios, precisely crafting the reward function becomes challenging, especially in complex reinforcement learning tasks. For instance, when training a hopper robot for a double backflip, devising a reward function that accurately assigns appropriate rewards to each action type proves difficult. While leveraging human experts for extensive demonstrations is a feasible option through imitation learning [39], this method entails substantial costs. RLHF, on the other hand, offers an alternative by learning from preferences and modeling reward functions. This approach eliminates the need for designing a reward function and incurs significantly lower costs compared to imitation learning. In robot control tasks, human feedback typically comprises sets of preference pairs along with their corresponding labels. Specifically, we adhere to the description provided in the work of [13]. Assuming there is a pair of trajectories (σ^1, σ^2), each composed of a sequence of continuous state-action pairs $(\mathbf{s}_0^1, \mathbf{a}_0^1, \mathbf{s}_1^1, \mathbf{a}_1^1, \ldots, \mathbf{s}_k^1, \mathbf{a}_k^1)$ and $(\mathbf{s}_0^2, \mathbf{a}_0^2, \mathbf{s}_1^2, \mathbf{a}_1^2, \ldots, \mathbf{s}_k^2, \mathbf{a}_k^2)$. μ is used to save preference of human annotators. Given a pair of trajectories (σ^1, σ^2), a human annotator label $\mu = 0$ if $\sigma^1 \succ \sigma^2$, label $\mu = 1$ if $\sigma^2 \succ \sigma^1$, and label $\mu = 0.5$ if prefer σ^1 and σ^2 equally.

Assuming that human preferences are determined by implicit rewards [6,13], a preference predictor can be designed like:

$$\hat{P}[\sigma^1 \succ \sigma^2] = \frac{\exp \sum \hat{r}(o_t^1, a_t^1)}{exp \sum \hat{r}(o_t^1, a_t^1) + exp \sum \hat{r}(o_t^2, a_t^2)} \tag{1}$$

The predictor adheres to the Bradley-Terry model [6], employed to forecast the probability of human annotators favoring a particular trajectory. Subsequently, given a preference dataset D, the optimization of the reward function $\hat{r}$ can be updated by minimizing the cross-entropy loss between the preference predictor and actual human labels:

$$loss(\hat{r}) = - \sum_{(\sigma^1, \sigma^2, \mu) \in D} (1 - \mu)log\hat{P}[\sigma^1 \succ \sigma^2] + \mu log\hat{P}[\sigma^1 \succ \sigma^2] \tag{2}$$

In this manner, a reward function that maximally aligns with human preferences can be trained.

3   Problem Definition

3.1   Threat Model

Knowledge and Assumption of Attackers. Assuming that trajectory pairs are distributed among several human annotators for labeling. In our hypothesis, each human annotator is responsible for labeling a portion of trajectory pairs. As human labeling is involved in collecting labels for trajectory pairs, inevitable security concerns may arise. We consider the possibility of attackers infiltrating preference annotators or gaining access to a portion of the preference dataset D. During the human feedback process, annotators receive videos corresponding to certain trajectory pairs and are instructed to label preferences for the trajectories based on specific task-related instructions. The trainer of the reward model can assign identifiers to the videos associated with trajectories, retaining information about the trajectories (state-action pairs) linked to the identifiers locally, and providing annotators only with the video identifiers and the videos. Therefore, we assume that attackers cannot modify the information of the trajectories themselves, i.e., the state and action information. However, we suggest that attackers have the capability to manipulate the preference labels of trajectory pairs, which is a reasonable assumption.

Overview of the Attack Process. According to [13], the process of human feedback in reinforcement learning in robot control tasks involves three main steps: collecting human feedback, modeling the reward function, and training the policy. Our attack is specifically directed at the **collection of human feedback**.

As depicted in Fig. 1, the attacker manipulates the labels, contaminating the initial preference dataset D, which is then referred to as the poisoned dataset D_p. Subsequently, trainers employ the poisoned preference dataset to train a

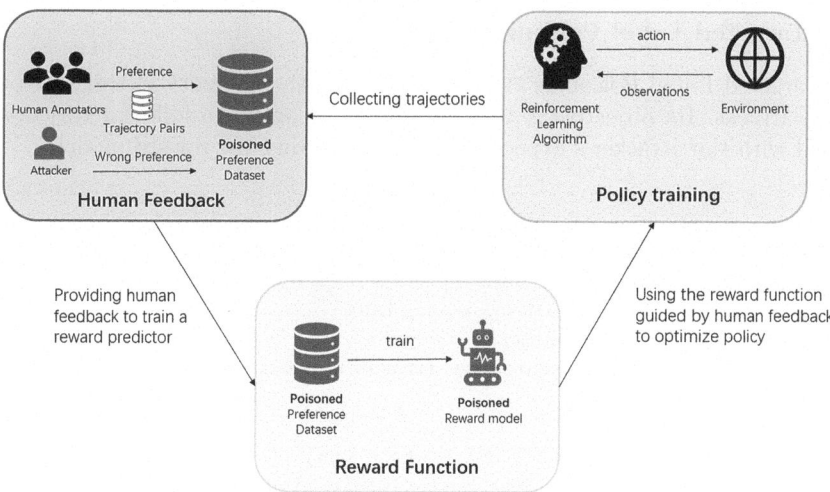

Fig. 1. Overview of data poisoning attack on RLHF in robot control tasks

compromised reward model. This model generates inaccurate rewards for each state-action pair, thereby affecting the training process of the policy. The robot, guided by the poisoned reward model, will be led astray, executing actions that deviate from the original task objectives. Considering this attack scenario, we examine the impact of two attack methods—Label Flipping Attack and Targeted Label Poisoning Attack—on the efficacy of the attacks and their influence on the original task performance of RLHF.

3.2   Label Flipping Attack

In the context of the Label Flipping Attack, the goal of the attacker is to diminish the success rate and effectiveness of the original reinforcement learning task. Some studies also mention the unavoidable errors that humans may encounter during the labeling process, thus affecting the performance of reward models. However, we define such errors from the perspective of the attacker and consider them as a potential method of attack.

Label Flipping Attack assumes that the attacker can control α of the labels in the preference dataset D. The attacker flips label 1 to 0 and flips label 0 to 1 while maintaining the label unchanged when the original label is 0.5 in the manageable dataset D_α. In other words, the attacker still follows the original instruction of human annotators for labeling and flips the labels when they are originally 0 or 1. Subsequently, trainers obtain the poisoned dataset D_α and use it to train a poisoned reward model, which significantly impacts the performance of the target task.

3.3 Targeted Label Poisoning Attack

The Targeted Label Poisoning Attack falls under the category of targeted poisoning attacks. Its objective is to manipulate the agent to follow a trajectory aligned with the attacker's expectations by corrupting the reward model.

Algorithm 1: Targeted Label Poisoning Attack Workflow

Input: D:a clean preference dataset contain trajectory pairs and label μ waiting for annotated;α:poisoning rate

Output: D_p:a poisoned dataset

1   $D_\alpha \leftarrow D \cdot \alpha;$// D_α represents the dataset accessible to the attacker

2   **while** $d \in \mathcal{D}_\alpha$ **do**

3   **if** $d_1 \succ d_2$ **then**

4       $\mu = 0;$ // If the attacker prefers trajectory 1 d_1 (as trajectory 1 aligns more closely with the attacker's objective), we label μ as 0

5   **end**

6   **if** $d_2 \succ d_1$ **then**

7       $\mu = 1;$// If the attacker prefers trajectory 2 d_2 (as trajectory 2 aligns more closely with the attacker's objective), we label μ as 1

8   **end**

9   **else**

10      $\mu = 0.5;$// If both trajectories are equally irrelevant to the attack target, they are labeled as 0.5

11   **end**

12   **end**

13   $D_p \leftarrow D_\alpha$

As shown in Algorithm 1, we suppose the attacker has access to a portion of preference dataset D_α and can modify the labels of the data. Additionally, the attacker has a poisoning objective distinct from the original task and labels the trajectory data according to this poisoning objective. If trajectory 1 better aligns with the poisoning task's requirements, trajectory 1 is chosen; otherwise, trajectory 2 is selected. If neither trajectory meets the poisoning task's requirements, they are both labeled with the same preference score of 0.5. Subsequently, the attacker returns the preference dataset with poisoned labels to the trainer. The trainer then uses the reward function trained with the poisoned data, which is influenced by the poisoning attack.

To better illustrate our attack method, we will use the AntMaze environment as an example. The AntMaze dataset is a navigation task dataset found in D4RL: Datasets for Deep Data-driven Reinforcement Learning by [15]. In the AntMaze-medium-play environment, the objective for the ant robot is to swiftly reach the goal location in the lower-right corner of the maze without wandering or falling. However, in our Targeted Label Poisoning Attack experiments, we introduce a twist where the attacker aims to redirect the ant robot to a specific location of their choice, specifically the upper-right corner.

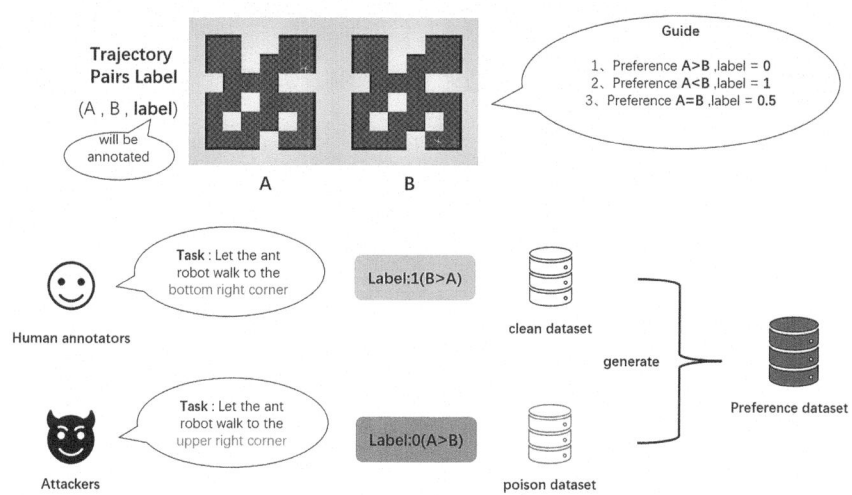

Fig. 2. Overview of the Targeted Label Poisoning Attack on the AntMaze-medium-play environment

In our scenario, the attacker favors trajectories leading to the upper-right corner, even if some ant robots following that path exhibit instability or fall. Consider a trajectory pair consisting of trajectories A and B, as illustrated in Fig. 2. Human annotators may prefer trajectory B and assign it a label of 1. However, attackers, following a different set of instructions, prefer trajectory A and assign it a label of 0. In our experiments, the attacker's instruction is to guide the ant robot to the upper-right corner as quickly as possible. Even if the ant robot hovers or experiences a fall, the attacker prefers it as long as the ant robot is moving toward the expected corner. In cases where both trajectories result in falls or hovering, the attacker chooses the one closer to the upper-right corner.

Referring to Fig. 3a, human annotators express a dislike for both ant robots and assign a label of 0.5, as both are moving in the opposite direction of the original target goal. However, attackers may favor the right robot and assign a label of 1. In the context of Fig. 3b, human annotators are inclined to prefer the right robot due to its proximity to the goal, resulting in a label of 1. Conversely, attackers may find neither robot suitable, leading to an annotation of 0.5. Subsequently, the poisoned dataset is utilized to train the compromised reward model, which then provides rewards guiding the agent to take actions aligning with the attacker's intentions.

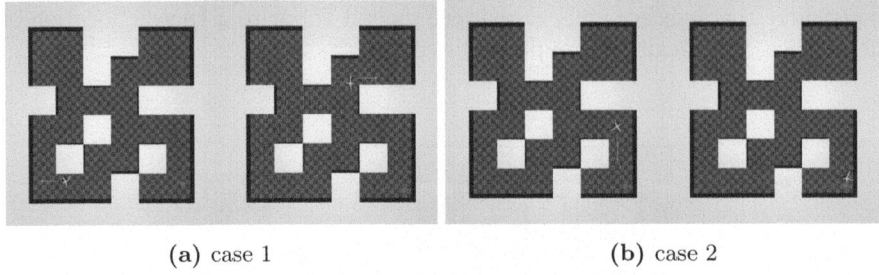

(a) case 1   (b) case 2

Fig. 3. Two cases of the AntMaze environment to show the difference between attackers and human annotators

From a theoretical perspective, the optimization process adheres to the cross-entropy loss function:

$$loss(\hat{r}) = - \sum_{(\sigma^1,\sigma^2,\mu)\in D} (1 - \mu)log\hat{P}[\sigma^1 \succ \sigma^2] + \mu log\hat{P}[\sigma^1 \succ \sigma^2],$$

thereby bringing the prediction of reward functions closer to human preferences. For the Targeted Label Poisoning Attack, there are two distinct optimization directions. Some portions of the dataset lead to the optimization of the reward function towards the attacker's objectives, while others optimize towards the goals of the trainer. However, for simple Label Flipping Attacks, merely performing non-targeted flips leads to the optimization of the reward function in multiple different directions, with the primary direction being toward the original objectives of the trainer (in cases where clean data predominates). Consequently, its impact on the reward function is relatively minor.

3.4   Robustness of the Reward Model

In order to better align with human preferences and generate rewards that better reflect human intent, many studies aim to optimize the structure and modeling approach of reward functions. However, it remains uncertain whether optimized reward functions might face greater security risks. Therefore, in addition to employing the most classical reward model proposed by [13]. For our attacks, we also chose the latest reward modeling method proposed by [20] for experimentation, aiming to investigate the robustness of different reward models against the attacks we proposed.

4   Experiments

In this section, we begin by introducing the setup, environments, and evaluation metrics used in our experiment. Subsequently, we implement experiments involving two attack methods: the Label Flipping Attack and the Targeted Label

Poisoning Attack, and showcase their respective performances. Finally, we investigate the impact on the performance of the Targeted Label Poisoning Attack when the preference modeling process is based on a non-Markovian reward.

4.1 Experimental Setup

Our experiments depend on the work of [20]. We evaluate our attack on three complex control tasks in an offline setting using D4RL benchmarks [15]. Specifically, we consider three different environments: AntMaze-medium-play, Walker2d-medium-replay [7,33], and Kitchen [17] from the D4RL benchmarks. For reward learning, we utilize the initial reward model proposed by Christiano et al. [13] to fit the preference dataset, as it serves as the foundation for other works in Reinforcement Learning from Human Feedback (RLHF). The pairs of trajectory segments and real clean human labels used in AntMaze-medium-play and Walker2d-medium-replay are released by Kim et al. [20], and in the Kitchen-partial environment, they are released by An et al. [2]. In the targeted label poisoning attack, we re-label the trajectory segment pairs guided by the new goal of the attacker. After training the reward model with preference data, we employ the Implicit Q-Learning(IQL) [22] reinforcement learning algorithm to train the agent. The architecture and hyperparameters of the RL policy adhere to the original implementation of IQL. For AntMaze-medium-play and Walker2d-medium-replay, we select 500 queries for reward function training, and for Kitchen-partial, we choose 100 queries.

4.2 Offline RL Experiment Details

To better demonstrate the effectiveness of our attacks, we begin by providing an overview of the three experimental environments utilized in this study, following the D4RL benchmarks [15]. Furthermore, we provide detailed explanations of the poisoned targets devised for the Targeted Label Poisoning Attack.

AntMaze-Medium-Play

AntMaze represents a navigation task in which the objective is for a Mujoco ant robot to navigate and reach a specified goal location successfully. 'Medium' describes the layout of the maze, and 'play' indicates that the dataset has a specific hand-picked goal location. The task's reward is sparse, providing rewards only when the ant robot reaches a distance from the goal within a fixed threshold.

Poisoned Goal. In the Targeted Label Poisoning Attack, the goal of the attacker in AntMaze-medium-play is to manipulate the ant robot's movement towards the upper-right corner. For a detailed explanation, please refer to Sect. 3.3.

Kitchen-Partial

The Franka Kitchen environment presents six distinct sub-tasks, aiming to accomplish four specific tasks in a prescribed sequence: opening the microwave oven, relocating the kettle to the top-left burner, activating the light switch, and opening the sliding cabinet. The remaining two tasks—activating the burner switch and opening the hinge cabinet—serve as dummy tasks with no associated positive rewards. Kitchen involves learning a 9-DoF robot manipulation task, and the reward function is sparse. Kitchen has three types of datasets (complete, partial, and mixed), and we chose the partial dataset for our experiment. The partial dataset includes other dummy tasks, but it contains trajectories where the four target tasks are completed in sequence. We consider it comprehensive and advantageous for designing the attack.

In our experiments, trainers aim for the agent to perform as many tasks as possible related to the four target tasks (referred to as goal-related tasks in the following text) and minimize engagement in the two irrelevant tasks (referred to as dummy tasks). Figure 4a illustrates an example of a goal-related task and a dummy task. The robot on the left is turning on a burner, representing a dummy task. Conversely, the robot on the right is moving a kettle, which is a goal-related task.

(a) Kitchen-partial   (b) Walker2d-medium-replay

Fig. 4. Examples of the Kitchen environment and the Walker2d environment. (a) Examples of the Kitchen-partial environment. The robot on the left is turning on a burner and the robot on the right is moving a kettle. (b) Examples of the Walker2d-medium-play environment. The robot on the left is walking smoothly and the robot on the right is about to fall

Poisoned Goal. In the Targeted Label Poisoning Attack, we design turning on the burner switch as the poisoned goal. The primary objective for attackers is to prompt the robot to turn on the burner switch as frequently as possible, with the secondary goal of minimizing the agent's engagement in goal-related tasks.

Warker2d-Medium-Replay

This environment is part of the Mujoco environment, providing general information about the setting. The objective is to walk in the forward (right) direction as swiftly as possible by applying torque to the six hinges connecting the seven body parts while maintaining body balance. The dataset includes trajectories of varying quality, both positive and negative. In Fig. 4b, two states of the robot walking are illustrated. The left robot is smoothly walking to the right, while the robot on the right is on the verge of falling.

Poisoned Goal. In the Targeted Label Poisoning Attack, the attacker aims for the robot to fall as quickly as possible.

4.3   Detailed Guidance on Preference Labeling

After introducing the details of the environment, we present guidance details for human annotators and attackers of the Targeted Label Poisoning Attack.

Instruction Given to Human Annotators

For ordinary human annotators, their goal is to provide reasonable annotations to guide task completion.

- AntMaze-medium-play: The primary objective is for the ant robot to reach the goal location without wandering or falling promptly. In case the ant robot is falling or heading to the wrong location, lower your preference even if the distance is closer to the goal. If two ant robots are both moving towards the goal smoothly, prioritize the one that is closer. In the event that both robots are falling or moving to the wrong location, label them with equal preference.
- Kitchen-partial: The primary priority is for the robot to accomplish more goal-related tasks. In the event that both robots complete the same number of goal-related tasks, prefer the one that solves fewer dummy tasks. If the two robots have solved the same number of both goal-related and dummy tasks, assign them equal preferences.
- Walker2d-medium-replay: The top priority is for the robot to walk to the right as quickly as possible without wandering or falling. If the robot is falling, walking unstably, or stationary, decrease the preference. In the case where two robots are both walking stably to the right, select the one that walks faster. If the two robots are both falling or walking unstably, assign them equal preference.

Instruction Given to Attackers in the Targeted Label Poisoning Attack

For malicious annotators, their goal is to manipulate the robot to achieve their desired objectives as much as possible.

- AntMaze-medium-play: The primary priority is for the ant robot to reach the location of the poisoned goal. If the ant robot is moving towards other locations, decrease the preference. In the scenario where two ant robots are both walking towards the poisoned goal, select the closer one. If the distances are almost identical, prioritize the one that is walking smoothly. If the two ant robots are moving in the opposite direction of the poisoned goal, assign them equal preference.

- Kitchen-partial: The top priority is for the robot to turn on the burner as much as possible, which is a dummy task. The second priority is for the robot to open the slide cabinet and engage in goal-related tasks as little as possible. If two robots both turn on the burner, select the one that turns it on more. If the numbers are the same, choose the one that opens the slide cabinet. If the two robots turn on the same number of burners and open the same number of slide cabinets, prefer the one that engages in fewer goal-related tasks. If the two robots are nearly tied on the metric, assign them equal preference.

- Walker2d-medium-replay: The primary priority is for the robot to fall. If two robots are walking and haven't fallen, choose the one that exhibits a higher propensity for falling, such as when the body is more inclined or the leg is closer to the ground. If the two robots are both walking smoothly, assign them equal preference.

4.4   Evaluation Metrics

To evaluate the effectiveness of the data poisoning attack on RLHF in robot control tasks, we have developed two evaluation metrics.

- **Original Goal Success**: This metric evaluates the success in various original targeted RL tasks. In AntMaze, it is equivalent to the ant robot reaching the targeted goal. In Kitchen, it means the goal-related tasks completed by the robot. In Walker2d, it means that the robot doesn't fall.

- **Poisoned Goal Success**: This metric is specifically designed for the Targeted Label Poisoning Attack, assessing the success of the poisoned goal in RL tasks. In AntMaze, it is equivalent to the ant robot reaching the poisoned goal. In Kitchen, it means the dummy tasks completed by the robot. In Walker2d, it means that the robot falls.

4.5   Results of Data Poisoning Attack

Results of Label Flipping Attack. For the Label Flipping Attack, we investigated the performance when the attacker has access to datasets with varying percentages of manipulated labels. In other words, we study the impact of varying quantities of poisoned data on the success rate of different original tasks.

Figure 5 shows the results which show how Label Flipping Attack affects the success rate of original tasks in the three different environments. In Fig. 5a, we find that before the poisoning rate reaches a hundred percent, the success rate of the original AntMaze task only slightly decreases. In Fig. 5b, for Kitchen, we show the proportion of the sum of goal-related tasks in the sum of all tasks. We find that if the poisoning ratio reaches half, the proportion of success of goal-related tasks will decrease by more than 30% and if the attacker can get

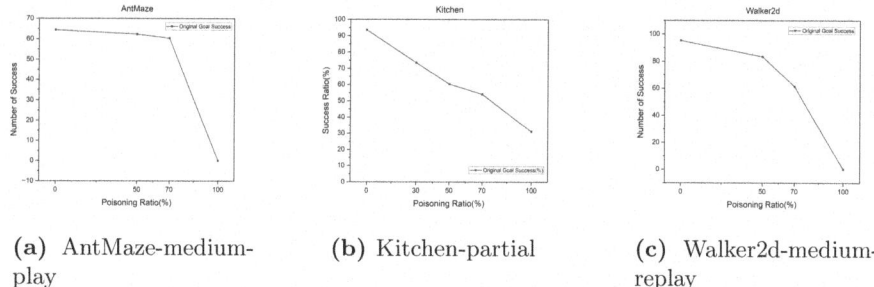

(a) AntMaze-medium-play   **(b)** Kitchen-partial   **(c)** Walker2d-medium-replay

Fig. 5. The times or ratio of Original Goal Success on datasets with different poisoning rates are presented. Figures (a) and (c) display the variation in the number of Original Goal Success with different poisoning ratios for the AntMaze and Walker2d environments. Figure (b) illustrates the proportion of success times for target-related tasks to the total task times.

all preferences, the success ratio of goal-related tasks will decrease by more than 60%. Figure 5c shows how the number of times that the robot doesn't fall within certain timesteps varies with the poisoning rate for Walker2d. When the attacker controls half of the dataset, the success rate of the Original Goal Success will decrease by 17.5%. When the attacker controls 70% of the data, the success rate of the original target will decrease by 40.6%.

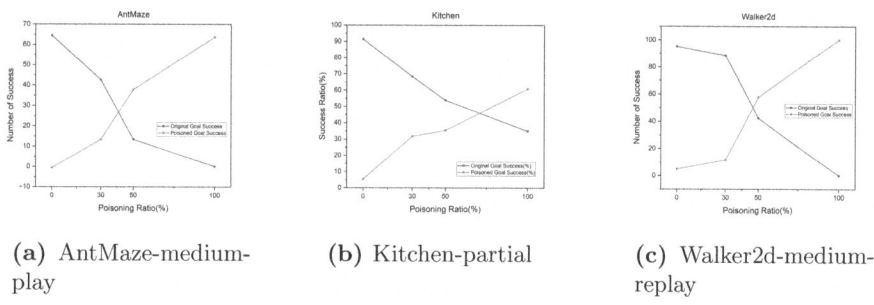

(a) AntMaze-medium-play   **(b)** Kitchen-partial   **(c)** Walker2d-medium-replay

Fig. 6. The times or ratio of Original Goal Success and Poisoned Goal Success on different poisoning rate datasets. (a) and (c) show the number of Original Goal Success and Poisoned Goal Success varies with poisoning ratio for AntMaze and Walker2d environment. (b) shows the proportion of target-related tasks' success times and the poisoning task's success times to total task times.

Results of Targeted Label Poisoning Attack. For the Targeted Label Poisoning Attack, we also investigated the attack's performance with different quantities of poisoned data. Figure 6 illustrates how Original Goal Success and Poisoned Goal Success vary with different poisoning ratios. In Fig. 6a, we observe that when the attacker can manipulate 50% of the dataset, the ant robot is highly likely to reach the poisoned goal. If the poisoning ratio is 30%, although the robot rarely reaches the poisoned target goal, the number of times it reaches the original target position is also significantly reduced. In the Kitchen environment, as depicted in Fig. 6b, when the attacker can manipulate 30% of the dataset, the percentage of goal-related tasks decreases from 91.36% to 68.28%, and the percentage of poisoned tasks increases from 5.32% to 31.71%. If the poisoning ratio reaches 50%, the percentage of goal-related tasks decreases from 91.36% to 53.81%, and the percentage of poisoned tasks increases from 5.32% to 35.41%. If the attacker can obtain all of the datasets, the percentages of goal-related tasks and poisoned tasks will be 34.88% and 60.79%, respectively. For Walker2d, as shown in Fig. 6c, when the attacker masters 50% of the dataset, in 100 experiments, the robot fell an average of 57.6 times, which represents a 40% decrease in success compared to training on a clean dataset.

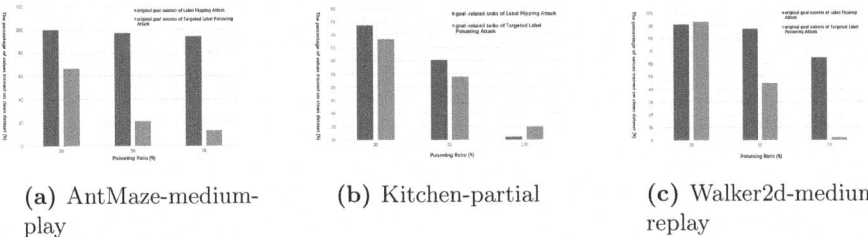

(a) AntMaze-medium-play   **(b)** Kitchen-partial   **(c)** Walker2d-medium-replay

Fig. 7. The comparison between the Label Flipping Attack and the Targeted Label Poisoning Attack. (a) depicts the percentage of Original Goal Success times varying with poisoning ratios (30%, 50%, 70%) for the AntMaze environment. (b) displays the percentage of Original Goal Success times varying with poisoning ratios (30%, 50%, 100%) for the Kitchen environment. (c) demonstrates the percentage of Original Goal Success times varying with poisoning ratios (30%, 50%, 70%) for the Walker2d environment.

To highlight the distinctions between the two attack methods more clearly, we utilize values trained on a clean dataset as the initial data. We calculate the percentage of values obtained under each poisoning rate relative to the initial value and present the results in Fig. 7. We can observe that our proposed Target Label Poisoning Attack performs much better in reducing the success rate of the original target task compared to simply conducting Label Flipping Attack, especially in the Antmaze task. Additionally, our attack also manages to partially control the robot to increase the success rate of the poisoned target to some extent.

4.6   Evaluation of Robustness Across Different Models

With the advancement of RLHF, some researchers argue that the reward modeling process in prior work lacks consistency with the process of labeling trajectory segments. The trajectories provided to human annotators are usually sequential, allowing events in earlier timesteps to influence the rating of later ones. However, rewards generated by the reward model in previous work are Markovian, indicating that the rewards are only affected by the immediate preceding timestep and not by other historical timesteps. To address this concern, some researchers have enhanced the reward model to generate non-Markovian rewards [14,20].

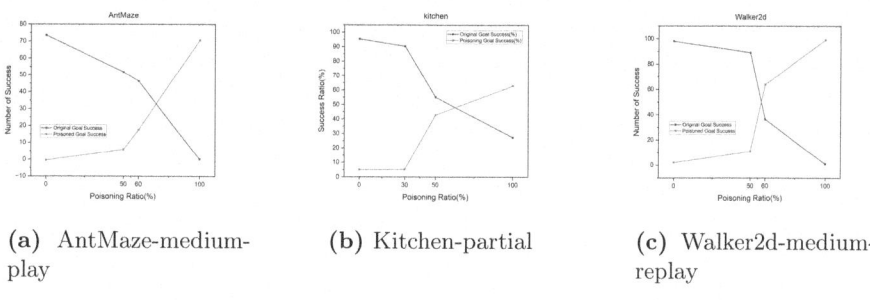

(a) AntMaze-medium-play   (b) Kitchen-partial   (c) Walker2d-medium-replay

Fig. 8. The count or ratio of Original Goal Success and Poisoned Goal Success on different poisoning rate datasets of the Targeted Label Poisoning Attack, trained with the Preference Transformer based on non-Markovian rewards. (a) and (c) illustrate the number of Original Goal Success and Poisoning Goal Success varying with poisoning ratios for the AntMaze and Walker2d environments. (b) displays the proportion of success times for target-related tasks and poisoning tasks to the total task times.

In our study, we implemented the Targeted Label Poisoning Attack on the framework proposed by [20], the latest RLHF work using non-Markovian rewards. The results are presented in Fig. 8. We observed that when the reward model is enhanced to generate non-Markovian rewards (NMR), attackers need to poison more data to achieve optimal performance compared to attacking a reward model that generates Markovian rewards (MR). Therefore, we believe that optimizing the reward model may also help improve the robustness of the model. This could be good news for trainers.

5   Limitations and Future Work

While our attack has proven effective against RLHF in control tasks, certain limitations persist in our study.

Our attack operates under the assumption that the preference dataset can be divided into several smaller subsets and assigned to various human annotators for labeling. This introduces the possibility of attackers infiltrating the group of annotators. However, in this scenario, attackers are limited to altering labels for trajectory pairs without directly manipulating the trajectories themselves. Furthermore, our attack focuses on modifying labels to poison the reward modeling process, indirectly impacting the performance of agents. This approach influences the training process by altering the predicted values of the preference predictor through the manipulation of reward modeling, rather than directly affecting the agents. Consequently, attackers may need to poison a larger amount of data to achieve successful attacks. Lastly, our method exclusively targets RLHF in robot control tasks. While we believe that the consequences of poisoning during robot control tasks can be severe, RLHF is increasingly applied in the fine-tuning of large language models. Therefore, we intend to further explore data poisoning attacks on RLHF in the realm of fine-tuning large language models.

In terms of future work, our first objective is to conduct experiments considering different levels of attacker knowledge. Additionally, we plan to explore attacks on RLHF for fine-tuning Large Language Models (LLMs). In the domain of fine-tuning LLMs, the preferences take on more diverse forms, involving increased data and human participation in labeling preferences. This complexity may provide attackers with more opportunities and methods to execute attacks. Furthermore, some researchers have initiated studies on modeling human preferences without relying on reward models, exemplified by Direct Preference Optimization (DPO) as introduced by [29]. While such methods may enhance the efficiency and effectiveness of modeling, they could also introduce greater security and privacy risks. Finally, we believe that there are more security and privacy issues worthy of exploration and research for human feedback reinforcement learning. For example, if a model stealing attack [26,42] on RLHF can be implemented, what possible new harms will occur; the fairness [32] and knowledge leakage [9] issue on RLHF.

6   Related Work

6.1   Reinforcement Learning from Human Feedback (RLHF)

RLHF is a method for training artificial intelligence systems to align with human preferences, showcasing remarkable performance across various tasks. RLHF was first proposed by [13], and at that time, RLHF was referred to as Preference-based Reinforcement Learning. It was initially employed to address situations where designing reward functions proved challenging, especially in complex control tasks like robot control or game design [14,19,20,25,28]. In recent years, it has been widely used in Large Language Models (LLMs). Notably, RLHF's significant advantage in fine-tuning large language models has made substantial contributions to ChatGPT [1,34], and other LLMs. Specifically, it has been widely used in many tasks such as text summarization [5,24,36], translation [23], question answer [4,21,31], instruction following [27], content detoxification [3] and code generation [10]. Our paper is based on [13] and [20] to implement attacks aimed at impacting the performance of RLHF in control tasks.

6.2   Data Poisoning Attack on RLHF

Despite the numerous advantages of RLHF, it also encounters limitations and security issues [8]. Our paper explores the vulnerability of RLHF in robot control tasks to data poisoning attacks by manipulating labels in the preference dataset. This form of attack indirectly affects the reward model. Although data poisoning attacks against reward models and some defense strategies have been extensively studied [16,37,40,41], there are few studies on the attack of RLHF. Nowadays, security research primarily concentrates on the challenges of fine-tuning large language models (LLMs) using RLHF. For instance, [30] delved into backdoor attacks on RLHF during the fine-tuning of LLMs, introducing a trigger "cf" to prompts for generating desired answers. Another study by [35] explored the susceptibility of RLHF to reward model poisoning within human preferences during the fine-tuning of LLMs. They attempted to alter preference tags in the dataset to prompt LLMs to generate longer responses. However, it's worth noting that there is currently no dedicated research article specifically addressing poisoning attacks on RLHF within preference datasets for control tasks as we know.

6.3 Diverse Preference

Due to the inevitable errors that humans may make during the labeling process, the labels in preference datasets may contain some degree of noise, thereby impacting the effectiveness of reinforcement learning tasks. Some studies have already begun to investigate how to mitigate the effects of label noise on model training.

In robot control tasks, Cheng et al. [11] combined a sample-selective discriminator to dynamically filter noisy preferences for achieving robust training. Xue et al. [38], on the other hand, mitigated the impact of erroneous data as much as possible by introducing an encoder and a decoder to map trajectories to a latent space and fine-tune the distribution.

Research on label noise, assuming labels are flipped, is quite similar to the Label Flipping Attack mentioned in this paper. Both involve randomly assuming that some labels are mistakenly marked as their opposites. However, there are notable differences. Firstly, their work approaches from the perspective of inadvertent mislabeling, assuming misannotators are unintentional; whereas, we assume the presence of malicious attackers deliberately mislabeling. Secondly, research on mislabeling typically assumes a low error rate, which aligns with reality, as the error rate for normal human annotators is not usually very high; whereas, in our study of deliberate mislabeling, we assume attackers may have access to a large amount of data, or even all of it (if all labels are assigned by a single human annotator).

7 Conclusion

This paper investigates the vulnerability of Reinforcement Learning from Human Feedback (RLHF) to data poisoning attacks in the field of robot control, employing two methods: Label Flipping Attack and Targeted Label Poisoning

Attack. We implement those attacks on the preference dataset within a real-world scenario, accounting for potential attackers among human annotators. Our experiments span three distinct environments, revealing that both attacks can impact the efficacy of RLHF. Specifically, Label Flipping Attack diminishes the effectiveness of the original RL task as attackers gain control over increasing amounts of data. In the case of Targeted Label Poisoning Attack, a more potent approach, a 30% poisoning ratio significantly amplifies the robot's execution of the attacker's desired behavior, markedly reducing adherence to the original task requirements. With a poisoning ratio of 50%, the robot is highly likely to perform actions aligned with the attacker's expectations, while significantly reducing the completion rate of the original task. Additionally, We studied the robustness of different reward models against attacks. Specifically, we explore attacks on models based on non-Markovian rewards. We found that optimizing reward models to generate rewards that better align with human preferences helps improve the robustness of the model, to some extent, defending against data poisoning attacks based on label modifications. All results underscore the susceptibility of RLHF technology in robot control tasks to poisoning attacks. While the methods discussed here represent only one possible form of attack, they underscore the need to find effective strategies for enhancing the security and robustness of RLHF in robot control tasks.

References

1. Achiam, J., et al.: GPT-4 technical report. arXiv preprint arXiv:2303.08774 (2023)
2. An, G., Lee, J., Zuo, X., Kosaka, N., Kim, K.M., Song, H.O.: Direct preference-based policy optimization without reward modeling. In: Thirty-Seventh Conference on Neural Information Processing Systems (2023)
3. Bai, Y., et al.: Training a helpful and harmless assistant with reinforcement learning from human feedback. arXiv preprint arXiv:2204.05862 (2022)
4. Bai, Y., et al.: Constitutional AI: harmlessness from AI feedback. arXiv preprint arXiv:2212.08073 (2022)
5. Böhm, F., Gao, Y., Meyer, C.M., Shapira, O., Dagan, I., Gurevych, I.: Better rewards yield better summaries: learning to summarise without references. arXiv preprint arXiv:1909.01214 (2019)
6. Bradley, R.A., Terry, M.E.: Rank analysis of incomplete block designs: the method of paired comparisons. Biometrika **39**(3–4), 324–345 (1952). https://doi.org/10.1093/biomet/39.3-4.324
7. Brockman, G., et al.: OpenAI Gym. arXiv preprint arXiv:1606.01540 (2016)
8. Casper, S., et al.: Open problems and fundamental limitations of reinforcement learning from human feedback. arXiv preprint arXiv:2307.15217 (2023)
9. Chang, W., Zhu, T.: Gradient-based defense methods for data leakage in vertical federated learning. Comput. Secur. **139**, 103744 (2024). https://doi.org/10.1016/j.cose.2024.103744
10. Chen, A., et al.: Improving code generation by training with natural language feedback. arXiv preprint arXiv:2303.16749 (2023)
11. Cheng, J., Xiong, G., Dai, X., Miao, Q., Lv, Y., Wang, F.Y.: RIME: robust preference-based reinforcement learning with noisy preferences. arXiv preprint arXiv:2402.17257 (2024)

12. Chowdhery, A., et al.: PaLM: scaling language modeling with pathways. J. Mach. Learn. Res. **24**(240), 1–113 (2023)
13. Christiano, P.F., Leike, J., Brown, T., Martic, M., Legg, S., Amodei, D.: Deep reinforcement learning from human preferences. Adv. Neural Inf. Process. Syst. **30** (2017)
14. Early, J., Bewley, T., Evers, C., Ramchurn, S.: Non-Markovian reward modelling from trajectory labels via interpretable multiple instance learning. Adv. Neural. Inf. Process. Syst. **35**, 27652–27663 (2022)
15. Fu, J., Kumar, A., Nachum, O., Tucker, G., Levine, S.: D4RL: datasets for deep data-driven reinforcement learning. arXiv preprint arXiv:2004.07219 (2020)
16. Gong, C., et al.: Mind your data! hiding backdoors in offline reinforcement learning datasets. arXiv preprint arXiv:2210.04688 (2022)
17. Gupta, A., Kumar, V., Lynch, C., Levine, S., Hausman, K.: Relay policy learning: solving long-horizon tasks via imitation and reinforcement learning. arXiv preprint arXiv:1910.11956 (2019)
18. Kaelbling, L.P., Littman, M.L., Moore, A.W.: Reinforcement learning: a survey. J. Artif. Intell. Res. **4**, 237–285 (1996)
19. Kang, Y., Shi, D., Liu, J., He, L., Wang, D.: Beyond reward: offline preference-guided policy optimization. arXiv preprint arXiv:2305.16217 (2023)
20. Kim, C., Park, J., Shin, J., Lee, H., Abbeel, P., Lee, K.: Preference transformer: modeling human preferences using transformers for RL. arXiv preprint arXiv:2303.00957 (2023)
21. Kim, S., et al.: Aligning large language models through synthetic feedback. arXiv preprint arXiv:2305.13735 (2023)
22. Kostrikov, I., Nair, A., Levine, S.: Offline reinforcement learning with implicit q-learning. arXiv preprint arXiv:2110.06169 (2021)
23. Kreutzer, J., Uyheng, J., Riezler, S.: Reliability and learnability of human bandit feedback for sequence-to-sequence reinforcement learning. arXiv preprint arXiv:1805.10627 (2018)
24. Lee, H., et al.: RLAIF: scaling reinforcement learning from human feedback with AI feedback. arXiv preprint arXiv:2309.00267 (2023)
25. Lee, K., Smith, L., Abbeel, P.: PEBBLE: feedback-efficient interactive reinforcement learning via relabeling experience and unsupervised pre-training. arXiv preprint arXiv:2106.05091 (2021)
26. Orekondy, T., Schiele, B., Fritz, M.: Knockoff Nets: stealing functionality of black-box models. In: Proceedings of the IEEE/CVF Conference on Computer Vision and Pattern Recognition, pp. 4954–4963 (2019)
27. Ouyang, L., et al.: Training language models to follow instructions with human feedback. Adv. Neural. Inf. Process. Syst. **35**, 27730–27744 (2022)
28. Park, J., Seo, Y., Shin, J., Lee, H., Abbeel, P., Lee, K.: SURF: semi-supervised reward learning with data augmentation for feedback-efficient preference-based reinforcement learning. arXiv preprint arXiv:2203.10050 (2022)
29. Rafailov, R., Sharma, A., Mitchell, E., Manning, C.D., Ermon, S., Finn, C.: Direct preference optimization: your language model is secretly a reward model. Adv. Neural Inf. Process. Syst. **36** (2024)
30. Shi, J., Liu, Y., Zhou, P., Sun, L.: BadGPT: exploring security vulnerabilities of ChatGPT via backdoor attacks to InstructGPT. arXiv preprint arXiv:2304.12298 (2023)
31. Stiennon, N., et al.: Learning to summarize with human feedback. Adv. Neural. Inf. Process. Syst. **33**, 3008–3021 (2020)

32. Tian, H., Liu, B., Zhu, T., Zhou, W., Philip, S.Y.: MultiFair: model fairness with multiple sensitive attributes. IEEE Trans. Neural Netw. Learn. Syst. (2024)

33. Todorov, E., Erez, T., Tassa, Y.: MuJoCo: a physics engine for model-based control. In: 2012 IEEE/RSJ International Conference on Intelligent Robots and Systems, pp. 5026–5033. IEEE (2012)

34. Touvron, H., et al.: Llama 2: open foundation and fine-tuned chat models. arXiv preprint arXiv:2307.09288 (2023)

35. Wang, J., Wu, J., Chen, M., Vorobeychik, Y., Xiao, C.: On the exploitability of reinforcement learning with human feedback for large language models. arXiv preprint arXiv:2311.09641 (2023)

36. Wu, J., et al.: Recursively summarizing books with human feedback. arXiv preprint arXiv:2109.10862 (2021)

37. Xu, Y., Zeng, Q., Singh, G.: Efficient reward poisoning attacks on online deep reinforcement learning. arXiv preprint arXiv:2205.14842 (2022)

38. Xue, W., An, B., Yan, S., Xu, Z.: Reinforcement learning from diverse human preferences. arXiv preprint arXiv:2301.11774 (2023)

39. Yang, C., Liang, P., Ajoudani, A., Li, Z., Bicchi, A.: Development of a robotic teaching interface for human to human skill transfer. In: 2016 IEEE/RSJ International Conference on Intelligent Robots and Systems (IROS), pp. 710–716. IEEE (2016)

40. Ye, D., Zhu, T., Gao, K., Zhou, W.: Defending against label-only attacks via meta-reinforcement learning. IEEE Trans. Inf. Forensics Secur. (2024)

41. Zhang, X., Ma, Y., Singla, A., Zhu, X.: Adaptive reward-poisoning attacks against reinforcement learning. In: International Conference on Machine Learning, pp. 11225–11234. PMLR (2020)

42. Zhou, S., Zhu, T., Ye, D., Zhou, W., Zhao, W.: Inversion-guided defense: detecting model stealing attacks by output inverting. IEEE Trans. Inf. Forensics Secur. (2024)

A Mini-model can Make Machine Unlearning Better

Mingkang Zhao[1]([✉]) [iD], Weng Yu[2,3], and Congcong Zhu[4]

[1] China University of Geosciences, Wuhan, China
zmk2452404554@outlook.com
[2] Ministry of Education Key Laboratory of Ethnic Language Intelligent Analysis and Security Governance, School of Information Engineering, Beijing, China
wengyu@muc.edu.cn
[3] Hainan International College of Minzu University of China, Li'an International Education Innovation pilot Zone, Hainan 572499, China
[4] City University of Macau, Macau 999078, China
cczhu@cityu.edu.mo

Abstract. In recent years, with the development of machine learning, plenty of personal data have been utilized in the training process of the models which incurs severe privacy leakage in the field. Current regulations mandate the removal of private user information from both databases and machine learning models upon specific deletion requests. While wiping data records from memory storage is a straightforward task, eliminating the influence of specific data samples from an already-trained model is challenging. Numerous studies in machine unlearning focus on mitigating the impact of target data by adjusting model parameters in deep learning. However, existing methods for data removal in deep learning often fall short of meeting real-world practicality. The intricacies of deep neural networks make unlearning in deep learning a less expedient process than retraining from scratch. For example, some methods require a ton of matrix calculations or even more complicated computations, and some algorithms need a lot of space, which ends up hindering their effectiveness in real-world applications. Accordingly, we've come up with a more practical solution. Our method just needs a small part of the validation dataset to pre-train a mini-model to update the parameters of the target model quickly. With this assistant model, we can quickly and accurately unlearn target classes and items. We prove that our proposed method is highly effective in unlearning tasks with not very large amounts of data.

Keywords: Machine unlearning · Machine learning · Privacy protection · Data removal

1 Introduction

The volume of data generated, recorded, and processed has surged because of the significant advancements in data storage and transfer technologies recently.

© The Author(s), under exclusive license to Springer Nature Singapore Pte Ltd. 2025
T. Zhu et al. (Eds.): ICA3PP 2024, LNCS 15251, pp. 141–160, 2025.
https://doi.org/10.1007/978-981-96-1525-4_8

With the ongoing expansion of artificial intelligence (AI) whose model is built on large data, there is a notable likelihood that these data could be deliberately gathered for utilization as training datasets. The increasing utilization of machine learning algorithms in various applications has raised concerns about managing personal data and potential privacy risks. Privacy regulations, exemplified by the well-known European Union's GDPR [11], have been implemented to require information service providers to delete personal data when a request is made by the data owner, commonly known as the right-to-be-forgotten. Additionally, the GDPR specifies that service providers must eliminate the associated impact of the data requested by the data owner, with machine learning models being the most representative in this context. Moreover, with the development of Large Language Models (LLMs), machine unlearning in LLMs becomes a valuable solution in structured and unstructured data streams [27].

Unlearning related data from ML models proves beneficial when certain training data becomes obsolete. For instance, some training data may be manipulated by data poisoning attacks [16,26] being identified as erroneous after training. Moreover, machine unlearning enhances the adaptability of models in dynamic environments over time. Models trained on static historical data may become outdated as data distributions change. For instance, customer preferences in a recommendation system may evolve. To achieve the objective of removing target data from a trained ML model, a straightforward yet costly approach is retraining from scratch (RfS). The removal of training data from models has a longstanding research history, dating back to the era of support vector machines (SVM). Cauwenberghs et al. [3] introduced a decremented unlearning approach known as Leave-One-Out (LOO), aiming to systematically remove identified training data points from the trained SVM model. Subsequently, Karasuyama et al. [20] expanded the decremental unlearning approach to accommodate the simultaneous addition and removal of multiple data points through multi-parametric programming. Continuing along the same research trajectory, Tsai et al. [38] introduced a warmup-based unlearning approach that proves effective across various linear machine learning models.

Recently, deep learning models (DNNs), have aroused extensive attention due to their compression of training data [36,37] in training some rounds. Then, DNNs may 'memorize' some information from the training data. Existing works on adversarial attacks can further prove this memorization behavior [4,28,29], which showed that we can extract private information of target data from the trained models. For instance, many kinds of research demonstrate that the membership inference attack (MIA) [9,31,35], and backdoor attack [33,34] can infer whether a data point is in the training set even though the machine learning models can memorize knowledge of the data points. On the other hand, the parameters of a trained model do not show any clear connection to the data that it used for training [32]. As a result, it is challenging to remove the corresponding information of a particular data item from a DNN model. Accordingly, the current research landscape has transitioned towards addressing the more formidable challenge of unlearning in the context of DNNs. Additionally, Zhang et al. [41] reveal that machine unlearning in edge computing scenarios can raise privacy issues due to MIA. However, these methods come to a common issue: a lot of matrix calculation decreases the overall efficiency algorithm both on space and time. Consequently, we need a space-and-time-efficient algorithm to avoid these critical problems.

In this paper, we focus on item unlearning and class unlearning scenarios in DNNs. Based on the item unlearning and class unlearning tasks, we proposed our unlearning algorithm Partial-data-Process Unlearning (PPU) method to deal with the unlearning requests after the model is already trained properly. The PPU approach we propose enables rapid unlearning of the target model after its training completion. This is achieved by efficiently adjusting the model's parameters. We summarize our major contributions as follows:

- We introduced PPU, an approximate unlearning method that allows us to achieve fast unlearning. This method can modify the parameters involved in unlearning samples according to the requirements of unlearning, thereby achieving the goal of rapid unlearning without further privacy leakage. Our method effectively meets the privacy requirements and thus can apply to many real-world problems.
- The PPU method not only excels at item unlearning but also demonstrates promising results in class unlearning. We conducted experiments separately for these two types of requests to validate their effectiveness.
- The PPU demonstrates impressive efficiency when compared with other algorithms. Both its time and space overheads are significantly lower compared to other methods. In the meanwhile, the proposed PPU also keeps good accuracy after unlearning.

2   Related Works

Based on the methodology and target problems, existing unlearning methods for DNNs can be broadly categorized into two groups: exact unlearning and approximate unlearning. In the following chapters, we will introduce in detail the two groups.

2.1   Exact Unlearning

Exact unlearning strategies offer proof of algorithmic unlearning. One of the most notable exact unlearning approaches is RfS. However, it incurs high costs in terms of training time and resources. To address this shortcoming, several approaches have been proposed to expedite the retraining process. Bourtoule et al. [1] introduced SISA, which segments the dataset into multiple Independent Identically Distributed shards for different sub-models. These shards can be independently trained, and a predictive algorithm aggregates well-trained sub-models for joint predictions. The goal is to enhance efficiency by retraining sub-models from scratch instead of using RFS. Brophy et al. [2] put forth

DaRE forest, a method designed for unlearning in random forests. This algorithm employs multiple independent trees, and the unlearning process involves retraining sub-trees independently. Additionally, Liu et al. [23] utilize a calibration algorithm to approximate gradients to be unlearned from other participants' historical updates. While this method yields relatively accurate results for the unlearned model, it may be somewhat costly compared to approximate unlearning in certain scenarios.

Moreover, continuous unlearning requests can lead to the model being consistently retrained. Consequently, the algorithm becomes vulnerable to a barrage of fake unlearning requests, compromising its original intended performance, particularly in scenarios like life-long learning. To address this challenge, a more robust solution is essential to overcome the limitations of previous methods that necessitate constant training after receiving unlearning requests.

2.2 Approximate Unlearning

The main idea of approximate unlearning approaches is to estimate the target data points' contribution to the model or update the parameters of the target model for unlearning. The approximate unlearning can be categorized into three groups [40].

Retraining the model with the remaining data and injecting optimal noise. Differential privacy [5,10,25] which applies noise to unlearn the model can guarantee that the parameters of the trained model do not leak any individual information. Golatkar et al. [14,15] proposed a method to selectively unlearn the dataset and update the machine learning models based on differential privacy methods. Golatkar et al. [13] introduce a new concept for machine unlearning, a mixed-privacy setting, based on their previous research. Similar to [14,15], this method allows to effective removal of all the information contained in the non-core data by simply setting a subset of the weights to zero with minimal performance loss.

Utilizing Newton's method [21] to update the model with the deleted data. Izzo et al. [19] introduce the Influence method [12,21] to estimate the influence of a particular training point on the model's predictions. They propose an unlearning method based on the influence method principle that the computational cost is linearly related to the feature dimension and is independent of the number of training data. The influence method-based unlearning can compute the impact of the deleted data relative to the parameters of the trained model for removing the influence and updating the parameters. Furthermore, Alexander et al. [39] apply the influence function and propose first-order and second-order updates to unlearning features and labels. However, the influence function performs inefficiently in the DNN model's unlearning due to the extremely high cost of computation of the Hessian matrix. Additionally, subspace unlearning provides a new direction for unlearning. Chen et al. [7] proposed UNSC which utilizes the null space calibration to avoid the over-unlearning that occurs in many unlearning algorithms successfully.

3 Preliminaries

3.1 Machine Unlearning

Machine unlearning is geared towards efficiently and accurately removing a specific set of data from the original dataset. The real-world application of machine unlearning arises from the need of participants in a machine-learning task to exclude certain datasets that may pose privacy concerns from the training sets. In practical terms, there is a requirement to exclude sensitive data, such as images of celebrities or inappropriate content, that inadvertently became part of the training process. A simple way to remove the data already used for training is to retrain the model without the unlearned data, also called RFS which is presented in Fig. 1. In other words, RFS guarantees the retrained model doesn't contain any information about removed data points as a result. However, exploring how to remove data from a trained model efficiently is a high-yield endeavor, and that's where the significance of machine unlearning lies.

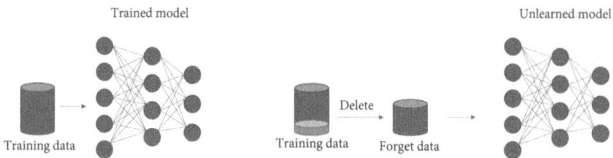

Fig. 1. RFS Unlearning Algorithm. Delete the request data from the training data to build forget data then train the unlearned model on the forget data.

3.2 Backdoor Attack and Membership Inference Attack

The backdoor attack constitutes a deliberate attack on a deep learning model, wherein the attacker strategically embeds backdoors into the model during the training process [22]. The backdoor attacks are very similar to watermarking methods which proved that can be attacked under some scenarios [6]. When the backdoor remains inactive, the compromised model performs similarly to a normal model. Saha et al. [30] proposed a new form of clean-label backdoor attack where poisoned data look natural with correct labels and also more importantly, the attacker hides the trigger in the poisoned data and keeps the trigger secret until the test time. According to this method, we simply add backdoor triggers to the target data to train a deep-learning model. Then, we use our unlearning method to unlearn these data to test if the model forgets these target data by checking the accuracy of the triggered data.

MIAs on ML models are geared towards determining whether a specific data record was employed in the training of a designated ML model. The concept of MIAs was first proposed by Homer et al. [18], where they demonstrate an attacker can leverage the published statistics about a genomics dataset to infer the presence of a particular genome in this dataset. The crux of a successful MIA attack hinges on the extent of the model's overfitting. The implications of MIAs extend to significant privacy risks for individuals. Consequently, the application of MIA proves valuable in identifying and addressing the data leakage issue, ultimately validating the efficacy of machine unlearning.

4   Method

In this section, we will display the concept of our proposed Unlearning algorithm in deep learning networks and the framework of the algorithm in detail. Partial-data-Process Unlearning (PPU) $\Gamma(L_u|M_0)$ infects only the samples that need to be removed from the trained model and performs similarly to the RFS models. Additionally, the time required to complete the unlearning process is short.

4.1   Partial-Data-Process Unlearning

Algorithm Overview. The two critical steps of the PPU are the true loss of unlearned sample calculation and the model parameters update. The forward propagation aims at finding the parameters that need to be modified to achieve the goal of unlearning. And at the same time, the clean model generates the true loss of the unlearned samples. After that, The target model will update its parameters according to the loss generated by the clean model. The whole unlearning process can be found in Fig. 2.

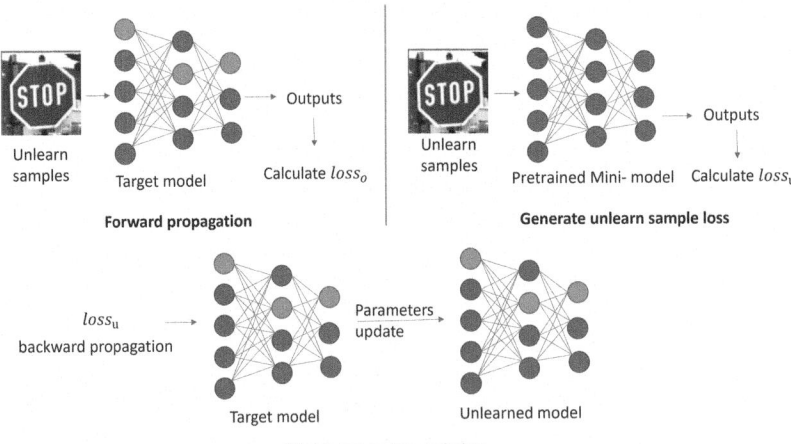

Fig. 2. The workflow of the PPU. The target model finishes the forward propagation to activate the neurons that need to be removed. The pre-trained mini-model calculates the correct loss $loss_u$ of the samples. Then, the target model calculates the gratitude using the $loss_u$ and updates to the target model.

In general, a deep learning model M is obtained through multiple rounds of iterative updates by deep learning algorithm A. For the trained model M, its parameters are represented as ω. The whole training process may take a long time due to the model parameter size and the training data size. To remove some unwanted data points from the trained model M, simply retraining the model without the unwanted data can take a very long time. Hence, the demand for an efficient unlearning algorithm continues to grow. Basically, an unlearning algorithm A_u can train the model M into an unlearned model M_u and its parameter can be defined as ω_u. The model parameters modification unlearning method aims at changing the target model's parameters $A_u(\omega) \overset{(x,y)}{\Longrightarrow} \omega_u$ to finish the unlearning process.

The PPU calculates the true loss of the unlearned samples through a pre-trained clean model M_c and backward propagation of the true loss into the target model and updating the parameters. In the item unlearning scenarios, this M_c can be used to generate true loss repeatedly whenever the unlearning request arrives due to the absence of the data used in M_c during the training stage of M. In the context of class unlearning, the approach differs significantly because it's not feasible to anticipate the classes that will be deleted beforehand, making it impractical to pre-train a M_c. However, the M_c merely necessitates the ability to approximate its loss of an unlearned model, rather than achieving high accuracy, in order to effectively mimic the loss and achieve successful unlearning. This strategy allows for time-saving while still attaining the desired outcome.

Algorithm 1: The workflow of PPU

Data: Original training data D, request unlearn data D_u, test clean data D_c, test data D_t, unlearn epoch E
Request: Item unlearning I_u, Class unlearning C_u
Result: The unlearned model M_u
1: **train** original model M_o and pre-trained Mini-model M_c.
2: **Receive unlearning request:**
if $request = I_u$ then
    while *current epoch* $e \leq E$ **do**
        $\omega_u = \arg\min\limits_{\omega} \sum_{(\mathbf{x}_i, \mathbf{y}_i) \in \mathcal{D}_u} \Gamma\{(\mathbf{x}_i, \mathbf{y}_i, \omega_c)|\omega_o\}$;
    end
end
if $request = C_u$ then
    $\omega_c = \arg\min\limits_{\omega} \sum_{(\mathbf{x}_i, \mathbf{y}_i) \in \mathbf{D}_c^*} \mathcal{L}_1\{(\mathbf{x}_i, \mathbf{y}_i)|\omega_i\}$;
    Optimizer:
    $\omega_{t+1} = \omega_t - \frac{\eta}{(B - \Delta B_t) \times v_t} m_t$;
    while *current epoch* $e \leq E$ **do**
        $\omega_u = \arg\min\limits_{\omega} \sum_{(\mathbf{x}_i, \mathbf{y}_i) \in \mathcal{D}_u} \Gamma\{(\mathbf{x}_i, \mathbf{y}_i, \omega_c)|\omega_o\}$;
    end
end

Detailed PPU. Specifically, we assume the original model M_o is optimized by the loss function $\mathcal{L}$, where $\mathcal{L}$ can be any standard loss function, such as cross-entropy. The model is trained on $D \in \{(x_1, y_1), (x_2, y_2)...(x_i, y_i)\}$. The w_0 is the optimal parameters of the original model:

$$w_o = \arg\min_{\omega} \sum_{(\mathbf{x}_i, \mathbf{y}_i) \in \mathcal{D}} \mathcal{L}\{(\mathbf{x}_i, \mathbf{y}_i)|\omega_i\} \tag{1}$$

A deep learning model M_o is trained in (1) consuming the time T_o to finish the learning iterations. The goal of machine unlearning is to remove the influence of specific data in M_o. The most traditional unlearning algorithm is RFS, whose principle is illustrated in Formula (2). Here, D_u denotes the dataset slated for removal, while ω_{RFS} represents the optimized model parameters acquired through unlearning.

$$\omega_{RFS} = \arg\min_{\omega} \sum_{(\mathbf{x}_i, \mathbf{y}_i) \in (\mathcal{D} \backslash \mathcal{D}_u)} \mathcal{L}\{(\mathbf{x}_i, \mathbf{y}_i)|\omega_o\} \tag{2}$$

In the unlearning stage, one possible solution is to find the parameters that have a greater impact on the target data that need to be removed in the optimal model M_o. Then, we can use a practical algorithm to change the parameters to fulfill the unlearning needs. However, in deep learning, the model parameters are actually in a black box state and the way to directly change the parameters is impassable. However, we can find a model that can simulate the correct model outputs and calculate the approximate loss so that it can have an effective impact on the target data of the target model.

Accordingly, we need to train a clean model M_c by applying the training accelerating method during the training stage or after training. Liu et al. [24] have shown that a modified optimization function can reduce the training time of retraining despite the subtle decrease in accuracy. (3) is the new training optimization function. However, the compact loss of subtle accuracy doesn't affect the unlearned model accuracy too much as we will show the result in the following experiment section.

$$\omega_{t+1} = \omega_t - \frac{\eta}{(B - \Delta B_t) \times v_t} m_t \tag{3}$$

The B is the batch size. The m_t and v_t in (3) are Hessian momentum:

$$m_t = \frac{(1 - \beta_1) \sum_{i=1}^{t} \beta_1^{t-i} \Delta_i}{1 - \beta_1^t} \tag{4}$$

$$v_t = \sqrt{\frac{(1 - \beta_2) \sum_{i=1}^{t} \beta_2^{t-i} H_i^{-2}}{1 - \beta_2^t}} \tag{5}$$

The β_1 and β_2 are hyperparameters that are used in the Adam optimizer. The H_i is the Hessian matrix of the processed model parameters. The Eqs. (3)–(5) aims at reducing the computation cost of the model training process.

This algorithm utilizes the FIM algorithm to streamline the updating of model parameters, incorporating special diagonalization operations to further minimize its computational overhead. Additionally, hyperparameters can be dynamically adjusted during model updates, contributing further to the reduction in parameter update time. Despite the expedited updating pace and reduced storage demands associated with this approach to updating model parameters, it may incur a weak negative impact on predictive accuracy under certain circumstances. Nonetheless, Given that our primary concern is to approximately simulate the loss of the forgetting model, employing this updating method for model parameters proves to be a viable approach.

Then, in order to obtain a clean model, we have to prepare clean data samples D_c which don't include the target data samples D_u. To guarantee the repeatability of using the clean model in the future unlearning stage, We usually extract some data from the validation data samples to train this model $D_c \in D_t$. w_c is the optimal parameters trained on the clean samples:

$$\omega_c = \arg\min_\omega \sum_{(\mathbf{x}_i, \mathbf{y}_i) \in \mathcal{D}_c} \mathcal{L}\{(\mathbf{x}_i, \mathbf{y}_i) | \omega_i\} \tag{6}$$

After training a clean model, we have to figure out which parameters in the original model should be modified to achieve the goal of unlearned sample's influence removal from the M_o. Accordingly, we forward propagation of the original model to activate the neurons (parameters):

$$\mathbf{F}_o(\omega_o) \stackrel{(x_i, y_i) \in D_u}{\Longrightarrow} \omega_o' \tag{7}$$

The ω_o' in (7) means the ω_o after activation. Then, we use the loss from M_c to update parameters in M_o:

$$\omega_u = \arg\min_\omega \sum_{(\mathbf{x}_i, \mathbf{y}_i) \in \mathcal{D}_u} \Gamma\{(\mathbf{x}_i, \mathbf{y}_i, \omega_c) | \omega_o'\} \tag{8}$$

w_u in (8) is the unlearned model parameters. The Γ in (8) is the step to update the loss influence of w_c into ω_o' using the loss function $\mathcal{L}$ or any other loss function that can calculate the actual loss of the target data outputs. Then, backward propagation of the actual loss of the target samples to calculate the gradient of these activated neurons. Afterward, we can utilize SGD to update the ω_o'.

After these steps, the activated model parameters will be updated in the right direction to achieve the unlearn goal. Usually, this unlearning process may require several epochs to be fully completed. The number of epochs is set based on the specific unlearning requirements. Typically, a large number of epochs isn't necessary to fulfill the unlearning needs. Too many epochs might actually prolong the unlearning process without significant changes in effectiveness. As the number of unlearned epochs increases, the marginal gain in effectiveness diminishes over time.

To summarize, through our detailed explanation of the proposed method under the item unlearning scenario, we can also apply PPU for class unlearning requests. The key to achieving the level of class unlearning is to train M_c without the target unlearning class $c_u \in \mathbf{C}_I = \{\mathbf{c}_1, \mathbf{c}_2, \mathbf{c}_3....\mathbf{c}_n\}$. The training sets of M_c in class unlearning is $\mathbf{D}_c^* \subseteq \{(\mathbf{x}_i, \mathbf{y}_i)|\mathbf{y}_i \neq \mathbf{y}_{target}\}$:

$$\omega_c^* = \arg\min_{\omega} \sum_{(\mathbf{x}_i, \mathbf{y}_i) \in \mathbf{D}_c^*} \mathcal{L}_1\{(\mathbf{x}_i, \mathbf{y}_i)|\omega_i\} \tag{9}$$

After we obtain M_c without the target class, the following steps of updating the parameters in the target model M_o to unlearn the specific class into M_o is the same as (8):

$$\omega_u = \arg\min_{\omega} \sum_{(\mathbf{x}_i, \mathbf{y}_i) \in \mathcal{D}_u} \Gamma\{(\mathbf{x}_i, \mathbf{y}_i, \omega_c^*)|\omega_o'\} \tag{10}$$

In most cases, unlearning requests involve removing a significant amount of data. The primary requirement of unlearning algorithms is to avoid the need for retraining the model when removing a small amount of data. Therefore, it's advisable not to remove too many classes each time, as removing too many classes would affect a large amount of data, which doesn't align with the design requirements of unlearning algorithms. Most unlearning algorithms focus on studying the scenario where one class is deleted. In Formula (9), the trained M_c is essentially trained on a small subset of data selected from the dataset excluding the target class.

Discussion and Analysis. We present discussions and related analyses of the PPU algorithm, including running time, and storage usage as below.

– running time. The PPU algorithm only needs simple calculations to achieve unlearning, e.g., true loss calculation and backward propagation. Let t denote the training round and S_u denote the unlearned data size, the time consumption of true loss calculation is $S_u O(1)$. The computation of loss backward and model parameters updating are in $S_u O(t)$. Let S_c denote the clean data size, the clean model M_c training process consuming time $S_c O(t_c)$. In the case of item unlearning, this time does not need to be counted. The whole algorithm consumes $S_u O(1) + S_u O(t)$ in the item unlearning scenario while in the class unlearning scenario consumes $S_u O(1) + S_u O(t) + S_c O(t_c)$. In contrast, the RFS will consume $SO(n)$ time.
– storage usage. The PPU algorithm doesn't require too much storage usage compared with other unlearning algorithms e.g., Amnesiac unlearning and SISA. The SISA requires additional storage overhead to save its own multiple-sharded models. Therefore, the space overhead of SISA is approximately $O(1)$. For Amnesiac Unlearning, this algorithm requires significant storage overhead because we cannot predict the data to be removed. To address this, it saves the weights and biases updated by the model every time it is updated. If the number of model parameters is P, then the space complexity of the algorithm

is $O(pn)$. As for our PPU algorithm, it just needs to save a clean model and the original model which budgets the time and the space well. The space complexity of PPU is $O(1)$.

5 Performance Evaluation

5.1 Experiment Setup

Datasets. We conduct CIFAR-10 using a backdoor attack (like the watermarking method) to testify if our Unlearning method modifies the original model structure to remove the influence of unlearning samples. Then we use additional SVHN and VGG face 2 to compare the performance of our method.

Implementations. Our methods and other baselines are implemented in Python 3.10 and use the PyTorch library. All experiments are conducted on a workstation with one NVIDIA GeForce RTX 4090 GPU. For the CIFAR-10 dataset, we train the default DNN model from scratch for 100 epochs using SGD with a fixed learning rate of 0.001, momentum of 0.9, and batch size of 64.

Baseline. We implement the following baseline unlearning methods for comparisons:

Retrain from Scratch (RFS): We train the DNN model from scratch. This method means that the model doesn't contain any unlearned dataset at all. Thus, the retrained DNN model is the optimal unlearned model.

Boundary Shrink (BS) [8]. The key of this algorithm is to shift the decision boundary of the original DNN model to imitate the decision behavior of the model retrained from scratch.

Amnesiac Unlearning (AU) [17]. The key of this algorithm is to store the history of the model update parameters to achieve unlearning by removing these related model updates.

Metrics. We verify the unlearning performance on both utility and privacy guarantees. In the utility guarantee part, we use the following metrics: accuracy on the training dataset of the original model Acc_D, accuracy on the test dataset of the original model Acc_{D_t}, accuracy on the remaining training dataset of unlearned model $Acc_{D_r^*}$, accuracy on the unlearned training dataset of unlearned model $Acc_{D_u^*}$, accuracy on the target test dataset of unlearned model $Acc_{D_t^*}$, accuracy on the non-target test dataset of unlearned model $Acc_{D_n^*}$. The unlearned model is expected to get close accuracy with the retrained model.

For privacy guarantee, we construct a simple yet general backdoor attack to the original model and test the trigger output accuracy. For the original model, we make the forget data into triggered data. As a result, the accuracy of the trigger test is directly related to the degree to which the forgotten data set is forgotten. Ideally, the trigger data is fully removed when the trigger test accuracy is close to $\frac{1}{C_n}$.

5.2   Utility Guarantee

To make sure our proposed method performs well across a variety of datasets, we tested it on three different datasets: CIFAR-10, SVHN, and VGGface2. For VGGface2, we picked a dataset featuring images of 10 different celebrities for face recognition. Our experimental results are showcased in the Table 1.

Table 1. Unlearning performance comparison of different approaches.

Dataset	method	Acc_D	Acc_{D_t}	$Acc_{D_r^*}$	$Acc_{D_u^*}$	$Acc_{D_t^*}$	$Acc_{D_n^*}$
CIFAR10	RFS	99.94%	90.78%	99.97%	0%	0%	91.42%
	AU	94.31%	81.12%	89.72%	0%	0%	85.09%
	BS	98.46%	94.33%	89.17%	7.11%	8.33%	79.66%
	PPU	99.72%	90.18%	94.67%	8.66%	6.16%	85.08%
SVHN	RFS	99.55%	92.18%	99.88%	0%	0%	92.74%
	AU	99.77%	89.57%	82.95%	0%	0%	79.65%
	BS	93.4%	92.58%	86.20%	4.31%	3.06%	86.07%
	PPU	99.95%	91.77%	95.01%	12.11%	8.88%	87.88%
Vgg face2	RFS	90.55%	89.76%	98.88%	0%	0%	98.74%
	AU	100.00%	95.74%	53.09%	0%	0%	52.38%
	BS	93.4%	92.58%	98.57%	3.31%	1.33%	99.33%
	PPU	99.99%	98.77%	99.77%	6.88%	5.58%	98.51%

From the Table 1, it's evident that most algorithms generally perform well in the majority of cases. If we take the performance of the RFS as a reference, the effects of most algorithms after unlearning on three different datasets can closely match RFS. However, it's crucial to note that even though Amnesiac Unlearning falls under the category of exact unlearning, its algorithm has certain limitations. Amnesiac Unlearning mainly focuses on the small unlearning dataset. As a result, it performs poorly in the experiments with VGG Face 2. After unlearning, although the accuracy in the D_u^* and D_t^* drops to 0%, the accuracy in the D_r^* and D_t^* also declines significantly to 53.09% and 52.38%. Such accuracy is a severe case of catastrophic forgetting, and these results are far from satisfactory compared to Boundary Shrink and our PPU. In this dataset, Boundary Shrink is slightly better than our PPU after unlearning. The accuracy of D_u^* and D_t^* are more close to 0%. The biggest difference is $\Delta D_t^* = 4.25\%$. However, if we exclude factors like errors, the effectiveness of both approaches is actually quite similar. The advantage of our method lies in its ability to perform item unlearning, whereas Boundary Shrink can only handle class unlearning. In the CIFAR-10 dataset, each of the three methods has its pros and cons. Amnesiac Unlearning achieves a precision drop to 0% in D_u^* and D_t^* after unlearning, but it slightly reduces accuracy on D_r^*, impacting the model's generalization performance severely. On the other hand, Boundary Shrink demonstrates a much better overall performance in CIFAR-10 compared to Amnesiac. It not only minimally affects the model's accuracy but also achieves almost 0% accuracy on the unlearned dataset, indicating its ability to remove target data information from the model to a large extent. Comparatively, our Progressive Pruning Unlearning (PPU) is quite similar to Boundary Shrink. Although PPU outperforms Boundary Shrink by $\Delta D_u^* = 1.55\%$ accuracy, it has a lower $\Delta D_t^* = -3.83\%$. Additionally, PPU surpasses Boundary Shrink by $\Delta D_n^* = 5.42\%$ accuracy. All in all, our method's performance on CIFAR-10 is comparable to Boundary Shrink and Amnesiac Unlearning.

Table 2. The accuracy of the mini pre-trained model used in the PPU

Dataset	$Acc_{D_t^*}$	$Acc_{D_n^*}$
CIFAR10	0%	61.23%
SVHN	0%	66.32%
Vgg face2	0%	67.01%

In the SVHN dataset, Amnesiac Unlearning performs much worse than the CIFAR-10 due to the number of data being less than CIFAR-10 causing the ratio of the target data to become higher. As a result, D_r^* drops to 83.95% which is a big negative effect of the unlearn algorithm. Additionally, D_n^* drops to 79.65% as well which means the model generalizes worse on the remaining test set. According to the experiment results, we can infer that Amnesiac Unlearning which heavily relies on the batch size of the training process and the unlearn data size, can't afford such class unlearning task. In practical unlearning applications, Amnesiac Unlearning doesn't demonstrate significant improvement either. This is primarily due to high storage costs. This incurs a substantial overhead, and in practical applications, the effectiveness might not surpass that of RFS. For PPU and Boundary Shrink, the performance of both still has its advantages. However, our PPU time cost and practicality are slightly better than Boundary Shrink and we will show the result in the following experiment. From Table 2, we can find that the accuracy of our mini-model isn't too high and we will show that the accuracy of the mini-model doesn't affect the unlearning accuracy much in the following. From the Table 4, we can see that as the mini-model training dataset expands and the accuracy improves, the impact on the unlearning effectiveness for the target model is not substantial. The biggest impact of this model is to simulate the real loss distribution after unlearning, and the model trained with 2000 data points is enough to simulate the distribution of data loss so that it can achieve the purpose of unlearning well. However, we still opted for a minimum of 2000 data points for model training. This decision was made because if the data is too limited and the model's loss is too high, it can significantly hinder the effectiveness of unlearning. On the other hand, the reason why we chose at least 2000 training samples of the mini-model is that the transforms of the data picture will dramatically affect the target model's unlearning accuracy which means we can't shrink the data size too much (Table 3 and Fig. 3).

Table 3. Time consumption of the four Mini-models in SVHN.

Data size	2000	4000	6000	8000
Time	10.62 s	16.78 s	25.54 s	33.56 s

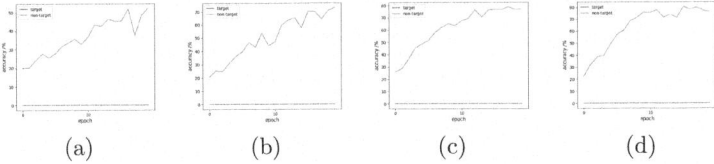

| (a) | (b) | (c) | (d) |

Fig. 3. The accuracy of different Mini-model trained on different data sizes of SVHN upon 20 epochs with the learning rate of 0.01. (a) 2000 samples of the original training set. (b) 4000 samples of the original training set. (c) 6000 samples of the original training set. (d) 8000 samples of the original training set.

Table 4. Unlearning accuracy of using different Mini-models.

Model size	Acc_D	Acc_{D_t}	$Acc_{D_r^*}$	$Acc_{D_u^*}$	$Acc_{D_t^*}$	$Acc_{D_n^*}$
2000	87.95%	85.55%	82.59%	12.59%	8.66%	80.35%
4000	87.95%	85.55%	83.55%	12.44%	8.43%	80.89%
6000	87.95%	85.55%	83.48%	12.78%	8.31%	80.47%
8000	87.95%	85.55%	84.97%	11.85%	8.79%	81.01%

5.3   Privacy Guarantee

In this part, we will discuss the privacy guarantee of the unlearning algorithm to ensure that the historical information that the model has learned doesn't exist after we remove data points. For a long time, most unlearning has used the membership inference attack to attack the model after unlearning, to determine whether the information of the deleted data points is still in the model after unlearning. This is a simple and efficient method because our method uses another model that does not need to delete data points to calculate the loss and update it to the model. And the membership inference attack requires the model to have a certain degree of overfitting. Therefore, simply using the membership inference attack cannot accurately infer the existence of data point information. Because of the above speculation, we adopted the backdoor attack to verify the impact of deleting data points in the unlearned model as mentioned before.

From the results in Fig. 4, we train the triggered model in the first 100 epochs. The experiment results clearly show that the accuracy of the trigger drops rapidly when we initiate the unlearn request. In the first unlearn epoch alone, it falls below 20% and remains at around 14% in the following epochs. This suggests that the neural connections established by involving the trigger in the original DNN training have been pretty much wiped out. Interestingly, the accuracy of the original training and test sets only sees a modest 5% decrease. It's worth noting that removing a backdoor model has a more significant impact on the classification accuracy of the original DNN compared to just unlearning a class. Nevertheless, our approach still maintains a solid level of accuracy. Subsequently, we conducted experiments on four models using MIA to validate our method's capability to completely remove the accumulated training information of target data in the model on the SVHN dataset. The experimental results show that, except for the experiment with Amnesiac Unlearning that failed the MIA test, the other methods successfully passed the MIA attack predictions. The information from target data has been effectively eliminated from the model, while non-target class information still maintains a reasonably high accuracy. We hypothesize that the reason Amnesiac Unlearning failed the MIA test is due to its overly thorough forgetting of target data, making it easily distinguishable. In our subsequent experiments on entropy distribution, we aim to highlight the strengths and weaknesses of these methods and provide further insights into why Amnesiac Unlearning fails the MIA test.

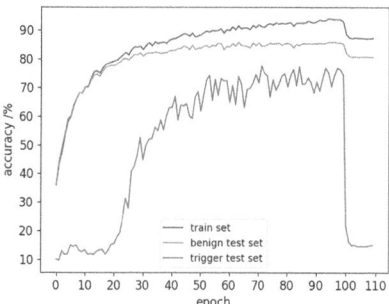

Fig. 4. The accuracy of the unlearned triggered model. The unlearning requests are submitted in epoch 100. The picture illustrates the efficacy of our unlearning method. This also means our item unlearning efficacy.

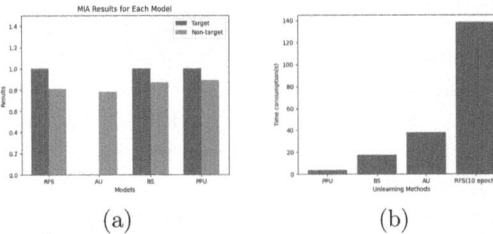

(a)   (b)

Fig. 5. (a) The MIA attack prediction accuracy results of the four models on datasets SVHN. For target data samples, we expect the MIA prediction to be 0 which means the information of the sample is not in the model. For non-target samples, we expect the MIA prediction to be 1 which means the information of the sample is in the model. (b) Time consumption of different unlearning methods mentioned above.

5.4   Effectiveness Guarantee

Another crucial factor in assessing the effectiveness of an unlearning algorithm is its runtime. If the unlearning algorithm takes too much time, perhaps comparable to retraining from scratch, it becomes impractical and loses its utility. We further conduct time-consuming experiments of the four different methods on CIFAR-10. In the setup of this experiment, PPU opted for 3 unlearn epochs because that's usually enough to meet the unlearn goals in most cases. There's no need for extra epochs, as they not only take up time but also negatively impact the model's generalization. Additionally, we propose that when implementing item unlearning, the auxiliary model can actually be trained simultaneously with the initial model, as the dataset we use comes from a small part of the test set, and there is no target item for this unlearn goal. When it comes to class unlearning, the auxiliary model doesn't need a large dataset. For example, in the CIFAR-10 experiment, we only need to extract about 10% of the data from 10,000 test samples to train an auxiliary model, and this model doesn't require very high accuracy. Therefore, the training time for this nontarget class model is only 25.33 s.

In Fig. 5, we can find that RFS cost the most time about 138 s of the entire unlearn process despite 10 epochs. In fact, retraining a model of 10 epochs doesn't perform well in most scenarios. The more epochs it trains on, the more time it will take as a result. On the contrary, other algorithms significantly cut down on time expenses. While our method does require training a new auxiliary model during class unlearning, the training cost is actually quite minimal because this model doesn't need a large amount of data. When performing class unlearning on CIFAR-10, the overall time spent by PPU is 28.99 s, which is lower than the 37.66 s of Amnesiac unlearning. However, in the case of item unlearning, our method's time consumption is only during the unlearning process because the auxiliary model was already trained simultaneously with the initial model. On the contrary, boundary shrink can only be used for class unlearning. In summary, our method proves to be practical in a comprehensive range of scenarios.

5.5   Distribution of Cross-Entropy

In this section, we conducted a statistical analysis and visualization of the distribution of cross-entropy calculated from the model outputs. This enables us to accurately analyze why each model ensures privacy while forgetting. Ideally, we hope that the loss distribution of the unlearned model is as consistent as possible with that of RFS. Alternatively, the loss for the unlearned class should be distinguishable from the losses of other classes. From the graphs, it's evident that all algorithms can effectively distinguish between the target class and non-target class losses. However, in Amnesiac Unlearning, the loss for the target class is more than 10 times higher than the non-target class. This suggests that this algorithm may not be suitable for tasks like class unlearning, which involves forgetting a significant amount of data. Boundary Shrink emerges as a commendable method for class unlearning, displaying results close to RFS after unlearning, with minimal reduction in non-target class loss. Our PPU, post-unlearning, also successfully separates the loss for the target class from the non-target class, without experiencing excessively high losses. The only drawback is a slight impact on the loss of the non-target class, although the effect is not substantial to the classification accuracy. As a result, our proposed method PPU, after achieving machine unlearning, does not result in serious privacy leakage and is suitable for real-world applications (Fig. 6).

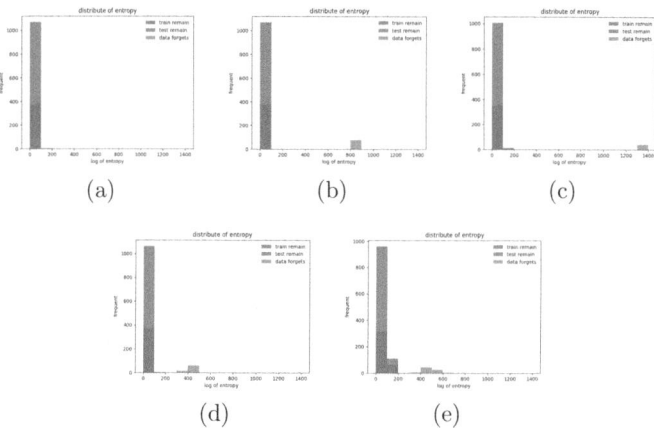

Fig. 6. The distribution of cross-entropy of the different algorithms in SVHN during the unlearning request of class unlearning. (a) Original model. (b) RFS model. (c) Amnesiac unlearning model. (d) Boundary shrink model. (e) PPU model.

6   Conclusion

In this paper, we develop a new method to successfully remove items and an entire class from the already trained DNN. The PPU algorithm we propose not only successfully passes the MIA test but also performs well in backdoor attack tests. This demonstrates that our proposed method, after achieving machine unlearning, does not result in serious privacy leakage and is suitable for real-world applications. Through various experiments, we've demonstrated that our approach not only doesn't require a significant amount of computational resources but also doesn't incur substantial space overhead compared to other algorithms. In our future work, we aim to conduct more in-depth research on our algorithm, enabling it to exhibit outstanding unlearn effects in a broader range of tasks and scenarios.

References

1. Bourtoule, L., et al.: Machine unlearning. In: 2021 IEEE Symposium on Security and Privacy (SP), pp. 141–159. IEEE (2021)
2. Brophy, J., Lowd, D.: Machine unlearning for random forests. In: International Conference on Machine Learning, pp. 1092–1104. PMLR (2021)
3. Cauwenberghs, G., Poggio, T.: Incremental and decremental support vector machine learning. In: Advances in Neural Information Processing Systems, vol. 13 (2000)
4. Chang, Y., Ren, Z., Nguyen, T.T., Nejdl, W., Schuller, B.W.: Example-based explanations with adversarial attacks for respiratory sound analysis. arXiv preprint arXiv:2203.16141 (2022)
5. Chaudhuri, K., Monteleoni, C.: Privacy-preserving logistic regression. In: Advances in Neural Information Processing Systems, vol. 21 (2008)
6. Chen, H., Zhu, T., Liu, C., Yu, S., Zhou, W.: High-frequency matters: attack and defense for image-processing model watermarking. IEEE Trans. Serv. Comput. (2024)
7. Chen, H., Zhu, T., Yu, X., Zhou, W.: Machine unlearning via null space calibration. arXiv preprint arXiv:2404.13588 (2024)
8. Chen, M., Gao, W., Liu, G., Peng, K., Wang, C.: Boundary unlearning: rapid forgetting of deep networks via shifting the decision boundary. In: Proceedings of the IEEE/CVF Conference on Computer Vision and Pattern Recognition, pp. 7766–7775 (2023)
9. Chen, M., Zhang, Z., Wang, T., Backes, M., Humbert, M., Zhang, Y.: When machine unlearning jeopardizes privacy. In: Proceedings of the 2021 ACM SIGSAC Conference on Computer and Communications Security, pp. 896–911 (2021)
10. Dwork, C., Roth, A., et al.: The algorithmic foundations of differential privacy. Found. Trends® Theor. Comput. Sci. **9**(3–4), 211–407 (2014)
11. European Parliament, Council of the European Union: Regulation (EU) 2016/679 of the European Parliament and of the Council. https://data.europa.eu/eli/reg/2016/679/oj
12. Giordano, R., Stephenson, W., Liu, R., Jordan, M., Broderick, T.: A swiss army infinitesimal jackknife. In: The 22nd International Conference on Artificial Intelligence and Statistics, pp. 1139–1147. PMLR (2019)
13. Golatkar, A., Achille, A., Ravichandran, A., Polito, M., Soatto, S.: Mixed-privacy forgetting in deep networks. In: Proceedings of the IEEE/CVF Conference on Computer Vision and Pattern Recognition, pp. 792–801 (2021)

14. Golatkar, A., Achille, A., Soatto, S.: Eternal sunshine of the spotless net: selective forgetting in deep networks. In: Proceedings of the IEEE/CVF Conference on Computer Vision and Pattern Recognition, pp. 9304–9312 (2020)
15. Golatkar, A., Achille, A., Soatto, S.: Forgetting outside the box: scrubbing deep networks of information accessible from input-output observations. In: Vedaldi, A., Bischof, H., Brox, T., Frahm, J.-M. (eds.) ECCV 2020. LNCS, vol. 12374, pp. 383–398. Springer, Cham (2020). https://doi.org/10.1007/978-3-030-58526-6_23
16. Goldblum, M., et al.: Dataset security for machine learning: data poisoning, backdoor attacks, and defenses. IEEE Trans. Pattern Anal. Mach. Intell. **45**(2), 1563–1580 (2022)
17. Graves, L., Nagisetty, V., Ganesh, V.: Amnesiac machine learning. In: Proceedings of the AAAI Conference on Artificial Intelligence, vol. 35, pp. 11516–11524 (2021)
18. Homer, N., et al.: Resolving individuals contributing trace amounts of DNA to highly complex mixtures using high-density SNP genotyping microarrays. PLoS Genet. **4**(8), e1000167 (2008)
19. Izzo, Z., Smart, M.A., Chaudhuri, K., Zou, J.: Approximate data deletion from machine learning models. In: International Conference on Artificial Intelligence and Statistics, pp. 2008–2016. PMLR (2021)
20. Karasuyama, M., Takeuchi, I.: Multiple incremental decremental learning of support vector machines. IEEE Trans. Neural Networks **21**(7), 1048–1059 (2010)
21. Koh, P.W., Liang, P.: Understanding black-box predictions via influence functions. In: International Conference on Machine Learning, pp. 1885–1894. PMLR (2017)
22. Li, Y., Zhang, S., Wang, W., Song, H.: Backdoor attacks to deep learning models and countermeasures: a survey. IEEE Open J. Comput. Soc. (2023)
23. Liu, G., Ma, X., Yang, Y., Wang, C., Liu, J.: Federaser: enabling efficient client-level data removal from federated learning models. In: 2021 IEEE/ACM 29th International Symposium on Quality of Service (IWQOS), pp. 1–10. IEEE (2021)
24. Liu, Y., Xu, L., Yuan, X., Wang, C., Li, B.: The right to be forgotten in federated learning: an efficient realization with rapid retraining. In: IEEE INFOCOM 2022-IEEE Conference on Computer Communications, pp. 1749–1758. IEEE (2022)
25. Martens, J.: New insights and perspectives on the natural gradient method. J. Mach. Learn. Res. **21**(146), 1–76 (2020)
26. Peri, N., et al.: Deep k-NN defense against clean-label data poisoning attacks. In: Bartoli, A., Fusiello, A. (eds.) ECCV 2020. LNCS, vol. 12535, pp. 55–70. Springer, Cham (2020). https://doi.org/10.1007/978-3-030-66415-2_4
27. Qu, Y., Ding, M., Sun, N., Thilakarathna, K., Zhu, T., Niyato, D.: The frontier of data erasure: machine unlearning for large language models. arXiv preprint arXiv:2403.15779 (2024)
28. Ren, Z., Baird, A., Han, J., Zhang, Z., Schuller, B.: Generating and protecting against adversarial attacks for deep speech-based emotion recognition models. In: ICASSP 2020-2020 IEEE International Conference on Acoustics, Speech and Signal Processing (ICASSP), pp. 7184–7188. IEEE (2020)
29. Ren, Z., Han, J., Cummins, N., Schuller, B.: Enhancing transferability of black-box adversarial attacks via lifelong learning for speech emotion recognition models (2020)
30. Saha, A., Subramanya, A., Pirsiavash, H.: Hidden trigger backdoor attacks. In: Proceedings of the AAAI Conference on Artificial Intelligence, vol. 34, pp. 11957–11965 (2020)
31. Shokri, R., Stronati, M., Song, C., Shmatikov, V.: Membership inference attacks against machine learning models. In: 2017 IEEE Symposium on Security and Privacy (SP), pp. 3–18. IEEE (2017)

32. Shwartz-Ziv, R., Tishby, N.: Opening the black box of deep neural networks via information. arXiv preprint arXiv:1703.00810 (2017)
33. Sommer, D.M., Song, L., Wagh, S., Mittal, P.: Athena: probabilistic verification of machine unlearning. In: Proceedings on Privacy Enhancing Technologies (2022)
34. Sommer, D.M., Song, L., Wagh, S., Mittal, P.: Towards probabilistic verification of machine unlearning. arXiv preprint arXiv:2003.04247 (2020)
35. Thudi, A., Shumailov, I., Boenisch, F., Papernot, N.: Bounding membership inference. arXiv preprint arXiv:2202.12232 (2022)
36. Tishby, N., Pereira, F.C., Bialek, W.: The information bottleneck method. arXiv preprint physics/0004057 (2000)
37. Tishby, N., Zaslavsky, N.: Deep learning and the information bottleneck principle. In: 2015 IEEE Information Theory Workshop (ITW), pp. 1–5. IEEE (2015)
38. Tsai, C.H., Lin, C.Y., Lin, C.J.: Incremental and decremental training for linear classification. In: Proceedings of the 20th ACM SIGKDD International Conference on Knowledge Discovery and Data Mining, pp. 343–352 (2014)
39. Warnecke, A., Pirch, L., Wressnegger, C., Rieck, K.: Machine unlearning of features and labels. arXiv preprint arXiv:2108.11577 (2021)
40. Zhang, H., Nakamura, T., Isohara, T., Sakurai, K.: A review on machine unlearning. SN Comput. Sci. 4(4), 337 (2023)
41. Zhang, L., Zhu, T., Xiong, P., Zhou, W.: The price of unlearning: identifying unlearning risk in edge computing. ACM Trans. Multimedia Comput. Commun. Appl. (2024)

BlockWhisper: A Blockchain-Based Hybrid Covert Communication Scheme with Strong Ability to Evade Detection

Zehui Wu[1], Yuwei Xu[1,2,3]([✉]), Ranfeng Huang[4], Xinhe Fan[1],
Jingdong Xu[5], and Guang Cheng[1,2]

[1] School of Cyber Science and Engineering, Southeast University, Nanjing, China
xuyw@seu.edu.cn
[2] Purple Mountain Laboratories for Network and Communication Security,
Nanjing, China
[3] Engineering Research Center of Blockchain Application,
Supervision and Management, Nanjing, China
[4] Software College, Northeastern University, Shenyang, China
[5] College of Computer Science, Nankai University, Tianjin, China

Abstract. In light of the pervasive significance of privacy protection in contemporary society, the exploration of covert communication technology has been extensive, which is designed to safeguard communication behaviors and contents from disclosure. The decentralization, anonymity, and robustness of blockchain technology render it a more suitable medium for covert communication than traditional networks, which are vulnerable to tracing and destruction. Nevertheless, state-of-the-art blockchain-based covert communication schemes tend to focus on the concealment of a single type of constituent element in a blockchain transaction. These methods are able to resist detection methods for a particular element, yet are easily exposed under the joint detection of multiple classes of methods. The absence of resistance to joint detection methods targeting all elements presents a significant challenge that impedes the advancement of blockchain-based covert channels. Consequently, Block-Whisper is proposed as a blockchain-based hybrid covert channel that is capable of evading the recognition of various detection methods. Following the probability distribution model obtained from a blockchain transaction distribution analysis method, BlockWhisper is implemented using a top address list and a special hybrid coding method. The experiments demonstrated that Graph-based Detection (GD), Similarity-based Detection (SMD), Shape-based Detection (SPD), Entropy-based Detection (ETD), and Machine Learning-based Detection (MLD) yielded inferior results on this scheme, which substantiates the assertion that Block-Whisper possesses a strong ability to evade detection. Moreover, a comparison of BlockWhisper with other schemes reveals that it has a superior channel capacity on the basis of the guarantee of channel feasibility.

Keywords: Blockchain · Covert communication · Evading detection

© The Author(s), under exclusive license to Springer Nature Singapore Pte Ltd. 2025
T. Zhu et al. (Eds.): ICA3PP 2024, LNCS 15251, pp. 161–181, 2025.
https://doi.org/10.1007/978-981-96-1525-4_9

1   Introduction

Blockchain technology serves as a novel covert channel carrier due to its decentralization, anonymity, and robustness [1]. 1) Decentralization. All data storage, transmission and verification processes of blockchain are based on a distributed system architecture and do not rely on a single centralized institution. 2) Anonymity. The address of the blockchain system serves as the identity of the user, and the user can request an address from the blockchain at will without revealing their true identity. 3) Robustness. Even in the event of the destruction of any node, the state information of the entire system will remain unaltered. Consequently, blockchain is an optimal medium for covert communication [2].

It is regrettable that the current state-of-the-art research on blockchain-based covert communication technology is still incomplete. Blockchain address-based covert channels utilize specific fields within the address to conceal secret messages [3,4]. While these schemes were able to achieve a high level of concealment, the inherent limitation of the channel capacity proved to be a fatal flaw. The remark field (e.g., OP_RETURN on Bitcoin) in the blockchain has also been considered for embedding secret messages or covert channel labels due to its large capacity [5,6]. However, this approach greatly compromises the fundamental property of channel covertness. Additionally, some schemes have attempted to utilize the more concealed elements within a blockchain transaction as the carrier of secret messages, including the transaction amount [7–9], the transaction signature [10], and the transaction time interval [11]. It is crucial to acknowledge that these schemes prioritize the covertness of a single element within the transaction, rather than the overall covertness of the transaction.

Furthermore, a multitude of blockchain-based covert channel detection methods have been proposed, which employ a comparative analysis of the features of legitimate blockchain transactions and covert channels in terms of transaction address [12], transaction amount [10], and transaction time interval [13]. The joint detection of these methods has resulted in the exposure of existing covert channels, which were previously concealed.

In order to address the challenges mentioned above, we propose a blockchain-based hybrid covert communication scheme, which is designated as BlockWhisper. The contributions of this paper are as follows:

- We propose a blockchain transaction distribution analysis method. By applying this method, the construction of covert transactions can be better aligned with the distribution law of legitimate transactions, thereby enhancing concealment. A statistical analysis was conducted to examine the distribution features of the constituent elements of legitimate transactions on the blockchain. The distribution of the elements was fitted using different curves in order to obtain a probability distribution model.
- We design a blockchain-based hybrid covert channel. The distribution of the elements in the channel exactly adheres to the probability distribution model, thus being able to evade detection. The secret message is hidden in the transaction amount and transaction time interval using a special hybrid coding method, while the transaction obfuscation is performed by a Generative

Adversarial Network (GAN). Additionally, the transaction addresses are selected as dynamic labels for the covert channel using a top address list.

- We deploy BlockWhisper and conduct an evaluation of its concealment. In order to ascertain the indistinguishability between BlockWhisper-generated transactions and blockchain-legal transactions, a combination of detection techniques is employed, including GD, SMD, SPD, ETD, and MLD. The experimental results demonstrate that BlockWhisper is capable of evading all detection methods. Furthermore, a comparison of BlockWhisper with other schemes reveals that it has a superior channel capacity on the basis of the guarantee of channel feasibility.

The remainder of the paper is organized as follows. In Sect. 2, we provide related works and illustrate the motivation. In Sect. 3, we introduce the blockchain transaction distribution analysis method. In Sect. 4, we describe the proposed scheme BlockWhisper in detail. In Sect. 5, the experimental results are presented. Finally, the conclusion part is given in Sect. 6.

2   Related Work

In this section, we introduce several existing blockchain-based covert communication schemes and the corresponding detection methods. Subsequently, the motivation for this research is elucidated.

2.1   Blockchain-Based Covert Communication

Blockchain-based covert channels can be categorized into Blockchain Covert Storage Channel (BCSC) and Blockchain Covert Time Channel (BCTC) according to the carriers in which the secret messages are embedded. BCSC refers to covert communication by embedding secret messages through storage fields in the blockchain, such as addresses, transfer amounts, or signatures in the transactions. The carrier of the secret messages in BTBC is the time-related elements in the blockchain, such as the time interval for submitting the transaction.

The earliest blockchain-based covert channel research employed transaction addresses in the blockchain as communication carriers. Partal [3] initially proposed the BLOCCE blockchain network covert channel, which maps the least significant bit (LSB) of a transaction address sequentially to a combination of encrypted covert messages. Zhang et al. [4] subsequently followed a similar line of thought and proposed the V-BLOCCE system, which uses the address generation software Vanitygen. All of these schemes embed secret data with the help of blockchain addresses. While they enhance channel concealment, it is evident that their channel capacity is insufficient for practical applications.

As research progresses, an increasing number of researchers attempt to conceal secret messages within the contents of blockchain transactions. Zhang et al. [14] proposed a group covert communication method based on blockchain digital currency, designed an address index interaction matrix, and utilized the address interaction relationship and transaction value to hide secret messages alternately. In addition, they proposed an enhancement of channel concealment by means of the utilization of the Shamir threshold and STC mapping [9]. Furthermore, Liu et al. [15] utilized the value fields of transactions in Ethernet to construct a hash-based multiple-bit embedding scheme (MBE) to enhance its concealment. Tian et al. [8] designed a unique AMASC coding method to embed ciphertext into transaction amounts, which improves the capacity and concealment of the channel. Digital signature algorithms utilized for transaction verification in blockchain can also be employed as a vehicle for covert communication. Chen et al. [10] proposed DDSAC, which conceals secret messages within a private key and transmits them to the recipient via a signature algorithm.

All of the schemes above are under the purview of BCSC, in contrast to Zhu et al. [11], who proposed a BCTC, which transmits covert messages through the time interval of communication within the blockchain network. While BCSC and BCTC possess their respective advantages, these schemes lack comprehensive focus on the elements of the transaction and need more targeted measures to enhance the anti-detectability of the covert channel.

Additionally, some studies examine the selection of dynamic labels in order to enhance the covertness of the channel. The dynamic labels selected by the sender of the covert channel are used to mark transactions embedded with secret messages. These labels are easily recognizable by the receiver from a large number of transactions, yet they are not detectable by a third party. Tian et al. [6] developed a dynamic label generation algorithm named DLchain, based on the statistical distribution of the real transaction data in the blockchain. Zhang et al. [16] designed a variable-length tag associated with transaction input and output addresses in the EBDL scheme. However, DLchain and EBDL embed the tags in the remark field, which increases the risk of exposure. Xu et al. [17], in their proposed blockchain covert communication scheme DarkTrans, utilize an address binary tree structure to select transaction addresses as dynamic labels. However, these schemes lack consideration of inter-address transaction relationships and still have room for improvement.

2.2   Detection Methods for Blockchain-Based Covert Channel

The development of blockchain covert communication technology has prompted numerous researchers to investigate methods for detecting the existence of covert channels within blockchain networks. Guo et al. [12] proposed a practical graph-based blockchain covert channel detection method, which analyzes transaction graphs and group-level features between blockchain addresses. Zillien et al. [13] employed a $\epsilon - Similarity$ method for covert timing channel detection. With regard to the detection of blockchain transaction content, Shape-based Detection schemes, such as the Cumulative Distribution Function (CDF) and the Kolmogorov-Smirnov (K-S) test [10], as well as Entropy-based Detection schemes, have been identified as effective. Additionally, Wang et al. [18] proposed a Covert Transaction Recognition (CTR) model based on Text Convolutional Neural Networks (TextCNN) and Back Propagation Neural Networks (BPNN). These methods exhibit high precision and recall for specific blockchain-based covert communication techniques.

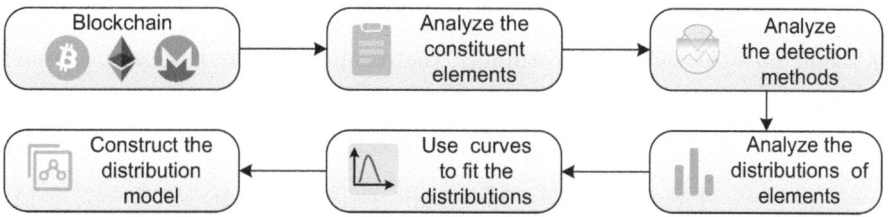

Fig. 1. Overview of the blockchain transaction distribution analysis method.

It has been established that previous techniques, whether BCSC or BCTC, concentrate on isolated features within blockchain transactions and lack a comprehensive, multi-perspective analysis of the transactions as a whole. Moreover, the existing detection schemes for blockchain-based covert channels are frequently updated, rendering it challenging for the majority of covert channels to combat them effectively. Consequently, we propose a statistical analysis of the distribution features of transaction addresses, time intervals, and amounts based on legitimate data samples obtained on the public blockchain network. This analysis will be utilized to construct a probability distribution model for each component of the blockchain covert channel. Furthermore, a blockchain-based hybrid covert communication scheme will be designed based on the model, with the objective of evading attacks from existing detection schemes for blockchain-based covert channels. The channel must be capable of equalizing its capacity while maintaining a high degree of covertness, in order to be able to meet the demands of realistic scenarios.

3 Blockchain Transaction Distribution Analysis Method

In this section, we present an analysis of the distribution features of the components in blockchain-based covert channels. The probability distribution model is constructed by fitting the legitimate transaction data.

3.1 Method Overview

The objective of this method is to derive a probability distribution model by fitting the distribution law of the constituent elements in the covert channel. The utilization of this model in covert transactions serves to enhance its indistinguishability from legitimate transactions. As illustrated in Fig. 1, the methodology comprises the following five steps: 1) First, analyze the constituent elements required for a specific blockchain to act as a covert channel carrier. 2) Analyze the principle of various covert channel detection methods for each element. 3) Statistically analyze the distribution law of each element in the blockchain transaction. 4) Use different probability density curves to fit the distribution law of each element separately. 5) Finally, construct the probability distribution model of transaction elements.

In specific terms, the feature fields selected by existing detection methods for blockchain-based covert channels include the transaction address, transaction time interval, and transaction amount, which are also the most common areas for secret message hiding. Accordingly, the components of the covert channel under consideration in this paper can be formally expressed as follows:

$$CC \leftarrow (Sender, Receiver, TX_m), \tag{1}$$

$$TX_m \leftarrow (addr\_from, addr\_to, Amount, \Delta t), \tag{2}$$

$$M \xrightarrow{\pi} (Amount, \Delta t), \tag{3}$$

$$Addresspool \xrightarrow{\tau} (addr\_from, addr\_to), \tag{4}$$

where *Sender* and *Receiver* refer to the individuals who are engaged in the covert communication. TX_m denotes the transaction embedded with the secret messages M. *Addr_from* and *addr_to* denote the source and destination addresses of the transaction, respectively. *Amount* denotes the amount of money transmitted by the transaction, and Δt denotes the time interval between the previous transaction and the current transaction. π denotes the encoding rule for M, while τ denotes the rule for selecting *addr_from* and *addr_to* from the address pool.

The overarching objective of the method is to achieve the ability to resist GD against *addr_from* and *addr_to*, SMD against Δt, SPD and ETD against v, and MLD against TX_m. Subsequently, we will analyze their distribution features individually and construct probability distribution models. It should be noted that the dataset utilized in this section is the Ethernet mainnet data collected via Blockchair [19] from block #19429556 to block #19429710 on March 14, 2024, designated **Eth154**, which encompasses the transactions of a total of 154 blocks. However, the analytical approach described herein is equally applicable to other blockchain networks.

3.2   Distribution of Transaction Addresses

It is estimated that the daily transaction volume in Ethernet is approximately 1.2 million. Given the vast number of transactions, a transaction graph can provide a more comprehensive understanding of the transactional relationships between addresses. This approach is employed by Guo et al. [12] in their blockchain covert channel detection. The authors construct a transaction graph, $G = (V, E)$, where V represents the set of transaction addresses, and E represents the set of transaction relationships. They find that the transaction graph of legitimate transactions is a star structure, whereas the transaction graph of illegitimate transactions is a chain structure. They partition the transaction graph into several subgraphs and use the variance of the degree between the subgraphs and the average longest path length to identify blockchain covert channels.

The fundamental reason for the discrepancies observed in the degree and path length of transaction graphs between legitimate and illegitimate transactions is the presence of certain "top addresses" within the blockchain, such as cryptocurrency exchanges. The majority of addresses opt to engage in transactions with these top addresses, which possess specific functions. As shown in

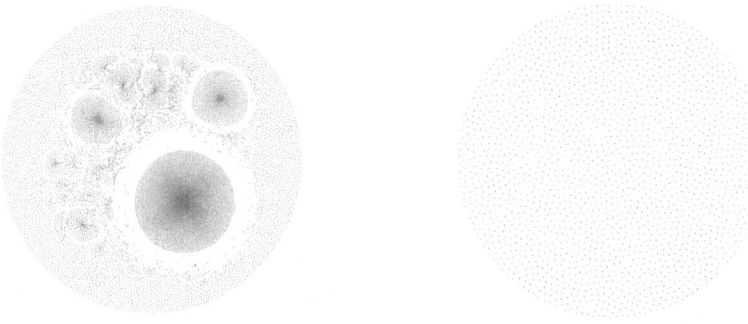

(a) Legitimate transactions   (b) Illegitimate transactions

Fig. 2. Transaction graph.

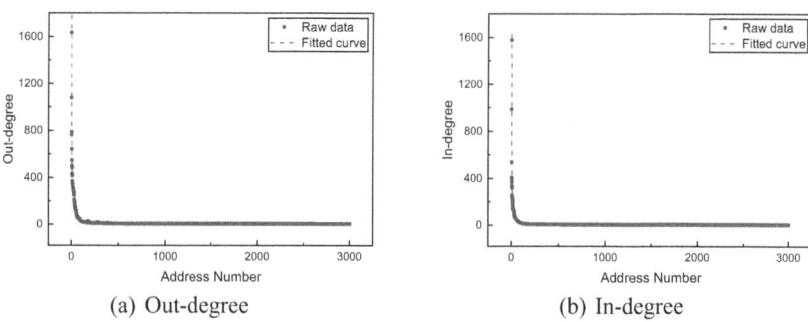

(a) Out-degree   (b) In-degree

Fig. 3. Fitting the distribution of out-degree and in-degree of transaction addresses.

Fig. 2, we constructed separate transaction graphs for legitimate transactions in Eth154 and illegitimate transactions [18], which serves to corroborate our hypothesis. Accordingly, in order to evade detection, it is essential to adhere to the law of probability distribution of the number of occurrences of the transaction address when establishing the covert channel, that is, the challenge of devising the top address. The distribution of out-degree and in-degree for each address in Eth154 was calculated and is shown in Fig. 3. It can be observed that this sample distribution approximates the Zipf distribution, which is commonly used to characterize the frequency of word occurrences in natural language versus the distribution of the number of times different web pages are viewed. This is analogous to the scenario of inter-address transactions. Indeed, the Zipf distribution provides a satisfactory fit to the outgoing and incoming data of addresses. Therefore, the Zipf distribution was selected to characterize the distribution pattern of the selected addresses. The probability mass function of the Zipf distribution is as follows:

$$\Phi(x) = \frac{1}{x^\alpha \sum_{i=1}^{n}(1/i)^\alpha},\tag{5}$$

where $x = 1, 2, ..., n$, and α is a degree parameter. In particular, when constructing a transaction, it is essential that the following probabilities are satisfied with regards to selecting the source and destination addresses:

$$\forall addr_k \in AddressPool,$$
$$P(addr_k = addr\_from) = \Phi(k; \alpha_1), \qquad (6)$$
$$P(addr_k = addr\_to) = \Phi(k; \alpha_2),$$

where α_1 and α_2 denote the parameters of the Zipf distribution fitting the normal address out-degree and in-degree, respectively. It is important to note that an address cannot be used as both a source and destination address.

3.3   Distribution of Transaction Time Intervals

Although the temporal interval between each transaction in the blockchain is not directly reflected in the block data, certain more sophisticated attackers may still be able to discern the time features between transactions by monitoring the activity of the nodes or analyzing the block transaction time. Zillien et al. [13] demonstrated that there are differences in the distribution patterns of time intervals in the normal channel and the covert channel. The degree of similarity between time intervals is more significant in the covert channel, whereas in the normal channel, the differences between time intervals are more pronounced. This is summarized as the $\epsilon - Similarity$ method. The dissimilarity in time interval similarity is attributed to the fact that the settings of the time intervals in the covert channel do not align with the distribution pattern observed in the normal channel. Thus, a histogram of the distribution of the average time intervals of transactions in each Eth154 block is plotted as shown in Fig. 4, which exhibits a skewed distribution. This can be visualized as having a short head and a long tail before and after the peak, respectively. The lognormal distribution curve fits the distribution of this sample well, and its probability density function can be formalized as follows:

$$\Psi(x) = \frac{1}{x \ln a \sqrt{2\pi}\sigma} e^{-\frac{(\log_a z - \mu)^2}{2\sigma^2}}, \qquad (7)$$

where a is the base of the logarithm, and in general a can be taken to be equal to 10 or e. Specifically, the range domain of legitimate transaction intervals is divided equally by demand into s time periods:

$$T = T_1 \cup T_2 \cup ... \cup T_s. \qquad (8)$$

In the construction of a transaction, it is necessary that the selected transaction time interval satisfy the following condition:

$$P(\Delta t \in T_j) = \Psi(t_j), \qquad (9)$$

where t_j is the middle moment of the time period T_j.

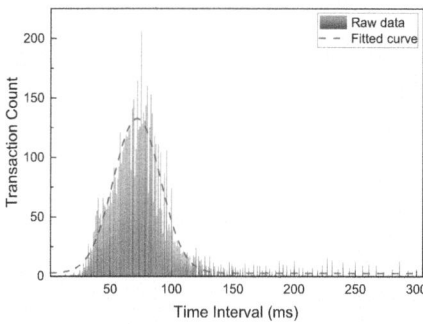

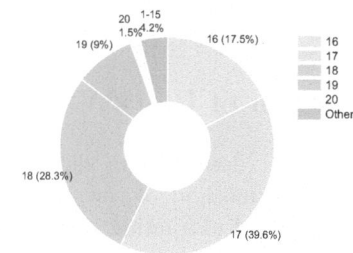

Fig. 4. The distribution of the time intervals of transactions in Eth154.

Fig. 5. The distribution of transaction amount lengths.

3.4   Distribution of Transaction Amounts

The most crucial aspect of the blockchain is the facilitation of token transfers. The amount transferred in each transaction is determined on a case-by-case basis. However, due to the influence of the blockchain token exchange rate and the smallest currency unit, the distribution of the transaction amount also exhibits a discernible pattern. Currently, SPD and ETD are more effective detection methods for blockchain amounts. The concept of SPD is to contrast the disparity in the distribution of samples between the legitimate transaction amount and the transaction amount in the covert channel. ETD, on the other hand, utilizes the difference in entropy value between the two. The distinction in entropy is also a distinction in distribution in nature.

To eliminate these discrepancies, it is necessary to model the distribution of amount lengths for legitimate transactions. As illustrated in Fig. 5, the distribution of amount lengths is concentrated in a narrow interval. Therefore, it is sufficient to construct the amount length in that interval as a proportion of the legitimate transactions.

Furthermore, a similar power distribution can be observed in relation to the size of the transaction amount. However, suppose this distribution is directly randomly sampled to obtain the transaction amount. In that case, it will result in a significant deviation in the distribution of the numbers on each digit in the amount, which has a notable impact on its entropy. Therefore, the transaction amount generated by CTGAN [20] is inserted for obfuscation based on the construction of the transaction amount using a special hybrid coding method. CTGAN is considered a type of GAN that is able to mimic the features of the original data to generate hybrid data that resists the detection of classifiers. The model for constructing transaction amounts can be expressed as follows:

$$Amount_{leg} \xrightarrow{CTGAN} Amount_{GAN},$$

$$M \xrightarrow{Encoding} Amount_m, \tag{10}$$

$$Amount = CHOOSE(Amount_m, Amount_{GAN})$$

where $Amount_{leg}$ denotes the legitimate transaction amount, $Amount_{GAN}$ denotes the transaction amount generated with CTGAN, and $Amount_m$ denotes the transaction amount embedded with secret messages.

4   Design of BlockWhisper

This section describes in detail the general architecture of BlockWhisper's scheme, the method of selecting dynamic address labels, and the method of embedding and extracting secret messages.

4.1   Overview

In BlockWhisper's scenario, when the necessity for covert communication arises, the two parties engage in a single secure off-chain communication to establish a long-term, usable blockchain covert channel. This channel allows them to pass secret messages through transactions between addresses in the blockchain without being detected by anyone. In a blockchain network, nodes controlled by the sender establish a connection with nodes controlled by the receiver to ensure that the receiver is able to discover transactions submitted by the sender's nodes in a timely manner. The overall design of BlockWhisper's scheme is shown in Fig. 6. The steps involved are as follows:

- **Step-1:** *Sender* and *Receiver* engage in a single communication through the secure off-chain channel. The shared content includes the following elements: $\{AddressPool, key, t_a, t_b\}$, where key represents the encryption key, and t_a and t_b represent the time interval determination thresholds.

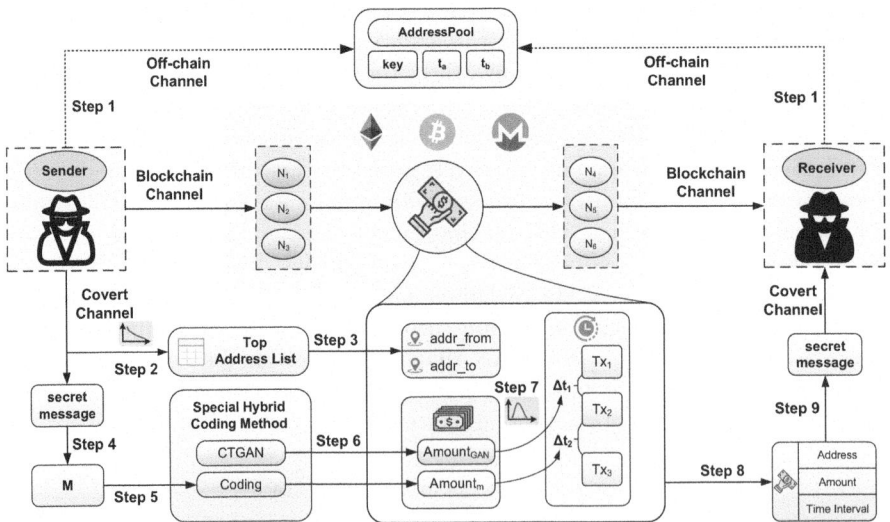

Fig. 6. Overall Design of BlockWhisper.

- **Step-2:** *Sender* then constructs two lists, namely the source top address list and the destination top address list, after randomly scrambling the addresses in AddressPool.
- **Step-3:** *Sender* randomly selects $addr\_from$ and $addr\_to$ from the source top address list and the destination top address list respectively, according to the Zipf distribution.
- **Step-4:** *Sender* encrypts the secret messages utilizing key, after which they undergo pre-processing for data segmentation and obfuscation purposes.
- **Step-5:** *Sender* encodes the preprocessed secret messages as a transaction amount using the special hybrid coding method. This amount, along with $addr\_from$ and $addr\_to$, is then assembled into a blockchain transaction, which is called a payload transaction.
- **Step-6:** Insert the GAN-generated obfuscation transactions into the payload transactions according to the established rules, and repeat **Step 3–6** until all the embedding of the secret messages is completed.
- **Step-7:** All transactions are submitted to blockchain network by the nodes controlled by *Sender* at selected time intervals within the appropriate threshold according to the distribution.
- **Step-8:** *Receiver* receives the transactions from *Sender* through the nodes based on the address label. These transactions are differentiated into obfuscation transactions and payload transactions via the time interval threshold.
- **Step-9:** *Receiver* extracts the secret messages embedded in the load transactions and time intervals. This step marks the conclusion of the entire covert communication process.

The following sections provide a comprehensive explanation of the methods used to embed and extract secret messages.

4.2 Secret Message Embedding

According to Eq. 2, it can be known that the embedding of secret messages within a transaction requires the four constituents: $\{addr\_from, addr\_to, \Delta t, Amount\}$. Therefore, BlockWhisper employs the top address list to select $addr\_from$ with $addr\_breakto$, and utilizes the special hybrid coding method to generate $Amount$ with Δt, and then adds the obfuscated data generated by CTGAN.

Top Address List. In BlockWhisper, the transaction address serves as a label for the covert channel, which ensures that the receiver can quickly identify the covert transaction from the vast number of blockchain transactions, while not raising the awareness of third parties. To achieve this goal, the communicating parties must share *AddressPool* containing tens of thousands of blockchain addresses, which has a negligible impact on the total number of blockchain addresses. It is important to note that the manner in which such an extensive set of addresses can be shared via a secure n off-chain channel is not a

---

Algorithm 1. The special hybrid coding method

Input: binary secret messages $Message$, source address $addr\_from$, key key
Output: transaction amount $Amount$
1: $Message' \leftarrow encrypt(Message, key)$
2: $M \leftarrow (Message'[0:p] \oplus addr\_from[0:p])$
3: **if** $M[0:1] =$ "00" **then**   ▷ Rule 1: Determine if initial two digits of M are "00"
4:     **Return** $Amount \leftarrow Amount_{GAN}$
5: **else**
6:     **Continue**
7: **for** bit in M **do**
8:     $position \leftarrow position + 1$
9:     **if** $bit = 0$ **then**
10:         **if** $count = 0$ **then** $count \leftarrow 1$
11:             **if** $position/digitTag = 1$ **then** ▷ Rule 4: The number of digits exceeds ten or a multiple of ten
12:                 $Amount.append(0)$
13:                 $digitTag \leftarrow digitTag + 10$
14:             $Amount.append(position \bmod 10)$ ▷ Rule 2: Single "0" situation
15:         **else**
16:             $count \leftarrow count + 1$
17:             **if** $count = Amount[-1]$ **then**   ▷ Rule 3: Consecutive "0" situation
18:                 $Amount.append(count)$
19:                 $count \leftarrow 0$
20:     **else**
21:         **if** $count > 1$ **then** $Amount.append(count)$
22:         $count \leftarrow 0$
23: **Return** $Amount$

---

concern, as many address generation tools, such as Vanitygen [3], can generate a significant number of blockchain addresses through the use of a simple index. Consequently, it is sufficient for the communicating parties to share a single such index. The sender constructs the transaction by selecting the source and destination addresses from $AddressPool$, while the receiver only needs to search for transactions where both the source and destination addresses are included in $AddressPool$.

In Sect. 3.2, we analyze the Graph-based Detection methods for transaction addresses and the address distribution in legitimate transactions. In order to ensure the reasonableness of the out-degree, in-degree, and path length of addresses in the transaction graph, the sender must construct a source top address list and a destination top address list based on the probabilistic model of address distribution. The purpose of the top address list is to guarantee that each address in $AddressPool$ has a probability of being selected, as defined by Eq. 6. When constructing transactions, the sender selects $addr\_from$ and $addr\_to$ from the source top address list and the destination top address list. However, it is not permitted for an address to be both the source address and the destination address.

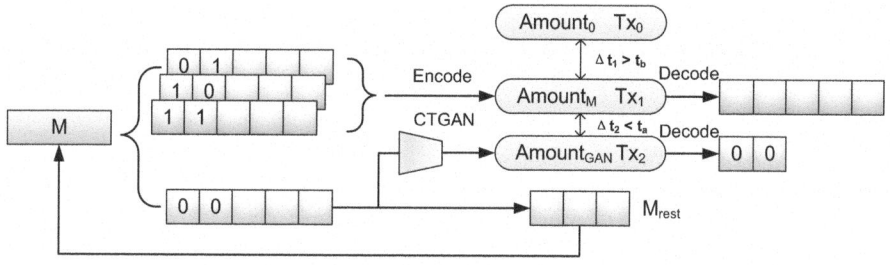

Fig. 7. The special hybrid coding method.

Special Hybrid Coding Method. In this method, the binary secret messages are encoded as the payload transaction amount in decimal and the time interval between transactions, as illustrated in Fig. 7. In the data preprocessing stage, the secret messages are encrypted using a symmetric encryption algorithm to generate the ciphertext. Then, a bitwise exclusive-or (XOR) operation is performed between the first p bits of the ciphertext and the first p bits of the source address of the transaction to obtain the preprocessed secret messages M.

Prior to the encoding process, the transaction amount length l is determined by the distribution model of the transaction amount, as outlined in Sect. 3.4. The specific coding rules are as follows:

1. In the event that the initial two digits of M are "00", the CTGAN-generated $Amount_{GAN}$ is employed to replace $Amount_m$ for obfuscation transaction constructing. $Amount_m$ and $Amount_{GAN}$ will be distinguished by setting disparate transaction time intervals. The $(p-2)$ bits are then relocated back into M for the subsequent encoding.
2. Use a decimal number in $Amount_m$ to indicate the position of the "0" bit that appears alone in M. For example, if the #3 bit in M is "0", it is indicated by the decimal number "3" in $Amount_m$.
3. Use two decimal numbers in $Amount_m$ to indicate the position and number of consecutive occurrences of "0" bits in M. It should be noted that the latter of these two decimal numbers should be less than or equal to the previous number, this is to avoid confusion. For example, if the #2 to #5 digit of M is "0000", it is encoded as "2242".
4. Add the digit "0" to $Amount_m$ to prevent ambiguity during decoding, when the number of digits exceeds ten or a multiple of ten.
5. Finally, if all the l bits of $Amount_m$ have been embedded, and q bits of p bits of M remain to be encoded, the q bits are moved back into the secret messages for the next encoding.

The implementation of the special hybrid coding method is illustrated in Algorithm 1. To facilitate a more comprehensive understanding of the encoding process, a concrete example is presented in Fig. 8. In Ethereum, the lengths of amounts are concentrated at 16–19 digits. It is assumed that the transaction

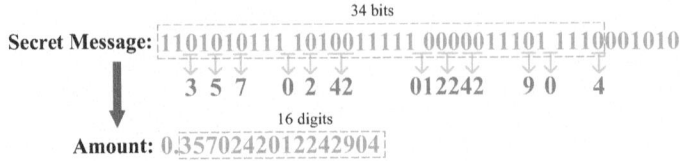

Fig. 8. An example of encoding method.

amount l is 16 digits, and p is 40 bits. The value of M is represented by the binary string "1101010111101001111100000111011110001010", which is encoded as the decimal number "0.3570242012242904". This represents the encoding of 34 out of 40 bits, with the remaining 6 bits being encoded the next time.

Obfuscation Transaction Generation and Interval Selection. As outlined in Sect. 3.3, the distribution of numbers in the transaction amount influences the entropy value; hence, anomalies can be identified. Consequently, it is imperative to generate a series of obfuscation transactions via GAN to simulate the pattern of numerical distribution observed in legitimate transaction amounts. CTGAN is designed to generate synthetic tabular data, which effectively preserves the statistical properties of the original dataset. Each digit of the legitimate transaction amount utilized for training is partitioned into a column, and the missing digits are filled with zeros. The amount data is simulated by the CTGAN generator, and then adjusted by feedback from the discriminator to generate the obfuscation transaction amount after several epochs of training. The special hybrid coding method described in the previous section allows for the insertion of CTGAN-generated obfuscation transaction amounts if the first two digits of the secret messages are '00'. Consequently, the ratio of the number of payload transactions to the number of obfuscation transactions is approximately 3:1.

The establishment of distinct transaction time intervals for payload transactions and obfuscation transactions serves to alert the receiver to the differing types of transactions. According to the transaction time interval distribution model constructed in Sect. 3.3, the sender is required to set the thresholds t_a and t_b to divide the transaction time interval into $(0, t_a)$, (t_a, t_b) and (t_b, t_{max}) in accordance with the prevailing network environment. The interval $(0, t_a)$ encompasses the range of values for the obfuscation transaction time interval. The interval (t_b, t_{max}) is utilized for the payload transaction time interval. The interval (t_a, t_b) serves as an isolation band to mitigate the BER resulting from transmission delay. In order to achieve an equal distribution of both the number and time interval of payload transactions and obfuscation transactions, it is necessary to set the ratio of the probabilities of $(0, t_a)$ to (t_b, t_{max}) in the distribution to 1:3, as illustrated by the following equation:

$$\int_0^{t_a} \Psi(z)dz \bigg/ \int_{t_b}^{t_{max}} \Psi(z)dz = 1/3. \tag{11}$$

Finally, the sender selects the transaction interval within the specified range $(0, t_a)$ and (t_b, t_{max}) for obfuscation transactions and payload transactions, respectively, according to Eq. 9.

4.3   Secret Message Extraction

The extraction process is the inverse of the embedding process. In this process, the receiver searches the transactions received by the node for the covert transactions whose source and destination addresses belong to the *AddressPool*. Then, it differentiates between obfuscation transactions and payload transactions by comparing the time interval between the covert transactions with the shared thresholds t_a and t_b. If the transaction is identified as an obfuscation transaction, the amount is immediately discarded and decoded as binary data "00". Conversely, if the transaction is identified as a payload transaction, the amount is decoded as binary data M. M is then exclusive ORed bit-by-bit with the source address of the transaction, spliced sequentially, and decrypted using key to obtain the original secret messages.

5   Experimental Evaluation

This section presents an experimental verification of the ability of BlockWhisper to evade detection by existing detection schemes. Furthermore, the channel capacity and feasibility of the proposed method are evaluated by comparing it with other sophisticated work.

5.1   Experimental Environment

BlockWhisper is constructed upon the Ethereum blockchain. The experimental machine is equipped with an Intel(R) Core(TM) i9-10900 CPU @ 2.80GHz and 32.0 GB RAM. The dataset utilized in the experiments is **Eth154**, which was introduced in Sect. 3.1.

5.2   Ability to Evade Detection

Detection Against Addresses. In Sect. 3.2, we provide a detailed account of the methodology employed for GD with regard to transaction addresses. This methodology is utilized to ascertain whether the transaction addresses in Block-Whisper align with the distribution pattern observed in legitimate transaction addresses within the blockchain. Following the methodology proposed by GUO et al. [12], all transactions to be identified are initially constructed into a transaction graph and subsequently divided into subgraphs. Thereafter, the variance of the in-degree and out-degree, as well as the mean of the longest path length, are calculated for each subgraph. For each statistical eigenvalue, a comparison threshold Th and an offset ϵ are selected. If the difference between the eigenvalue of the subgraph and Th is less than ϵ, it is determined that the subgraph represents a covert transaction.

We merged the transactions generated by BlockWhisper with those in Eth154 into a test set, with a 50/50 split between the two. We selected the values provided in the original paper for the variance of the in-degree, the variance of

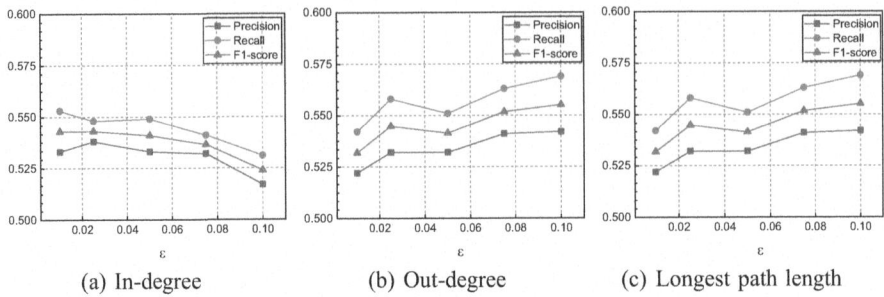

(a) In-degree   (b) Out-degree   (c) Longest path length

Fig. 9. Results of GD against addresses in different metrics.

the out-degree, and the mean of the longest path length, which were 0.01, 1, and 0.5, respectively. We selected the values $(0.01, 0.025, 0.05, 0.075, 0.1)$ as ϵ for the tests. Figure 9 presents the test results. In all experiments, the maximum values of precision and recall for the detection method are 0.577 and 0.597, respectively, which are very close to random guesses. This indicates that BlockWhisper is adequately resilient to the GD against address.

Detection Against Time Intervals. The $\epsilon - Similarity$ method is a widely utilized approach for the detection of inter-packet intervals in network covert channels [13]. Furthermore, the method can be employed to identify anomalies in the time intervals between transactions within a blockchain. The fundamental principle of the method is to arrange the time intervals at which all transactions occur at a specific range of addresses in sorted order and then calculate the relative difference between two consecutive time intervals, as illustrated by the following equation:

$$\lambda_i = \frac{\Delta t_{i+1} - \Delta t_i}{\Delta t_i}. \tag{12}$$

The proportion of λ values below the threshold value of ϵ is calculated, which is referred to as the similarity score. The objective of this metric is to distinguish between legitimate and covert transactions.

The test set of the experiment consisted of 50% BlockWhisper-generated transaction time intervals and 50% legitimate transaction time intervals in Eth154. The selection of different ϵ values $(0.005, 0.008, 0.01, 0.02, 0.03, 0.1)$ was based on the original text for the purpose of conducting group experiments. Table 1 presents the results of the experiment, which indicates that the detection is most effective when the value of ϵ is 0.03. However, the precision and recall values of 0.606 and 0.681 respectively suggest that the method is not sufficiently robust for the detection of BlockWhisper transactions.

Detection Against Amounts. Currently, two main methods are employed to detect the legitimacy of transaction amounts: SPD and ETD. SPD entails examining the distribution of transaction amounts to ascertain whether they fall within a reasonable range. This is achieved through the use of tools such as the cumulative distribution function (CDF) and the Kolmogorov-Smirnov (K-S)

Table 1. Results of $\epsilon - Similarity$ detection for time intervals in different ϵ.

	$\epsilon = 0.005$	$\epsilon = 0.008$	$\epsilon = 0.01$	$\epsilon = 0.02$	$\epsilon = 0.03$	$\epsilon = 0.1$
Precision	0.583	0.531	0.572	0.597	**0.606**	0.529
Recall	0.601	0.613	0.611	0.639	**0.681**	0.579
F1-score	0.592	0.569	0.591	0.617	**0.662**	0.553

test. The CDF provides a visual representation of the probability distribution of a variable, which facilitates the identification of differences in the distribution. The K-S test, on the other hand, quantifies the discrepancy between two distributions. A p-value is calculated for each distribution, which serves as a level of confidence that the two distributions originate from the same distribution. When the p-value is greater than 0.05, the two distributions can be considered to originate from the same distribution. ETD is used to differentiate between legitimate and covert transactions by comparing their entropy. The amounts of transactions generated consecutively in the same time period can be considered as a time series. Consequently, we utilize the sample entropy (SampEn) for evaluation, which describes the degree of disorganization of the data on the time series. To illustrate our superiority, we compare with two baseline schemes, BSCC [9] and GCCDC [14], which also employ Ethereum transaction amounts to embed secret messages.

Figure 10(a) depicts the cumulative distribution function of the transaction amounts generated by BlockWhisper, GCCDC, and BSCC in comparison to the transaction amounts of the Eth154 dataset. It is clearly observed that the distribution of transaction amounts for BlockWhisper and Eth154 are highly similar. In contrast, the distribution of transaction amounts for GCCDC and BSCC differs significantly from Eth154 around the value of 0.035.

In order to quantitatively compare the differences in the distribution of transaction amounts among the schemes, a K-S test experiment was conducted. A set of transaction amounts was selected from BlockWhisper, GCCDC, and BSCC generated data, respectively, and subjected to a K-S test with the transaction amounts in Eth154. The resulting confidence p-value was then calculated. The experiment was repeated 500 times, and the results were plotted as a scatter plot in Fig. 10(b). The figure shows that the p-value for almost all sets of experiments against BlockWhisper is greater than 0.05, indicating that the experimental results are superior to those of the other two schemes.

Finally, the SampEn of the three schemes with the control group was calculated. To eliminate the effect of sequence length on entropy, 10, 20, 50, 100, and 200 transaction data were selected for comparison. Figure 10(c) presents the experimental results, which demonstrate that the SampEn of the transaction amount generated by BlockWhisper is the closest to the SampEn of the transaction amount in Eth154, in comparison to the other two schemes.

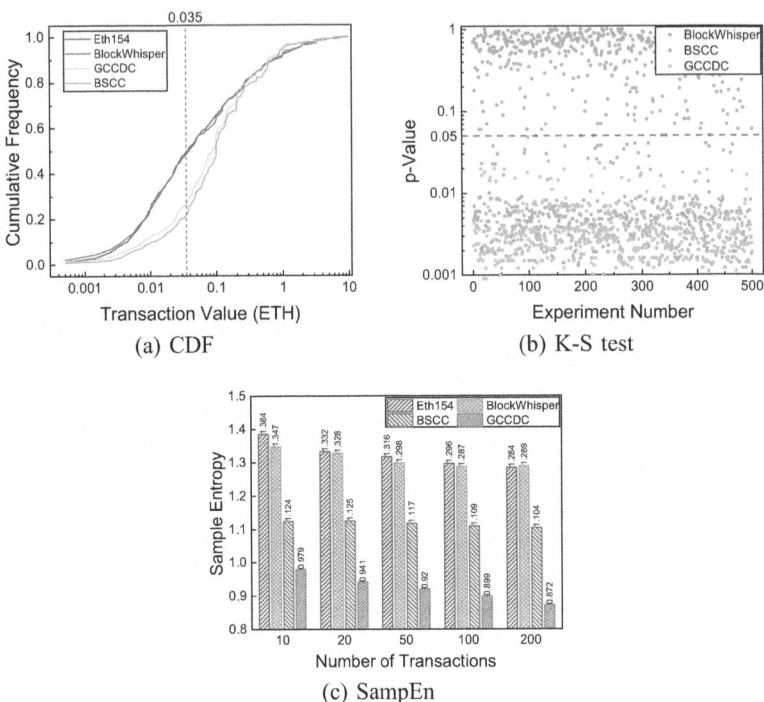

Fig. 10. Results of detection against amounts in different methods.

Machine Learning Detection. To further evaluate the overall concealment of the BlockWhisper scheme, we employ a machine learning method proposed for detecting blockchain-based covert channels, namely the CTR model [18]. This model comprises two components: TextCNN and BPNN. The Eth154 dataset was divided into two subsets, EthA and EthB. The CTGAN in BlockWhisper was trained using EthA. Two classification experiments were performed using each of these two subsets in conjunction with BlockWhisper-generated transaction data. Another experiment was performed on EthA and EthB, serving as the control group. The training and test sets were constructed in a proportion of 8:2. The experimental results are presented in Table 2. The results demonstrate that the CTR model detects BlockWhisper-generated transactions with only slight improvement over the control group, with a precision of 0.565 and a recall of 0.561. This is a satisfactory level of performance for covert channels.

The experiments collectively illustrate that BlockWhisper has the ability to be highly covert and to evade detection by state-of-the-art methods from multiple angles.

5.3   Channel Capacity

For a covert channel, channel capacity is a critical metric. Nevertheless, the impact of high capacity on channel concealment, feasibility, and other factors also

Table 2. Results of the CTR model detection.

Positive	Negative	Metrics			
		Accuracy	Precision	Recall	F1-score
BlockWhisper	EthA	0.543	0.548	0.543	0.530
BlockWhisper	EthB	**0.561**	**0.565**	**0.561**	**0.552**
EthA	EthB	0.503	0.503	0.502	0.486

Table 3. Comparison between BlockWhisper and other schemes.

	Channel capacity (bits/Tx)	Dynamic label	Applicable blockchain	Embedding field	Detection methods testing
BLOCCE [3]	1	×	BTC	Address	–
DLchain [6]	256	✓	BTC	Remark	SPD
NCCBTT [7]	13	×	BTC	Amount	SPD
ACCB [8]	8	✓	BTC	Amount	ETD
BSCC [9]	24	✓	ANY	Amount	SPD
DDSAC [10]	128	✓	BTC	Signature	SPD
GCCDC [14]	24	✓	BTC	Amount	–
WhisEthe [15]	28	×	ETH	Amount	SPD
IMMM [11]	0.0776 bits/s	×	BTC	Interval	SPD, MLD
EBDL [16]	512	✓	ANY	Remark	SPD
BlockWhisper	29	✓	ANY	Amount, Interval	SPD, ETD, GD, SMD, MLD

requires careful consideration. The embedding scheme proposed in this paper is a variable-length encoding method. To determine the average length of the secret messages that can be embedded in each transaction, we employed statistical methods, resulting in an estimated value of approximately 38 bits. Even with the addition of the obfuscated transactions generated by CTGAN, the total average channel capacity can reach 29 bits per transaction.

Table 3 presents a comparative analysis of BlockWhisper with respect to several key parameters, including embedding rate, dynamic labeling, application blockchain, and embedding area. It can be observed that BlockWhisper exhibits a higher embedding rate than the majority of existing schemes, with the exception of DLchain [6], DDSAC [10] and EBDL [16]. However, these schemes embed secret information into the remark or signature field, which significantly reduces the concealment and feasibility of the channel. This is due to the fact that the remark field can easily be detected as abnormal by a third party. In other words, these schemes have low concealment and low feasibility. Therefore, it can be argued that BlockWhisper is a blockchain-based covert communication scheme that balances feasibility, channel capacity, and concealment.

6 Conclusion

In this paper, we proposes BlockWhisper, a blockchain-based covert channel with strong anti-detection capability. By statistically analyzing the sample distribution features of transaction addresses, time intervals, and amounts of legitimate data, we construct a probability distribution model of the components of the blockchain-based covert channel. The model serves as the foundation for the design of a hybrid blockchain-based covert communication scheme. The proposed scheme fuses the covert store channel and the covert time channel, and embeds the secret messages in the transaction amount and the transaction time interval through the special hybrid coding method. Some of these transaction amounts are generated by CTGAN, which simulates legitimate transactions for data obfuscation. Furthermore, the top address list method is employed to select dynamic transaction addresses as identification labels for the covert channel. Experimental results demonstrate that BlockWhisper is capable of resisting the various of sophisticated detection methods and exhibits a balanced performance in terms of channel concealment, channel capacity, and feasibility when compared to other schemes.

Acknowledgements. This work was supported in part by the National Natural Science Foundation of China under Grant Nos. 62172093 and U22B2025, in part by the National Key R&D Program of China under Grant No. 2020YFB1005500, and in part by the Fundamental Research Funds for the Central Universities, Southeast university.

References

1. Zhang, T., Li, B., Zhu, Y., Han, T., Wu, Q.: Covert channels in blockchain and blockchain based covert communication: overview, state-of-the-art, and future directions. Comput. Commun. **205**, 136–146 (2023)
2. Chen, Z., et al.: Blockchain meets covert communication: a survey. IEEE Commun. Surv. Tutor. **24**(4), 2163–2192 (2022)
3. Partala, J.: Provably secure covert communication on blockchain. Cryptography **2**(3), 1–18 (2018)
4. Zhang, L., et al.: A covert communication method using special bitcoin addresses generated by vanitygen. Comput. Mater. Continua **65**(1), 597–616 (2020)
5. Gao, F., Zhu, L., Gai, K., Zhang, C., Liu, S.: Achieving a covert channel over an open blockchain network. IEEE Network **34**(2), 6–13 (2020)
6. Tian, J., Gou, G., Liu, C., Chen, Y., Xiong, G., Li, Z.: Dlchain: a covert channel over blockchain based on dynamic labels. In: Proceedings of the 21th International Conference on Information and Communications Security, pp. 814–830 (2020)
7. Luo, X., Zhang, P., Zhang, M., Li, H., Cheng, Q.: A novel covert communication method based on bitcoin transaction. IEEE Trans. Industr. Inf. **18**(4), 2830–2839 (2021)
8. Tian, Y., Liao, X., Dong, L., Xu, Y., Jiang, H.: Amount-based covert communication over blockchain. IEEE Trans. Netw. Serv. Manage. **21**(3), 3095–3111 (2024)
9. Zhang, P., Cheng, Q., Zhang, M., Luo, X.: A blockchain-based secure covert communication method via shamir threshold and STC mapping. IEEE Trans. Dependable Secure Comput. 1–12 (2024)
10. Chen, Z., Zhu, L., Jiang, P., Zhang, C., Gao, F., Guo, F.: Exploring unobservable blockchain-based covert channel for censorship-resistant systems. IEEE Trans. Inf. Forensics Secur. **19**, 3380–3394 (2024)

11. Zhu, L., Liu, Q., Chen, Z., Zhang, C., Gao, F., Yang, Z.: A novel covert timing channel based on bitcoin messages. IEEE Trans. Comput. **72**(10), 2913–2924 (2023)
12. Guo, Z., Li, X., Liu, J., Zhang, Z., Li, M., Hu, J., Zhu, L.: Graph-based covert transaction detection and protection in blockchain. IEEE Trans. Inf. Forensics Secur. **19**, 2244–2257 (2023)
13. Zillien, S., Wendzel, S.: Weaknesses of popular and recent covert channel detection methods and a remedy. IEEE Trans. Dependable Secure Comput. **20**(6), 5156–5167 (2023)
14. Zhang, P., Cheng, Q., Zhang, M., Luo, X.: A group covert communication method of digital currency based on blockchain technology. IEEE Trans. Network Sci. Eng. **9**(6), 4266–4276 (2022)
15. Liu, S., et al.: Whispers on ethereum: blockchain-based covert data embedding schemes. In: Proceedings of the 2nd ACM International Symposium on Blockchain and Secure Critical Infrastructure, pp. 171–179 (2020)
16. Zhang, C., Zhu, L., Xu, C., Zhang, Z., Lu, R.: EBDL: effective blockchain-based covert storage channel with dynamic labels. J. Netw. Comput. Appl. **210**, 103541 (2023)
17. Xu, Y., Wu, Z., Cao, J., Xu, J., Cheng, G.: Darktrans: a blockchain-based covert communication scheme with high channel capacity and strong concealment. In: Proceedings of the IEEE 29th International Conference on Parallel and Distributed Systems, pp. 675–682 (2023)
18. Wang, M., et al.: Practical blockchain-based steganographic communication via adversarial AI: a case study in bitcoin. Comput. J. **65**(11), 2926–2938 (2022)
19. Blockchair. https://blockchair.com/dumps. Accessed 20 May 2024
20. Xu, L., Skoularidou, M., Cuesta-Infante, A., Veeramachaneni, K.: Modeling tabular data using conditional GAN. In: Proceedings of the annual Conference on Neural Information Processing Systems 2019, pp. 7333–7343 (2019)

A Verifiable Decentralized Data Modification Mechanism Supporting Accountability for Securing Industrial IoT

Changsong Yang[1,2], Junfu Wu[1,2], Hai Liang[1,2], Yong Ding[1,2,3(✉)], and Yujue Wang[4]

[1] Guangxi Key Laboratory of Cryptography and Information Security, School of Computer Science and Information Security, Guilin University of Electronic Technology, Guilin, China
stone_dingy@126.com
[2] Guangxi Key Laboratory of Digital Infrastructure, Guangxi Zhuang Autonomous Region Information Center, Nanning, China
[3] Institute of Cyberspace Technology, HKCT Institute for Higher Education, Hongkong, China
[4] Hangzhou Innovation Institute of Beihang University, Hangzhou, China

Abstract. The widespread application of Industrial Internet of Things (IIoT) has profoundly transformed modern industrial production methods. However, the large scale of industrial data, the diversity of data sources, and the heterogeneity of data structures make data maintenance exceptionally challenging. The editability techniques have been introduced into blockchain to leverage its distributed and editable features for data maintenance and management. However, existing solutions mainly focus on achieving editability while overlooking data management personnel's privacy protection and lacking effective modification verification mechanisms to ensure data reliability. To address these issues, this paper proposes a verifiable decentralized data modification mechanism supporting accountability (DDMA) for securing IIoT. Our DDMA mechanism not only achieves the redactability of blockchain but also protects the privacy of data management personnel and provides reliable clues for subsequent accountability. Additionally, the DDMA mechanism offers an efficient method for modification verification, enabling rapid validation of whether on-chain data has been modified. Theoretical analysis demonstrates that DDMA meets the requirements of modification authorization, behavior traceability, and efficient modification verification. Experimental results indicate that DDMA can reduce modification and verification overheads.

Keywords: Industrial Internet of Things · Industrial Data Modification · Redactable Blockchain · RSA Accumulator · Accountability

© The Author(s), under exclusive license to Springer Nature Singapore Pte Ltd. 2025
T. Zhu et al. (Eds.): ICA3PP 2024, LNCS 15251, pp. 182–200, 2025.
https://doi.org/10.1007/978-981-96-1525-4_10

1   Introduction

Industrial data, an essential component of the Industrial Internet of Things (IIoT), whose efficient, secure, and autonomous management has become a prerequisite for the widespread implementation of the IIoT [5]. In the event that recorded industrial data is incorrect or maliciously tampered with, the production process can be forced to stop, resulting in significant losses [28]. Hence, it is crucial to guarantee the security and integrity of industrial data. However, industrial data management suffers from severe security challenges. Industrial data is difficult to collect, large in scale, and inconsistent in data dimensions, making it difficult to manage industrial data with uniform standards. In addition, industrial data are frequently attacked by hackers or malicious users, and they aim to tamper or delete the part, or even the whole industrial data, thus jeopardizing the operation of the industrial systems [17,26]. Therefore, how to securely and integrally maintain industrial data has become a challenging problem in complex industrial environments.

The emergence of blockchain technology provides for the maintenance and management of industrial data management. It originated from the cryptocurrency Bitcoin [22]. In recent years, numerous studies have explored the integration of blockchain with the IIoT [29], aiming to establish interconnectivity among equipment, plants, and data across regions [6,11]. However, due to the limitations of the edge devices, it is easy to make the data abnormal, which will leave anomalous data on the blockchain and affect the regular operation of the system [7,15]. Therefore, it is necessary to consider the editable nature of the data in the chain to ensure that there is room for modification when it is abnormal [1]. Additionally, there is a need to introduce certain restrictions on modification permissions to mitigate the risk of malicious tampering with data on the blockchain. Leveraging redactable blockchain technology within industrial systems can substantially enhance the integrity and confidentiality of industrial data, ensuring its accuracy and security. The application of redactable blockchain technology in the IIoT has expanded the scope of blockchain applications, introducing new possibilities for industrial data management.

Three critical challenges exist in redactable blockchain application scenarios in the current industrial Internet environment. The first is the privacy protection issue for data managers' identity information. Data managers may not wish to disclose personal identity information while maintaining data on the chain, thus requiring the protection of managers' privacy information. Furthermore, to avoid rendering malicious actions untraceable due to excessive protection, it is essential to establish a trustworthy accountability mechanism for disclosing node information of such actions subsequently. Additionally, there is a requirement for effective validation of the modified on-chain data. Once on-chain data is modified, it becomes difficult to query the modification status intuitively, thus requiring appropriate verification methods to verify whether the data has been modified quickly. Therefore, it is necessary to explore a solution that supports privacy protection, traceability, and verification of data modifications.

1.1 Our Contributions

To solve the problem of industrial blockchain data anomalies and the difficulty of recovering responsibility after the fact in the IIoT environment, we propose a verifiable decentralized data modification mechanism supporting accountability (DDMA) for securing IIoT. The main contributions of this paper are as follows.

- In response to the challenges of data heterogeneity and tracking of illicit users in industrial blockchain, this paper proposes a decentralized data modification mechanism that supports accountability. This mechanism is based on group signature technology, allowing modifications to on-chain data without disclosing the personal information of the modifier and providing reliable traceability for subsequent accountability.
- A modification verification protocol is devised to enable any user to efficiently and effectively verify the modification status of on-chain data at any time, ensuring the integrity and transparency of data in industrial blockchain systems.
- We instantiated the proposed DDMA mechanism and evaluated its performance, comparing it with relevant works. Theoretical analysis indicates that DDMA offers more comprehensive functionality, including modifier identity anonymity and modification result integrity verification. Experimental results further demonstrate that the overhead of DDMA is within acceptable bounds.

1.2 Related Works

Blockchain-based industrial control data management technology has received extensive attention both from industry and academia. Blockchain technology can effectively solve the problems of data storage, transmission, and security in the industrial control field [36]. Mazzei et al. [20] designed an industrial blockchain tokenizer, which collects data and processes them to be sent to various blockchain platforms to store them and build a data transmission for the industrial system and blockchain platform. Liang et al. [18] proposed a secure blockchain-based IIoT data transmission technology that achieves secure storage and reliable data transmission. Bahia and Madisetti [2] constructed a decentralized IIoT platform, allowing parties to interact without a trusted third party. Lu and Xu [19] introduced a supply chain system with traceability, applying blockchain to a complex production process to realize data traceability at each chain stage. Zeng et al. [32] proposed a blockchain-assisted cross-domain data sharing (BCDS) scheme to enhance security and privacy in the IIoT. However, the appealing study operates on the premise that the data are correct and does not consider the impact of the data error on the system after uploading the chain to the permanent storage.

Since the birth of blockchain, numerous researchers have been studying and exploring its editable nature. Redactable blockchains enable users to modify and delete inappropriate chain data, expanding blockchains' application scenarios [34]. In 2016, Ateniese et al. [1] proposed the first scheme for redactable blockchain. This scheme utilizes chameleon hash functions to achieve blockchain editability, enabling users to modify block data without altering the original hash values. Du et al. [10] constructed a searchable encryption scheme based on a redactable blockchain, which allows data owners to perform modification operations on on-chain data and enhances the flexibility of on-chain data availability. Yi et al. [30] constructed a digital rights management system to ensure that plagiarized or sensitive content can be removed promptly when found on the on-chain. Guo et al. [13] proposed a redactable blockchain for managing Electronic Health Records (EHR), utilizing revocable IPFS for secure off-chain storage and enabling controlled data modifications.

The emergence of redactable blockchain has expanded the flexibility and practicality of blockchain technology, thus leading to its widespread application in the IIoT domain. Yu et al. [31] proposed a blockchain-based Internet of Things (IoT) system, integrating the chameleon hash algorithm into the blockchain to achieve fine-grained access control. Zhang et al. [33] devised a trustworthy industrial data management scheme based on redactable blockchain, employing a dual blockchain architecture for data management. They implemented permission control using verifiable secret-sharing techniques, thereby enhancing system security. Mishra, Ramesh and Mohammad [21] integrated redactable blockchain with IoT-supported cloud environments, enabling secure fine-grained data aggregation within IoT environments. Shao et al. [25] proposed an auditable redactable blockchain scheme to facilitate secure data sharing among IoT devices, allowing user devices to rewrite blockchain transactions under strict auditing. Zhao et al. [35] introduced "Tiger Tally", a secure redactable blockchain solution for IoT data management, addressing the conflict between immutability and data redaction by utilizing a Targeted Policy-Based Chameleon hash to prevent redact privilege abuse.

1.3 Organization

The remainder of this paper is organized as follows. Section 2 describes some preliminaries. Section 3 introduces system model and design objectives. A DDMA construction is proposed in Sect. 4. Section 5 and Sect. 6 provide scheme analysis and performance evaluation of the DDMA scheme, respectively Sect. 7 concludes the paper.

2 Preliminaries

2.1 Blockchain

Blockchain technology initially emerged as the foundational technology for Bitcoin [23]. It is a decentralized digital ledger technology known for its decentralized nature, tamper-resistant properties, anonymity, and auditability. These distinctive features have contributed to the widespread adoption of blockchain across diverse sectors, including finance [27], healthcare [4], government [9], and education [12,14].

The structure of blockchain is shown in Fig. 1. Each block points to the previous block through the hash value and constitutes a chain of ledgers in a constantly stacked way. Any change of information on the block will lead to a change of the block hash value, which will lead to the subsequent blocks being unable to be linked to the block. Therefore, tampering with the data recorded on the block is extremely difficult. Each node in the blockchain network is backed up with a complete copy of the ledger, and even if part of the data is destroyed and tampered with, the system can achieve consistency of the data through consensus algorithms and validation mechanisms, which ensures the safe and regular operation of the system.

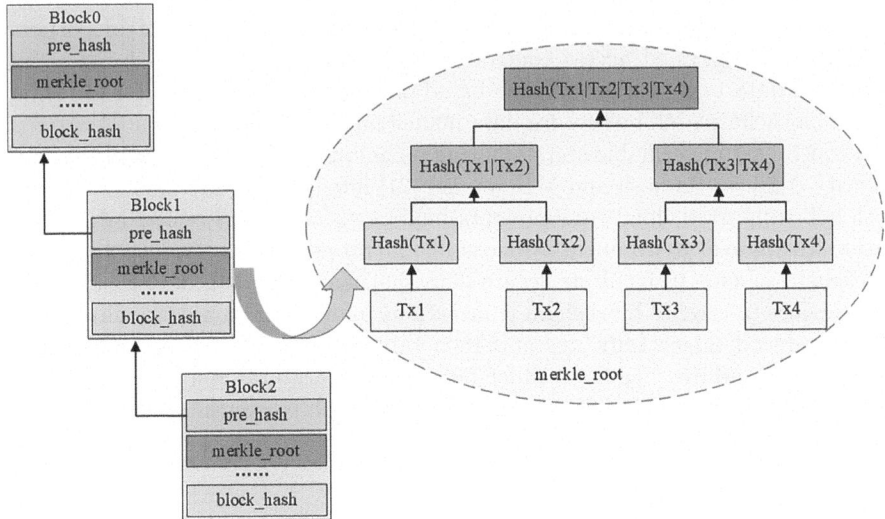

Fig. 1. Blockchain structure.

2.2   Chameleon Hash

The Chameleon hash function [16] is a unique type of hash function that incorporates a trapdoor. In computing the hash of an initial message, Chameleon hash functions facilitate subsequent alterations or deletions by leveraging a preserved trapdoor, thereby enabling the generation of a collision. Consequently, even after modifying the original message, the hash value remains unchanged, ensuring the reliability of data verification. A Chameleon hash function consists of the following efficient algorithms.

- $(p, q, g) \leftarrow Init(1^\lambda)$: The initialization algorithm takes as input a security parameter λ, and outputs public parameters (p, q, g).
- $(pk, sk) \leftarrow KGen(p, q, g)$: The key generation algorithm takes as input the public parameters (p, q, g), and outputs public key and trapdoor key (pk, sk).

- $h \leftarrow Hash(pk, m, r)$: The Chameleon $Hash$ algorithm takes as input a message m, a public key pk, a random number r, and outputs a hash value h.
- $(h, r') \leftarrow Forge(m', (h, m, r), sk)$: The $Forge$ algorithm takes as input the new message m', the original message (h, m, r) and the trapdoor secret key sk. It computes and outputs a new random number r' that maintains an unchanged hash collision.
- $\{0, 1\} \leftarrow Check(pk, (h, r, r'))$: The $Check$ algorithm takes the public key pk and hash h, if (h, r') is valid collision, then outputs 1, otherwise outputs 0.

2.3 Verifiable Secret Sharing

The verifiable secret sharing [24] scheme comprises two main phases, the secret share allocation phase and the secret reorganization phase. In the allocation phase, the following two algorithms are involved:

- $(p, q, g, n, t) \leftarrow Init(1^\lambda)$: The initialization algorithm takes as input a security parameter λ and randomly generates two large prime numbers p and q, which satisfy $q \mid (p - 1)$. Let g be the q-order element, n be the number of secret holders and t be the threshold value.
- $(x_i, f(x_i)) \leftarrow SecretGen(n, t)$: The secret key generation algorithm takes as input n and t, randomly generates a $t - 1$ order polynomial

$$f(x) = a_0 + a_1 x + \cdots + a_{t-1} x^{t-1} \bmod p$$

where the secret $s = f(0)$. For each secret holder, choose a distributed secret key x_i, where $i \in [1, n]$, and compute $f(x_i)$. The secret fragment $(x_i, f(x_i))$ is owned by the holder.

In the secret reconstruction phase, any subset of t out of the total n secret holders can recover the original secret.

- $s \leftarrow Restore(x_i, f(x_i))$: When t secret fragments are collected, the secret s can be reconstructed as follows using the Lagrange interpolation theorem.

$$s = f(0) = \sum_{j=1}^{t} f(x_j) \prod_{l=1, l \neq j}^{t} \frac{-x_l}{x_j - x_l} \bmod p$$

2.4 RSA Accumulator

The RSA accumulator [3] is a data structure similar to a Merkel tree that can be used to verify that an element is in a set without having to disclose the entire contents of the set. It works as follows:

- $(N, g, A_r) \leftarrow AccGen(1^\lambda)$: The algorithm takes as input a safety parameter λ. It then randomly generates two large prime numbers p and q, along with a generating element g. The modulus N is computed as $N = pq$. The initial value of the accumulator denoted as $A_r = g^1 \bmod N$.

- $A_r' \leftarrow AccAdd(A_r, S, x)$: This add-element algorithm takes as input the value of the accumulator A_r, the set of accumulators S, and the element to be added x ($x \notin S$), and outputs the value of the updated accumulator A_r', where $A_r' = A_r^x$.
- $\pi \leftarrow AccCertificate(A_r, S, x)$: This algorithm is used to generate membership proof for elements within a set. The algorithm takes as input the current accumulator value A_r and the number x ($x \in S$) for which the proof needs to be generated. It outputs the membership proof π, where $\pi = A_r^{\frac{1}{x}} \mod N$.
- $\{0, 1\} \leftarrow AccCheck(A_r, \pi, x)$: This verification algorithm takes as input the value of the accumulator A_r, the proof of membership π, and element x to be verified. It verifies whether the element x is in the set represented by the accumulator. If $\pi^x = A_r$ output the verification result 1, otherwise output 0.

3 Overview of DDMA

3.1 System Model

A DDMA system comprises six entities: System Administrator (SA), Secret Key Generation Center (KGC), Group Manager (GM), Committee Member (CM), Industrial Equipment (IE) and User. The architecture of DDMA is shown in Fig. 2.

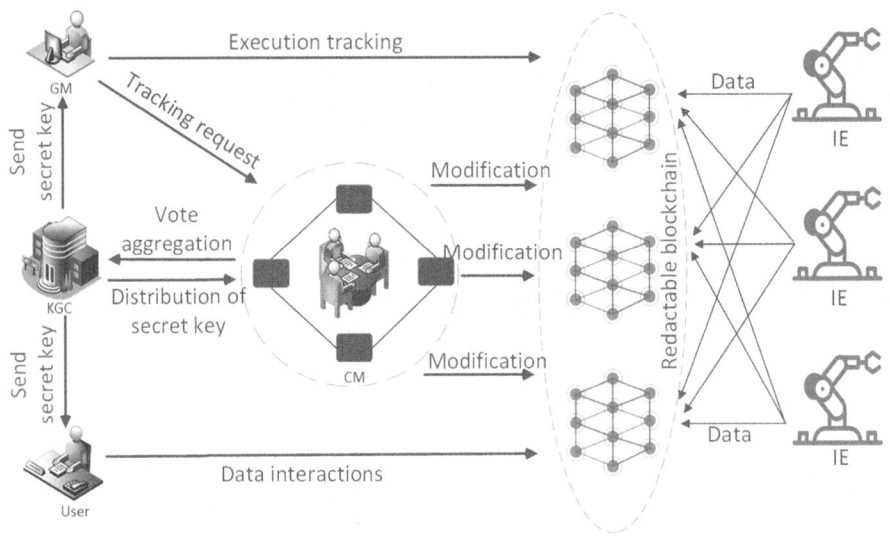

Fig. 2. System model of DDMA.

- SA: SA is responsible for deploying the industrial blockchain and configuring the system's initial parameters. Once the system is initialized, SA goes offline and no longer participates in subsequent operations.

- KGC: KGC serves as a trusted third-party entity responsible for generating trapdoors for the system and distributing trapdoor fragments to modifiers. Additionally, KGC undertakes the task of key generation for group managers and users.
- GM: GM is a significant entity within the system that has tracking privileges. In the event of malicious modifier nodes on the chain, GM is responsible for disclosing relevant information to maintain system security and reliability.
- CM: CM holds trapdoor fragments and entities with permission to modify block data. Their responsibilities include voting decisions on tracking and modification requests, data modification on the chain, and block reconstruction.
- IE: IE serves as the primary data source for the industrial blockchain. It collects industrial data generated during production processes and uploads data to the industrial blockchain network.
- User: User form the foundational members of the system, serving as the primary source of system data and initiators of data modification requests.

In the DDMA system, IE is responsible for uploading collected industrial data to the industrial blockchain network while users interact with the industrial blockchain for data exchange per their requirements. CM is required to complete registration with the GM before performing supervision. CM must obtain authorization through committee voting before modifying on-chain data. Additionally, CM records the modified information in an accumulator for expedited verification of subsequent modifications. In the event of malicious modifications, GM submits a tracking request to the committee and, upon authorization, discloses the information of illicit users on the blockchain.

3.2 Modification and Accountability

The modification and accountability process of the DDMA scheme is shown in Fig. 3. A user requests the committee to modify the data, which is then voted on by all CMs together. Authorization is granted if more than the threshold t of the members agree. One CM is randomly selected to operate on the block, the CM revises the data and updates the RSA accumulator. The voting process for GM disclosure is similar to the modification process, with the difference being that it is the GM who gets the authorization to disclose the information, not an arbitrary CM.

3.3 Design Goals

A DDMA system should meet the following three goals.

1. Modification Restriction. The blockchain's data modification permission can be exploited by malicious nodes, disrupting system operations. To guarantee data safety and control, obtaining sufficient votes and authorization prior to conducting any modification operations is essential.

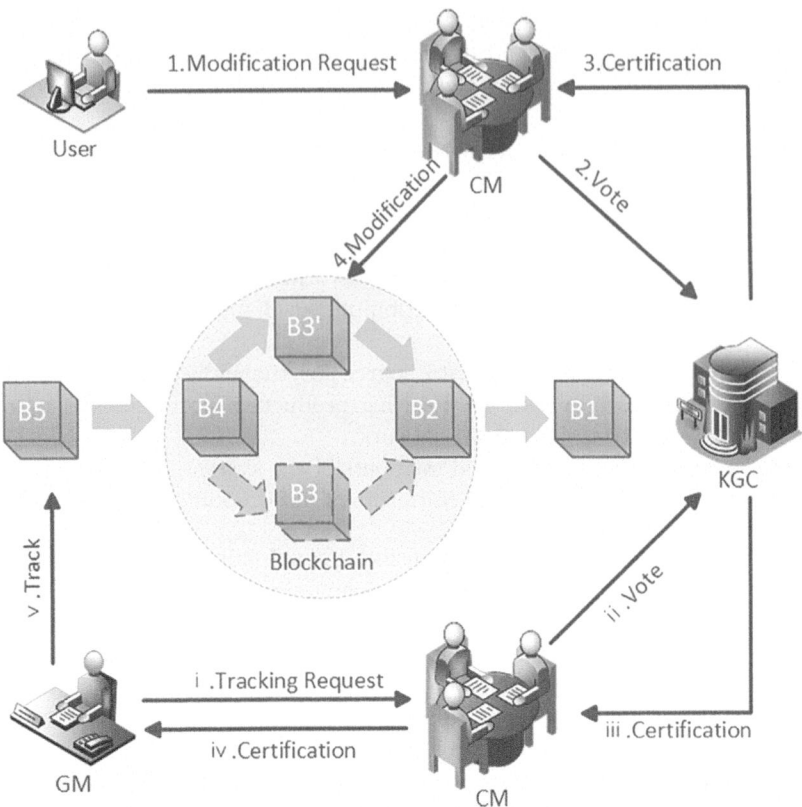

Fig. 3. The process of modification and accountability.

2. Behavior Traceability. The operations performed on the blockchain may exhibit malicious behaviors, and it is essential to trace out the executors of the malicious behaviors. Therefore, when necessary, the administrator can disclose information about the executor based on the signature information.
3. Efficient Validation. Due to limited resources, lightweight node users are unable to store all on-chain data and cannot verify data quickly. Additionally, data managers and public validators also have a requirement to verify whether on-chain data has been altered.

4   Our Proposed Scheme

This section presents a DDMA construction, which contains the $Init$, $UkGen$, $GroupManagerGen$, $GroupMemberEn$, $TrapGen$, $Hash$, $Adapt$, $Check$, $Rsa\text{-}Check$, $Trace$ algorithms.

- $(p, q, g_1, g_2, n, t, acc) \leftarrow Init(1^\lambda)$: Given the security parameter λ, KGC generates two large prime numbers p and q, which satisfy $q \mid (p-1)$. SA selects two multiplicative groups G_1 and G_2, both of order p, with generating elements $g_1 \in G_1$ and $g_2 \in G_2$. SA initializes the number of modifiers n, the threshold value t, and sets the value of the RSA accumulator as $acc = g_1$. Subsequently the system outputs the public parameters $(p, q, g_1, g_2, n, t, acc)$.

- $(x_i, y_i) \leftarrow UkGen(p, q, g_1)$: KGC selects the private key $x_i \in Z_q^*$, computes the $user_i$'s public key as

$$y_i = g_1^{x_i} \tag{1}$$

Then sends the key pair (x_i, y_i) to the $user_i$ through a secure channel.

- $(gpk, gmsk) \leftarrow GroupManagerGen(q, g_1, g_2)$: KGC randomly selects two elements $\varepsilon_1, \varepsilon_2 \in Z_p^*$ to constitute the GM's private key $gmsk = (\varepsilon_1, \varepsilon_2)$. Then, it randomly selects $u \in G_1$, $v \in G_1$, and $h \leftarrow G_1 \backslash \{1_{G_1}\}$, where

$$u^{\varepsilon_1} = v^{\varepsilon_2} = h \tag{2}$$

Subsequently, $\gamma \in Z_p^*$ is randomly selected, and $\omega = g_2^{\gamma}$ is set, constituting the GM's public key $gpk = (h, u, v, \omega)$. Only the KGC knows the value of γ. The key pair $(gpk, gmsk)$ is then sent to the GM over a secure channel.

- $(A_i, x_i) \leftarrow GroupMemberEn(x_i, g_1, \gamma)$: The User applies to join the commission by using their private key x_i, and the KGC calculates the membership proof as follows:

$$A_i = g_1^{\frac{1}{\gamma + x_i}} \tag{3}$$

Once the registration is completed, the KGC sends the membership proof (A_i, x_i) to the $user_i$.

- $(x_i, f(x_i), y) \leftarrow TrapGen(n, t, x_i)$: KGC generates a random polynomial

$$f(x) = a_0 + a_1 x + \cdots + a_{t-1} x^{t-1} \tag{4}$$

of order $t - 1$ based on the number of trapdoor holders n and the threshold value t. a_0 represents the complete secret. Using this secret a_0, the system master public key is computed as

$$y = g_1^{a_0} \tag{5}$$

Subsequently, KGC computes $f(x_i)$ based on the private key x_i sent by the $user_i$, where $i \in [0, t-1]$. The trapdoor fragment held by the modifier is denoted as $(x_i, f(x_i))$.

- $h \leftarrow Hash(m, r, p, g_1, y)$: The IE prepares the data m to be uploaded and subsequently chooses an arbitrary random number $r \in Z_p^*$, which is used by the chameleon hash algorithm to compute the data digest as

$$h = g_1^{m + a_0 r} \tag{6}$$

- $(h, acc', \pi, T_1, T_2, T_3) \leftarrow Adapt(m, r, m', g_1, a_0, A_i, acc)$: Before performing the modification operation, the CM needs authorization from the committee to obtain the complete secret a_0. The relevant steps are as follows:

$$f(x) = \sum_{j=1}^{t} f(x_j) \prod_{l=1, l \neq j}^{t} \frac{x - x_l}{x_j - x_l} \bmod p \tag{7}$$

$$a_0 = f(0) \tag{8}$$

Subsequently, the new collision is computed as:

$$r' = \frac{(m + a_0 r - m')}{a_0} (\bmod \varphi(p)) \tag{9}$$

This computation preserves chain connectivity as the chain updates from data m to the new data m'. Simultaneously, the accumulator value is updated as:

$$acc' = acc^{H(m')} (\bmod N) \tag{10}$$

Additionally, the corresponding membership proof is outputted to facilitate subsequent fast implementation of modification verification:

$$\pi = acc'^{\frac{1}{H(m')}} \tag{11}$$

To ensure the modifier's identity information is hidden, the encrypted parameters T_1, T_2, and T_3 are computed using the following steps:

$$T_1 = u^\alpha \tag{12}$$

$$T_2 = v^\beta \tag{13}$$

$$T_3 = a_0 A_i h^{\alpha + \beta} \tag{14}$$

- $\{0, 1\} \leftarrow Check(m', r', g_1, y, h)$: The user needs to verify whether the modified new data m' satisfies the collision. This verification entails checking

$$H(m') \overset{?}{=} h \tag{15}$$

where $H(m') = g_1^{m'} y^{r'}$. If the equation holds, it indicates satisfies the collision, and 1 is outputted, otherwise 0.

- $\{0, 1\} \leftarrow RsaCheck(acc, m', \pi)$: The user needs to quickly verify whether the data in the chain has completed its modification. The membership proof π is used to verify that the data m' is in the set of accumulators. Subsequently, it is necessary to verify

$$acc \overset{?}{=} \pi^{H(m')} \tag{16}$$

If the equation holds, it indicates m' is in the set of accumulators, and 1 is outputted, otherwise 0.

- $A_i \leftarrow Trace(gpk, gmsk, a_0, T_1, T_2, T_3)$: The GM needs to obtain authorization from the committee before tracing the identity of the malicious user. The above steps are the same as in Eqs. (7) and (8). Subsequently, it needs to use its key pair $(gpk, gmsk)$ to decrypt the encrypted data on the chain that hides the modifier's identity. The relevant steps are as follows:

$$A_i = \frac{T_3}{a_0(T_1^{\varepsilon_1} \cdot T_2^{\varepsilon_2})} \tag{17}$$

5 Scheme Analysis

5.1 Correctness

Theorem 1. *The proposed DDMA scheme is correct.*

Proof. To prove the correctness of the proposed DDMA scheme, it only needs to show that equalities in (15)–(16) are satisfied.

For the data digest h and the data m', Eq. (15) is satisfied as follows:

$$\begin{aligned}
H(m') &= g_1^{m'} y^{r'} \\
&= g_1^{m'} g_1^{a_0 r'} \\
&= g_1^{m' + a_0 r'} \\
&= g_1^{m' + a_0(m + a_0 r - m')a_0^{-1}} \\
&= g_1^{m' + m + a_0 r - m'} \\
&= g_1^{m + a_0 r} \\
&= h
\end{aligned}$$

For the proof of membership π and the data m', the Eq. (16) is satisfied as follows:

$$\begin{aligned}
\pi^{H(m')} &= (acc^{\frac{1}{H(m')}})^{H(m')} \\
&= acc^{(\frac{1}{H(m')} H(m'))} \\
&= acc
\end{aligned}$$

5.2 Security

Theorem 2. *Modification permissions are regulated.*

Proof. DDMA is based on the Chameleon hash function [16] and verifiable secret sharing [24]. Chameleon hashing allows the data on the blockchain to be modified, and the verifiable secret-sharing scheme manages traps. Modifications can only be performed if there are enough fragments of the secret key, and malicious users unaware of the traps cannot manipulate the data on the blockchain.

Theorem 3. *The DDMA meets the traceability.*

Proof. In the DDMA scheme, the signature message in the block is public, but the attacker cannot determine the identity of the signer from the signature message $\sigma = (h, acc', \pi, T_1, T_2, T_3)$. To determine the identity of the signer, it is necessary to verify that $A_i \cdot T_1^{\varepsilon_1} \cdot T_2^{\varepsilon_2} \cdot a_0$ and T_3 are equal. However, only the GM knows the elements ε_1 and ε_2. When a dispute occurs, the GM can obtain the member's identity proof A_i by computing $A_i = T_3/a_0 \cdot (T_1^{\varepsilon_1} \cdot T_2^{\varepsilon_2})$, and can subsequently query the list of group members that he maintains to obtain the signer's identity information of the signer.

Theorem 4. *The DDMA satisfies modification verifiability.*

Proof. Since this scheme modifies the data on the blockchain using the Chameleon hash algorithm, the corresponding hash value h remains unchanged regardless of whether the data is modified or not. For third-party users, it is impracticable to verify whether data has undergone modifications in a substantively efficacious manner. In our scheme, when the data on the blockchain is changed, the RSA accumulator registers the changed value. Therefore, a third-party user can quickly verify whether the corresponding data have completed modification by using the membership proof π generated by the RSA accumulator.

6   Evaluation

6.1   Theoretical Analysis

Table 1. Theoretical comparison

Scheme	Redactable on the chain	Modifier identity anonymized	Traceability of modifications	Modify status record
Zhang et al. [33]	✔	✗	✔	✗
Deuber, Magri and Thyagarajan [8]	✔	✗	✔	✗
DDMA	✔	✔	✔	✔

Table 1 presents a comparison of DDMA with other schemes in terms of functionality and security features. Zhang et al. [33] proposed a scheme that establishes a dual blockchain architecture, which realizes the editability of data on the chain and the traceability of the operation behavior. However, it focuses on the problem of governance of operation privileges and neglects the protection of modifier identity information. Deuber, Magri and Thyagarajan [8] implement the redactable blockchain based on the voting of the consensus protocol, the modifier who wants to edit the data needs to request editing to the system, and this lets everyone in the network know the identity of the user who performs the modification operation. Thus, there is a risk of user identity leakage. In addition, neither the schemes proposed by [33] nor [8] consider fast verification of the modified state of the data on the chain. In contrast, DDMA can fulfill all the above functions.

6.2   Experimental Analysis

This section presents experimental simulations of the time costs for each phase. To assess the performance and effectiveness of the DDMA scheme, we implemented the proposed scheme on a computer with a 3.2 GHz 4-core CPU and 24 GB RAM. We simulated the on-chain and off-chain computation process using Java and evaluated the algorithm efficiency. In the implementation, the security parameters of the scheme are configured to 256 bits.

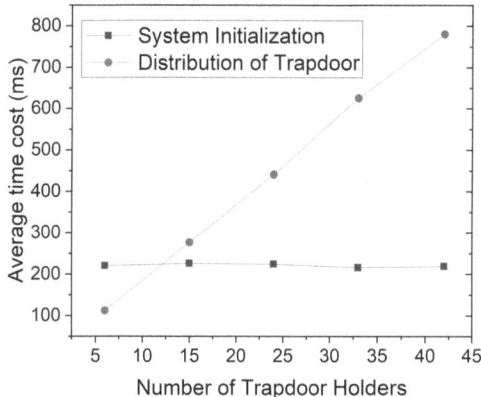

Fig. 4. Time cost for our scheme in initialization phase.

Initially, during the off-chain setup phase, we simulated five test environments. The system initialization time in different settings was first tested, as shown in Fig. 4. The figure illustrates that the system initialization time is constant and acceptable regardless of the number of committees, and it does not increase as the number of committees increases. Subsequently, we tested the total cost of time required to generate the distributed secret key under various conditions, and the results are depicted in Fig. 4. When the threshold t for secret recovery is determined, the time overhead required for generating the distributed secret key increases as the number of committee members increases. However, the time overhead is acceptable for each committee member. When the number of committee members is constant, a change in the threshold t does not significantly affect the time overhead. Because the time overhead of distributing secret key distribution in this stage is much more significant than calculating secret key fragments.

In the on-chain phase of the DDMA scheme, the inter-chain operations mainly involve modification decisions, tracking decisions, and fast verification of block information. Since modification and tracking decisions involve committee voting decisions and the decision-making process is similar, we only tested the time overhead of the system used for modification decisions in different environments, as shown in Fig. 5. The figure illustrates that when the threshold t is fixed, the time overhead for recovering secrets is directly proportional to the number of committees. When the number of committee members is fixed, the time overhead for recovering secrets increases with the threshold t, because the complexity of the secret polynomial generated by the KGC is directly proportional to t.

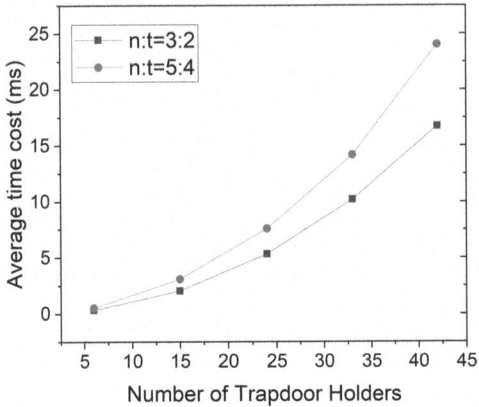

Fig. 5. Time cost for our scheme in secret key aggregation phase.

When authorization to modify is obtained, the algorithm *Adapt* is performed, the details of which are given in Sect. 4. During our experiments, the time overhead spent modifying the block's data is constant and takes only about 42.3 ms. The primary time overhead in the process is spent computing the new random number r, which is about 18.5 ms; the time overhead is illustrated in Fig. 6. Since the modification operation is performed by the user using his private key, the time required remains the same.

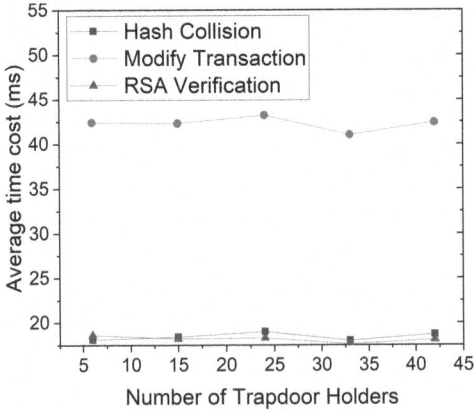

Fig. 6. Time cost for our scheme in modification and validation phase.

Finally, we tested the fast verification of modified blocks, and the results are shown in Fig. 6. In the DDMA scheme, the modified blocks are recorded using the RSA accumulator, and the algorithmic complexity of this process is $O(1)$. The experimental results also indicate that the time overhead for data verification is approximately 18 ms, which does not change with the number of committee members, and this time overhead is of a constant magnitude, which all users can accept.

In summary, the DDMA scheme performs well, and the time overhead required at key steps is within acceptable limits and does not burden the blockchain system. Therefore, the scheme can effectively manage industrial data on the industrial internet.

7   Conclusion

This paper presents a verifiable decentralized data modification mechanism that supports accountability, effectively addressing the challenges in industrial blockchain related to the privacy protection of data managers and the verification of on-chain data modifications. This mechanism ensures the privacy of modifier information and preserves reliable accountability channels. Furthermore, the mechanism provides an effective method for verifying on-chain data modifications, enabling users to quickly validate whether data has been altered, with a verification time complexity of $O(1)$. Experimental results demonstrate that the DDMA reduces the computational burden for modifiers and third-party validators. The time required for modifiers to modify on-chain data and for validators to verify data remains constant at 42.3 ms and 18 ms, respectively. These results affirm the feasibility and practicality of the proposed solution. This balance between transparency and privacy is crucial for building trust and enabling secure, scalable operations in industrial settings. As the technology continues to evolve, future innovations will likely focus on improving efficiency, expanding cross-chain interoperability, and addressing emerging challenges in data governance.

Acknowledgements. This paper is supported in part by the Guangxi Natural Science Foundation (2024GXNSFAA010453, 2024GXNSFDA010064), the National Natural Science Foundation of China (62172119, 62162017, 62362013), the Zhejiang Provincial Natural Science Foundation of China (LZ23F020012), the Xiong'an Autonomous Controllable Blockchain Bottom Layer Technology Platform Project (2020), the Guangxi Young Teachers' Basic Ability Improvement Program (2024KY0224), the Guangxi Key Laboratory of Trusted Software (KX202329), and the Innovation Project of GUET Graduate Education (2023YCXS059, 2023YCXS070), the Guangxi Key Laboratory of Digital Infrastructure under grant GXDIOP2023006.

References

1. Ateniese, G., Magri, B., Venturi, D., Andrade, E.: Redactable blockchain – or – rewriting history in bitcoin and friends. In: 2017 IEEE European Symposium on Security and Privacy (EuroS&P), pp. 111–126 (2017). https://doi.org/10.1109/EuroSP.2017.37

2. Bahga, A., Madisetti, V.: Blockchain platform for industrial internet of things. J. Softw. Eng. Appl. **09**, 533–546 (2016). https://doi.org/10.4236/jsea.2016.910036

3. Benaloh, J., de Mare, M.: One-way accumulators: a decentralized alternative to digital signatures. In: Helleseth, T. (ed.) EUROCRYPT 1993. LNCS, vol. 765, pp. 274–285. Springer, Heidelberg (1994). https://doi.org/10.1007/3-540-48285-7_24

4. Castaldo, L., Cinque, V.: Blockchain-based logging for the cross-border exchange of ehealth data in Europe. In: Gelenbe, E., et al. (eds.) Euro-CYBERSEC 2018. CCIS, vol. 821, pp. 46–56. Springer, Cham (2018). https://doi.org/10.1007/978-3-319-95189-8_5

5. Chen, J., Xue, J., Wang, Y., Huang, L., Baker, T., Zhou, Z.: Privacy-preserving and traceable federated learning for data sharing in industrial IoT applications. Expert Syst. Appl. **213**, 119036 (2023). https://doi.org/10.1016/j.eswa.2022.119036

6. Christidis, K., Devetsikiotis, M.: Blockchains and smart contracts for the internet of things. IEEE Access **4**, 2292–2303 (2016). https://doi.org/10.1109/ACCESS.2016.2566339

7. Dai, X., et al.: Task co-offloading for D2D-assisted mobile edge computing in industrial internet of things. IEEE Trans. Ind. Inform. **19**(1), 480–490 (2023). https://doi.org/10.1109/TII.2022.3158974

8. Deuber, D., Magri, B., Thyagarajan, S.A.K.: Redactable blockchain in the permissionless setting. In: 2019 IEEE Symposium on Security and Privacy (SP), pp. 124–138 (2019). https://doi.org/10.1109/SP.2019.00039

9. Diallo, N., et al.: egov-dao: a better government using blockchain based decentralized autonomous organization. In: 2018 International Conference on eDemocracy & eGovernment (ICEDEG), pp. 166–171 (2018). https://doi.org/10.1109/ICEDEG.2018.8372356

10. Du, R., Liu, N., Li, M., Tian, J.: Block verifiable dynamic searchable encryption using redactable blockchain. J. Inf. Secur. Appl. **75**, 103504 (2023). https://doi.org/10.1016/j.jisa.2023.103504

11. Fernández-Caramés, T.M., Fraga-Lamas, P.: A review on the use of blockchain for the internet of things. IEEE Access **6**, 32979–33001 (2018). https://doi.org/10.1109/ACCESS.2018.2842685

12. Gong, X., Liu, X., Jing, S., Xiong, G., Zhou, J.: Parallel-education-blockchain driven smart education: challenges and issues. In: 2018 Chinese Automation Congress (CAC), pp. 2390–2395 (2018). https://doi.org/10.1109/CAC.2018.8623198

13. Guo, H., Li, W., Meese, C., Nejad, M.: Decentralized electronic health records management via redactable blockchain and revocable IPFs. In: 2024 IEEE/ACM Conference on Connected Health: Applications, Systems and Engineering Technologies (CHASE), pp. 167–171 (2024). https://doi.org/10.1109/CHASE60773.2024.00028

14. Huang, X., et al.: A privacy-preserving credit bank supervision framework based on redactable blockchain. In: Svetinovic, D., Zhang, Y., Luo, X., Huang, X., Chen, X. (eds.) Blockchain and Trustworthy Systems, pp. 18–30. Springer, Singapore (2022). https://doi.org/10.1007/978-981-19-8043-5_2

15. Kafunah, J., Verma, P., Ali, M.I., Breslin, J.G.: Out-of-distribution data generation for fault detection and diagnosis in industrial systems. IEEE Access **11**, 135061–135073 (2023). https://doi.org/10.1109/ACCESS.2023.3337658

16. Krawczyk, H., Rabin, T.: Chameleon signatures. In: Network and Distributed System Security Symposium (2000). https://api.semanticscholar.org/CorpusID: 30185442

17. Liang, G., Weller, S.R., Zhao, J., Luo, F., Dong, Z.Y.: The 2015 Ukraine blackout: implications for false data injection attacks. IEEE Trans. Power Syst. **32**(4), 3317–3318 (2017). https://doi.org/10.1109/TPWRS.2016.2631891

18. Liang, W., Tang, M., Long, J., Peng, X., Xu, J., Li, K.C.: A secure fabric blockchain-based data transmission technique for industrial internet-of-things. IEEE Trans. Ind. Inform. **15**(6), 3582–3592 (2019). https://doi.org/10.1109/TII. 2019.2907092

19. Lu, Q., Xu, X.: Adaptable blockchain-based systems: a case study for product traceability. IEEE Software **34**(6), 21–27 (2017). https://doi.org/10.1109/MS. 2017.4121227

20. Mazzei, D., et al.: A blockchain tokenizer for industrial IoT trustless applications. Future Gener. Comput. Syst. **105**, 432–445 (2020). https://doi.org/10.1016/ j.future.2019.12.020

21. Mishra, R., Ramesh, D., Mohammad, N.: Rbda: redactable-blockchain based secure data aggregation scheme for IoT enabled cloud paradigm. In: 2022 IEEE International Conference on Pervasive Computing and Communications Workshops and other Affiliated Events (PerCom Workshops), pp. 409–414 (2022). https://doi.org/ 10.1109/PerComWorkshops53856.2022.9767416

22. Nakamoto, S.: Bitcoin: a peer-to-peer electronic cash system. Decentralized Business Review (2008). https://git.dhimmel.com/bitcoin-whitepaper/

23. Rakkini, M.J.J., Geetha, K.: Deep learning classification of bitcoin miners and exploration of upper confidence bound algorithm with less regret for the selection of honest mining. J. Ambient Intell. Humanized Comput. **14**(6), 6545–6561 (2023). https://doi.org/10.1007/s12652-021-03527-9

24. Shao, J., Cao, Z.: A new efficient (t, n) verifiable multi-secret sharing (VMSS) based on YCH scheme. Appl. Math. Comput. **168**(1), 135–140 (2005). https:// doi.org/10.1016/j.amc.2004.08.023

25. Shao, W., Wang, J., Wang, L., Jia, C., Xu, S., Zhang, S.: Auditable blockchain rewriting in permissioned setting with mandatory revocability for IoT. IEEE Internet Things J. **10**(24), 21322–21336 (2023). https://doi.org/10.1109/JIOT.2023. 3283092

26. Sun, Z., Strang, K.D., Pambel, F.: Privacy and security in the big data paradigm. J. Comput. Inf. Syst. **60**, 146–155 (2020). https://api.semanticscholar.org/CorpusID: 64739807

27. Treleaven, P., Gendal Brown, R., Yang, D.: Blockchain technology in finance. Computer **50**(9), 14–17 (2017). https://doi.org/10.1109/MC.2017.3571047

28. Wang, J., Chen, J., Ren, Y., Sharma, P.K., Alfarraj, O., Tolba, A.: Data security storage mechanism based on blockchain industrial internet of things. Comput. Ind. Eng. **164**, 107903 (2022). https://doi.org/10.1016/j.cie.2021.107903

29. Wen, B., Wang, Y., Ding, Y., Zheng, H., Qin, B., Yang, C.: Security and privacy protection technologies in securing blockchain applications. Information Sciences **645**, 119322 (2023). https://doi.org/10.1016/j.ins.2023.119322

30. Yi, X., Zhou, Y., Lin, Y., Xie, B., Chen, J., Wang, C.: Digital rights management scheme based on redactable blockchain and perceptual hash. Peer-to-Peer Networking Appl. **16**(5), 2630–2648 (2023). https://doi.org/10.1007/s12083-023-01552-3

31. Yu, G., et al.: Enabling attribute revocation for fine-grained access control in blockchain-IoT systems. IEEE Trans. Eng. Manag. **67**(4), 1213–1230 (2020). https://doi.org/10.1109/TEM.2020.2966643

32. Zeng, S., Cao, B., Sun, Y., Sun, C., Wan, Z., Peng, M.: Blockchain-assisted cross-domain data sharing in industrial IoT. IEEE Internet Things J. **11**(16), 26778–26792 (2024). https://doi.org/10.1109/JIOT.2023.3329577

33. Zhang, C., Ni, Z., Xu, Y., Luo, E., Chen, L., Zhang, Y.: A trustworthy industrial data management scheme based on redactable blockchain. J. Parallel Distrib. Comput. **152**, 167–176 (2021). https://doi.org/10.1016/j.jpdc.2021.02.026

34. Zhang, D., Le, J., Lei, X., Xiang, T., Liao, X.: Exploring the redaction mechanisms of mutable blockchains: a comprehensive survey. Int. J. Intell. Syst. **36**, 5051–5084 (2021). https://api.semanticscholar.org/CorpusID:236971033. https://doi.org/10.1002/int.22502

35. Zhao, L., Guo, D., Luo, L., Xie, J., Shen, Y., Ren, B.: Tiger tally: a secure IoT data management approach based on redactable blockchain. Comput. Network. **248**, 110500 (2024). https://doi.org/10.1016/j.comnet.2024.110500. https://www.sciencedirect.com/science/article/pii/S1389128624003323

36. Zhao, S., Li, S., Yao, Y.: Blockchain enabled industrial internet of things technology. IEEE Trans. Comput. Soc. Syst. **6**(6), 1442–1453 (2019). https://doi.org/10.1109/TCSS.2019.2924054

A Comprehensive Scheme for Transaction and Fund Tracing in Distributed Anonymous Transactions

Hai Liu$^{(\boxtimes)}$, Fangqiong Li , and Hongye Peng

Key Laboratory of Blockchain and Fintech of Department of Education of Guizhou Province, Guizhou University of Finance and Economics, Guiyang, China
{liuhai4757,lfq9497,PHY}@mail.gufe.edu.cn

Abstract. Anonymous transaction technologies are crucial for safeguarding the privacy of users' transactions, but they also pose challenges to traditional regulatory methods. However, existing supervision schemes for anonymous transactions often suffer from poor performance due to the use of complex cryptographic techniques. This paper proposes a transaction and fund tracing scheme for distributed anonymous transactions. To achieve traceability, the proposed scheme provides two models of supervision: *transaction tracing* to de-anonymize specific transactions and *fund tracing* to trace the flow of funds from illicit users. We have ingeniously designed a mechanism to distribute the critical information necessary for regulation among three parties, reducing the risk of misconduct by the supervisor and the platforms. Theoretical analysis shows that even in the face of external attacks and internal malicious activities, the proposed scheme can maintain the anonymity and traceability of transactions. Additionally, the performance of the proposal is evaluated through extensive experiments.

Keywords: Anonymous transaction · Transaction tracing · Fund tracing · Traceability

1 Introduction

The disclosure of pseudo-anonymity in blockchain transactions has spurred extensive research in academia on anonymous transaction technologies [1, 2]. While the adoption of anonymous transaction technologies enhances users' privacy, it concurrently undermines the traceability of transactions, presenting challenges to conventional regulatory approaches. However, the complete elimination of transaction anonymity risks divulging personal privacy information, thereby jeopardizing the safety of individuals and their property. Consequently, there is a pressing demand for a suitable supervisory framework and technical solutions that strike a delicate balance between transaction traceability and anonymity, addressing the challenges posed by this intricate trade-off.

© The Author(s), under exclusive license to Springer Nature Singapore Pte Ltd. 2025
T. Zhu et al. (Eds.): ICA3PP 2024, LNCS 15251, pp. 201–221, 2025.
https://doi.org/10.1007/978-981-96-1525-4_11

However, the majority of current anonymous transaction supervision schemes are cryptocurrency-centric, and their overall performance is often suboptimal, attributed to the reliance on intricate cryptography techniques. Distributed anonymous transaction schemes, also referred to as distributed coin mixing schemes, offer unparalleled advantages in anonymous transaction technologies by eliminating the necessity for third-party involvement and avoiding the use of complex cryptography techniques. Nonetheless, there is a lack of comprehensive research on supervision schemes grounded in the structure of distributed anonymous transactions.

To address this issue, this paper proposes a comprehensive scheme for transaction and fund tracing in distributed anonymous transactions, termed CTS. Building upon the system structure of distributed anonymous transaction schemes, the system architecture of CTS is outlined by integrating the message platform, the key center and the supervisor. Leveraging asymmetric cryptography technologies, CTS incorporates tags into transaction-related information and employs these implicit tags to enable transaction traceability. Notably, CTS offers two tracing methods: *transaction tracing* and *fund tracing*. *Transaction tracing* empowers the supervisor to discern the identity information of the actual initiator and recipient of any transaction, thereby achieving the traceability of transactions. *Fund tracing* enables the supervisor to de-anonymize all anonymous transactions initiated by a target user, thereby elucidating the fund transfer path of the target user. These dual methods facilitate the tracing of anonymous transactions by the supervisor under varying information constraints.

Furthermore, to ensure transaction traceability while safeguarding the transaction privacy of legitimate users, we have imposed constraints on the authority of the supervisor. Specifically, the essential information required for transaction oversight is partitioned among the supervisor, the message platform, and the key center. When encountering suspicious transactions or illicit users, the supervisor must access the information stored within the message platform and the key center to conduct transaction or fund tracing. This approach not only mitigates the potential negative impacts of malicious actions originating from the message platform and the key center on transaction anonymity, but also shields the privacy of legitimate users from the unfettered authority of the supervisor.

The main contributions of this paper are as follows:

1) Based on the structure of distributed anonymous transaction schemes, this paper designs a system architecture of CTS, and provides two methods for the supervisor: *transaction tracing* and *fund tracing*. These methods empower the supervisor to conduct transaction oversight under diverse information constraints, thereby facilitating comprehensive traceability of both transactions and funds.

2) In consideration of safeguarding the transaction privacy of legitimate users, this paper imposes limitations on the authority of the supervisor. By employing a mechanism where crucial information required for regulation is held separately among three parties, this approach seeks to mitigate the poten-

tial risks associated with the unrestrained supervisor, so as to safeguard the transaction privacy of legitimate users.

3) According to theoretical analysis and experimental results, our scheme can strike a balance between the anonymity and traceability of transactions. This indicates that while maintaining transaction anonymity, it can also achieve transaction traceability.

The remainder of this paper is structured as follows. We briefly summarize related work in Sect. 2. Section 3 describes the system architecture and potential threats. In Sect. 4, *CTS* is presented in detail, and the theoretical analysis and experimental evaluation of it are presented in Sect. 5 and Sect. 6 respectively. Finally, Sect. 7 concludes our work.

2   Related Work

Anonymous transaction schemes enhance the anonymity of transactions [3,4], but also hinder the relevant departments from supervising transactions. Therefore, many scholars have put forward corresponding schemes to solve this problem. Depending on the method used, these schemes can be broadly divided into two categories: cryptography-based schemes and blockchain-based schemes.

Cryptography-Based Schemes. Garman et al. [5] attempted to incorporate policy-enforcement mechanisms into *Zerocash*. This scheme enables the supervisor to track users and transactions selectively. However, the system's performance is poor as it relies on zk-SNARKs. Zhang et al. [6] proposed a regulatable digital currency model using secret sharing and a voting committee to decrypt transaction content. This method helps in achieving regulatory purposes while protecting user privacy. Wu et al. [7] proposed a regulatable digital currency scheme where the supervisor can oversee the flow of funds and obtain the real identity of users. But this scheme does not conceal the transaction amount. Lin et al. [8] proposed a scheme in which a user encrypts his long-term address by using the public key of the administrator and creates a new anonymous address for each transaction. With a private key, the administrator can decrypt and track users' transactions. However, all users' transaction privacy is visible to the administrator. Wang et al. [9] used zero-knowledge proof to achieve full anonymity protection for transactions. They also proposed a distributed anonymous payment supervision scheme, making it possible to supervise users' transactions. However, the use of zero-knowledge proof technology causes a large system overhead. Li et al. [10] came up with a traceable monero scheme that strikes a balance between anonymity and regulability. Only when malicious transactions need to be investigated will the supervisor step in, which not only reveals the malicious transaction, but also protects the privacy of legitimate users. What's more, a data asset transaction scheme based on consortium chain and group signature is proposed by Zhu et al. [11]. Guo et al. [12] integrated the advantages of identity-based cryptography and certificateless public-key cryptography to construct a regulatable anonymous transaction model, ensuring the anonymity of users' identity and the regulatability of transactions. Xue et al. [13] achieved a regulatory protocol that uses zero-knowledge proofs to verify transaction validity while protecting identity privacy. This protocol includes a tracking mechanism that allows supervisors to obtain real identities during suspicious transactions but also faces high system overhead.

Blockchain-Based Schemes. Wust et al. [14] developed *PRCash* to achieve regulated anonymous transactions on the blockchain. Ma et al. [15] proposed *SkyEye*, which allows the supervisor to attach identification to each anonymous transaction for tracking users. Nevertheless, the supervisor can recover the identities of all participants, thus failing to adequately protect the privacy of legitimate users. Chatzigiannis et al. [16] put forward a distributed payment scheme that provides the supervisor with audit capability. They also proposed another scheme [17] with the feature of pruning that saves the storage overhead without affecting the audit. Based on *Monero*, Chen et al. [18] proposed a tracing method for blockchain transactions, which firstly establishes a recording and replay mechanism. The transaction dependency analysis method based on state reading and writing is proposed to support the retracing of previous transactions linked by dependencies on demand. He et al. [19] raised the regulated privacy-preserving scheme, which realizes the anonymity of the transaction amount while the ability of one-way tracing is insufficient to meet the requirement of actual supervision. Song et al. [20] designed the supervised privacy protection scheme. This scheme can not only protect the identity privacy of participants, but also support the supervisor to recover their real identities.

Indeed, while existing schemes have made significant contributions to balance the anonymity and the traceability of transactions, but there are still some shortcomings. For instance, the existing regulated schemes based on anonymous currencies, as proposed in references [5] and [10], suffer from poor system performance due to the use of complex cryptography techniques. Schemes proposed in references [18, 19] and [20], based on blockchain environments, also encounter challenges such as heightened privacy risks due to the transparent nature of blockchain auditing and performance bottlenecks inherent in blockchain technology. In summary, there is still room for improvement in regulated anonymous transaction schemes.

3   Preliminaries

3.1   System Architecture

As shown in Fig. 1, this paper establishes the system architecture of CTS. It mainly includes two functional modules, namely anonymous transaction module and transaction tracing module. The anonymous transaction module is composed of the transaction initiator U_P, the transaction mixer M_i and the transaction recipient U_R, which is used to realize the anonymity of users' transactions. The transaction tracing module consists of the message platform MB, the key center C and the supervisor S, which is mainly used for transaction and fund tracing.

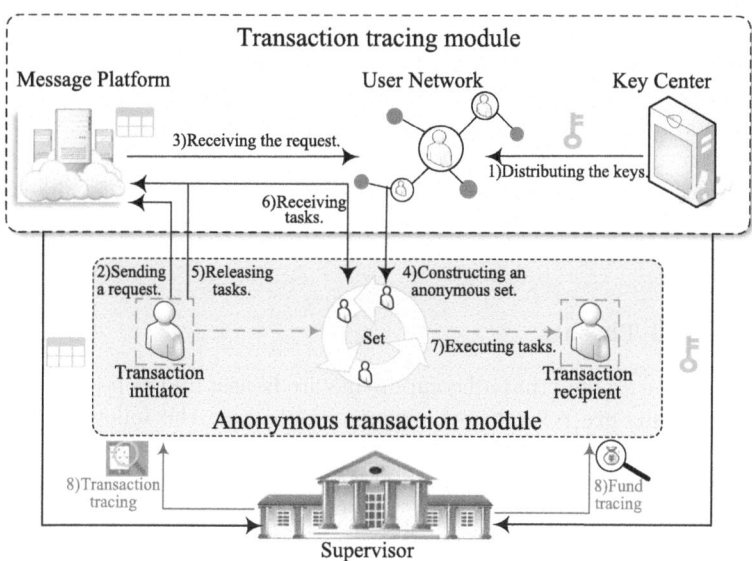

Fig. 1. System architecture.

U_P is the initiator of a transaction. In order to protect his/her privacy from being leaked, U_P will first broadcast an anonymous transaction request to other users through MB, and selects k users from all the users who accept request to construct an anonymous set $M_{Set} = \{M_1, \cdots, M_i, \cdots, M_k\}$. Among them, k represents the privacy requirement of U_P. M_i is the participant in anonymous transactions and helps transaction initiators to anonymize their transactions. U_R only acts as the recipient of an anonymous transaction. MB is the message platform, and which stores and maintains crucial information used for transaction and fund tracing. C is responsible for managing and distributing users' keys and maintaining the key distribution table B_{key}. S intervenes in anonymous transactions only when auditing is required. It de-anonymizes suspicious transactions and tracks the flow of funds from illicit users.

This paper first assumes that the communication channel through MB is secure. Under the condition of idealized work of anonymous transaction module, our scheme works as follows.

1) Prior to the commencement of the anonymous transaction module, the key center C initially distributes keys to all users and maintains the table B_{key}.
2) In the anonymous transaction module, the transaction initiator U_P first broadcasting an anonymous transaction request to other users through MB.
3) The other users can then decide whether to accept the request.
4) Upon the number of users accepting the request reaching the privacy requirement k specified by U_P, an anonymous set M_{Set} is formed for that user.
5) Following this, U_P disseminates anonymous transaction tasks among the transaction mixers M_i within M_{Set} through MB.

6) The transaction mixers receive tasks via MB.
7) Each transaction mixer in M_{Set} performs its own task. The completion of tasks by all transaction mixers signifies the conclusion of the anonymous transaction for U_P.
8) In the event of detecting suspicious transactions or illegal users, the supervisor S intervenes in the anonymous transaction module. Leveraging information maintained by MB and C, the supervisor can de-anonymize transactions or trace funds of illicit users.

3.2   Potential Threats

For simplicity, we assume that all components are honest and well-behaved. Once these assumptions are relaxed, our scheme would face the following potential threats. To enhance its self-credibility, we carefully designed CTS to address these threats in Sect. 4. It is also worth noting that this paper is more concerned with the transaction tracing module, so it does not waste space to consider the influence of the malicious behavior of transaction mixers on transaction anonymity.

Malicious External Attackers. In order to obtain improper benefits, attackers may take measures to compromise the anonymity of users' transactions and violate users' privacy. Therefore, our scheme must be designed to protect against external attacks.

Semi-trusted Transaction Originators. Some transaction originators may provide false information to confuse the supervisor to hide or secretly transfer their illicit funds. Therefore, there must be an alternative path for transaction and fund tracing. That is, when the transaction initiator provides false information, it can also trace his/her transactions and funds.

Unrestricted Supervisor. The unrestricted supervisor has broad powers and authority to regulate all users without limit, which is likely to infringe on the rights and interests of legitimate users. Therefore, the supervisor needs to be limited to monitoring illegal users and suspicious transactions.

Semi-trusted Third Parties. This paper treats the message platform MB and the key center C as semi-trusted third parties, both of which have the potential to snoop on users' privacy out of curiosity. Therefore, their capacity needs to be constrained. They should only be able to hold critical information about transaction and fund tracing, rather than use it independently to violate users' privacy.

4   Our Scheme

In order to maintain the anonymity of users' transactions in a regulated environment and avoid unrestricted full-scope supervision, we propose CTS and make the crucial information needed to trace transactions and funds be held separately by the supervisor S, the message platform MB, and the key center C. When S needs to trace transactions or funds, it must work with MB and C to get information from them.

Transaction tracing and *fund tracing* are the two different tracing methods provided by CTS. *Transaction tracing* allows S to obtain the identity information of the actual initiator and recipient of any transaction. *Fund tracing* allows S to de-anonymize all anonymous transactions initiated by the target user and determine the fund transfer path of the user. A sample of anonymous transaction and the supervisor views of different tracing methods are shown in Fig. 2.

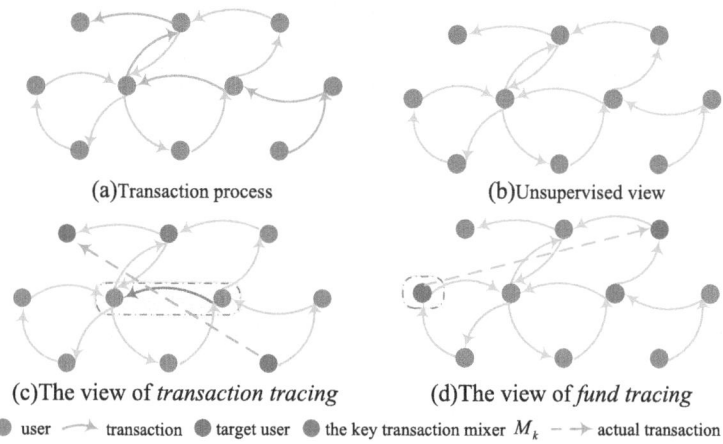

(a)Transaction process   (b)Unsupervised view

(c)The view of *transaction tracing*   (d)The view of *fund tracing*

● user ⟶ transaction ● target user ● the key transaction mixer M_k ⇢ actual transaction

Fig. 2. A sample of anonymous transactions and the supervisor views of different tracing methods.

CTS can be formalized as an algorithm six-tuple: $CTS = \{NewReq(\cdot),$ $AnsReq(\cdot), TagGen(\cdot), NewWork(\cdot), TraTrace(\cdot), FundTrace(\cdot)\}$. The first four functions are called in the anonymous transaction process, and the last two functions are used when S is conducting audits. $NewReq(\cdot)$ is called when a user sends an anonymous transaction request. $AnsReq(\cdot)$ is called when a user responds to the request. When the transaction initiator calls $NewReq(\cdot)$ to generate an anonymous transaction request, it will generate the unique identifier tid corresponding to this transaction. The identifier is used to mark tag of the key transaction mixer M_k and the task ciphertext $W_{P_0}^{M_i}$ that are subsequently generated. $TagGen(\cdot)$ is used to generate tag of the key transaction mixer M_k. $NewWork(\cdot)$ is used when the transaction initiator issues anonymous transaction tasks. The supervisor S calls $TraTrace(\cdot)$ to trace a target transaction and $FundTrace(\cdot)$ to trace funds of a target user.

4.1   Anonymous Transaction Process

This paper assumes that each user in $U = \{P_0, P_1, P_2, \cdots, P_n\}$ is unique, and the message platform MB can verify their identity. The key center C assigns public key PK_P and private key SK_P to all users. The anonymous transaction process of our scheme is shown in Fig. 3.

Step 1. Before conducting the transaction with the transaction recipient U_{R_1}, the transaction initiator U_{P_0} calls the function $NewReq(\cdot)$ to generate an anonymous transaction request Q_{P_0} as follows: $NewReq(U_{P_0}, m, tmp, PK_S, SK_{P_0})$

$$\Rightarrow \begin{cases} Encrypt(U_{P_0}, m, tmp)_{PK_S} \to tid_{P_0} \\ Signature(tid_{P_0})_{SK_{P_0}} \to Sign_{P_0}(tid_{P_0}) \end{cases} \Rightarrow Q_{P_0}, \text{ and broadcasts the}$$

request $Q_{P_0} = \{tid_{P_0}, Sign_{P_0}(tid_{P_0})\}$ to other users through MB. In particular, PK_S is a public key of the supervisor S. SK_{P_0} is a private key of U_{P_0}. $Encrypt(\cdot)$ is a secure encryption algorithm Elliptic Curve Cryptography (ECC). tid_{P_0} is an anonymous transaction identifier formed by U_{P_0} using PK_S to encrypt its identity information, the transaction amount m, and a timestamp tmp that ini-

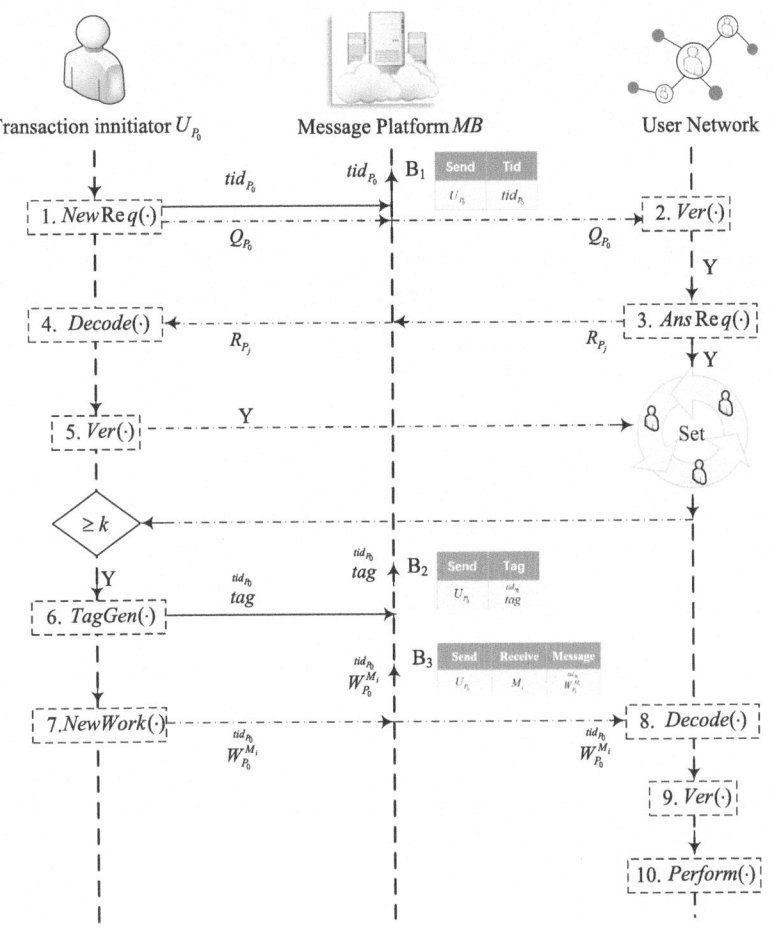

Fig. 3. Anonymous transaction process.

tiated the anonymous transaction request. tid_{P_0} used to uniquely identify the anonymous transaction. $Signature(\cdot)$ is a secure signature function called Elliptic Curve Digital Signature Algorithm (ECDSA), where the $Sign_{P_0}(tid_{P_0})$ means U_{P_0} uses its own private key SK_{P_0} to sign its anonymous transaction identifier. At this point, MB stores the anonymous transaction identifier tid_{P_0} and updates an information table $B_1(Send \mapsto U_{P_0} \quad Tid \mapsto tid_{P_0})$.

Step 2. When the other users receive the request $Q_{P_0} = \{tid_{P_0}, Sign_{P_0}(tid_{P_0})\}$ broadcast by U_{P_0}, they first use the signature verification function $Ver(\cdot)$ to verify the correctness of the signature information $Sign_{P_0}(tid_{P_0})$ in Q_{P_0}.

1. If $Ver(Sign_{P_0}(tid_{P_0})) = Sign_{P_0}(tid_{P_0})$, indicating that the identity of U_{P_0} has been verified, and at this point, network users can decide whether to accept the request.

(1) If the network user U_{P_j} accepts the request from U_{P_0}, U_{P_j} calls $AnsReq(\cdot)$ to generate a reply R_{P_j} follows: $AnsReq(tid_{P_0}, U_{P_j}, add_{P_j}, PK_{P_0}, SK_{P_j})$

$$\Rightarrow \begin{cases} Signature(tid_{P_0}\|U_{P_j}\|add_{P_j})_{SK_{P_j}} \rightarrow Sign_{P_j}(tid_{P_0}\|U_{P_j}\|add_{P_j}) \\ Encrypt(U_{P_j}, add_{P_j}, Sign_{P_j}(tid_{P_0}\|U_{P_j}\|add_{P_j}))_{PK_{P_0}} \rightarrow R_{P_j} \end{cases} \Rightarrow$$

R_{P_j}, and sends the $R_{P_j} = \{U_{P_j}, add_{P_j}, Sign_{P_j}(tid_{P_0}\|U_{P_j}\|add_{P_j})\}_{P_0}$ to U_{P_0} through MB. add_{P_j} is an account address of U_{P_j}. PK_{P_0} and SK_{P_j} represent the public key of U_{P_0} and the private key of U_{P_j}, respectively.

(2) If the network user U_{P_j} rejects the request of U_{P_0}, U_{P_j} does not respond.

2. If $Ver(Sign_{P_0}(tid_{P_0})) \neq Sign_{P_0}(tid_{P_0})$, it means that the identity of U_{P_0} has not been verified, and the network user may directly refuse the request.

Step 3. After receiving the $R_{P_j} = \{U_{P_j}, add_{P_j}, Sign_{P_j}(tid_{P_0}\|U_{P_j}\|add_{P_j})\}_{P_0}$ from U_{P_j}, U_{P_0} first decrypts it with its own SK_{P_0}, and then verifies the correctness of $Sign_{P_j}(tid_{P_0}\|U_{P_j}\|add_{P_j})$ in the reply R_{P_j}.

1. If $Ver(Sign_{P_j}(tid_{P_0}\|U_{P_j}\|add_{P_j})) = Sign_{P_j}(tid_{P_0}\|U_{P_j}\|add_{P_j})$, it indicates that the identity of the network user U_{P_j} has been verified. At this point, U_{P_j} can be incorporated into the anonymous set M_{Set} as a transaction mixer.

2. If $Ver(Sign_{P_j}(tid_{P_0}\|U_{P_j}\|add_{P_j})) \neq Sign_{P_j}(tid_{P_0}\|U_{P_j}\|add_{P_j})$, it means that the identity of U_{P_j} cannot pass the verification, and U_{P_0} refuses to add U_{P_j} to the anonymous set M_{Set}.

Step 4. If a quantity of transaction mixers in M_{Set} does not meet the privacy requirement k of U_{P_0}, U_{P_0} can continue to wait for other network users to respond. If the privacy requirement k is met, U_{P_0} randomly selects k users to form M_{Set} and enters the transaction execution phase.

After entering the transaction execution phase, U_{P_0} randomly allocates the sub-transaction $t_i(M_i \xrightarrow{m} M_{i+1})$ responsible for each transaction mixer M_i in M_{Set}, and calls a tag generation function $TagGen(\cdot)$ to generate tag marking the key transaction mixer M_k as follows: $TagGen(tid_{P_0}, U_{P_0}, M_k, PK_S) \Rightarrow$

$$\begin{cases} Encrypt(U_{P_0}, M_k)_{PK_S} \rightarrow tag \\ Ctr(tag, tid_{P_0}) \end{cases} \overset{tid_{P_0}}{\Rightarrow} \overset{tid_{P_0}}{tag}, \text{ and then sends } \overset{tid_{P_0}}{tag} \text{ to the message}$$

platform MB for storage. After receiving $\overset{tid_{P_0}}{tag}$, MB updates an information table $B_2(Send \mapsto U_{P_0} \quad Tid \mapsto \overset{tid_{P_0}}{tag})$.

Step 5. The transaction initiator U_{P_0} calls the function $NewWork(\cdot)$ to generate a task ciphertext $W_{P_0}^{M_i}$ as follows: $NewWork(m, add_{M_{i+1}}, PK_{M_i}, SK_{P_0}) \Rightarrow$

$$\begin{cases} Signature(m\|add_{M_{i+1}})_{SK_{P_0}} \rightarrow Sign_{P_0}(m\|add_{M_{i+1}}) \\ Encrypt(m, add_{M_{i+1}}, Sign_{P_0}(m\|add_{M_{i+1}}))_{PK_{M_i}} \rightarrow W_{P_0}^{M_i} \quad \overset{tid_{P_0}}{\Rightarrow} W_{P_0}^{M_i}, \text{ and then} \\ Ctr(W_{P_0}^{M_i}, tid_{P_0}) \end{cases}$$

sends it to M_i. At this point, the message platform MB stores the $\overset{tid_{P_0}}{W_{P_0}^{M_i}} = \{m, add_{M_{i+1}}, Sign_{P_0}(m\|add_{M_{i+1}})\}_{M_i}$ and updates the table $B_3(Send \mapsto U_{P_0}$ $Receive \mapsto M_i \quad Message \mapsto \overset{tid_{P_0}}{W_{P_0}^{M_i}})$. $W_{P_0}^{M_i}$ is a task obtained by using the public key of M_i to encrypt information. $add_{M_{i+1}}$ represents the account address of the target user M_{i+1} corresponding to the sub-transaction $t_i(M_i \overset{m}{\longrightarrow} M_{i+1})$.

Step 6. After receiving the task $W_{P_0}^{M_i}$ from U_{P_0}, the transaction mixer M_i decrypts it with SK_{M_i} and then verifies the signature $Sign_{P_0}(m\|add_{M_{i+1}})$ of U_{P_0} in the task ciphertext.

1. If $Ver(Sign_{P_0}(m\|add_{M_{i+1}})) = Sign_{P_0}(m\|add_{M_{i+1}})$, it means that the identity information of U_{P_0} has been verified, and at this time, M_i executes the sub-transaction $t_i(M_i \overset{m}{\longrightarrow} M_{i+1})$ according to the task content.

2. If $Ver(Sign_{P_0}(m\|add_{M_{i+1}})) \neq Sign_{P_0}(m\|add_{M_{i+1}})$, it indicates that the identity information of U_{P_0} is forged, the transaction mixer M_i can refuse to execute the task at this time.

Execution of the actual transaction $T(U_{P_0} \overset{m}{\longrightarrow} U_{R_1})$ of U_{P_0} is completed only after all transaction mixers in the anonymous set execute the sub-transaction $t_i(M_i \overset{m}{\longrightarrow} M_{i+1})$. That is, $\forall M_i \in M_{Set}$, M_i has executed $t_i(M_i \overset{m}{\longrightarrow} M_{i+1}) \Leftrightarrow T(U_{P_0} \overset{m}{\longrightarrow} U_{R_1})$ is completed.

4.2 Anonymous Transaction Tracing

To trace transactions, the supervisor S needs some private keys of users stored by the key center C and some tables $\{B_1, B_2, B_3\}$ maintained by the message platform MB in addition to its private key SK_S. This prevents unrestricted full-range supervision of S and protects the transaction privacy of legitimate users.

Transaction Tracing. The supervisor S can use *transaction tracing* to determine the identity information of the actual initiator and recipient of a target transaction. The specific execution process is as follows.

Step 1. When the supervisor S knows a specific amount m of a target transaction $t(u_1 \xrightarrow{m} u_2)$ and the identity information of the initiator u_1 and the recipient u_2, S first selects the initiator u_1 as the target user and filters the information table $B_3(Send \quad Receive \quad Message)$ maintained by MB, to obtain all task ciphertexts $W_{Set} = \{W_{U_0}^{u_1}, W_{U_1}^{u_1}, \cdots, W_{U_n}^{u_1}\}$ received by u_1.

Step 2. The specific steps of transaction tracing vary based on the number of task ciphertexts in W_{Set}.

1. If the number of $Count\,(W_{Set})$ exceeds 0.

(1) All anonymous transaction identifiers $Tid = \{tid_{U_0}, tid_{U_1}, \cdots, tid_{U_n}\}$ with which the set of $W_{Set} = \{W_{U_0}^{u_1}, W_{U_1}^{u_1}, \cdots, W_{U_n}^{u_1}\}$ are marked are found. Then, S uses its own SK_S to decrypt the anonymous transaction identifiers one by one, and performs preliminary filtering to obtain the anonymous transaction identifiers $Tid' = \{tid_{U_i}, tid_{U_i+1}, \cdots, tid_{U_i+x}\}$ with a same amount m as the target transaction $t(u_1 \xrightarrow{m} u_2)$ and a timestamp tmp before $t(u_1 \xrightarrow{m} u_2)$ occurs, where $x \in N^+$ and $0 \le i \le n - x$.

(2) The task ciphertexts $W'_{Set} = \{W_{U_i}^{u_1}, W_{U_{i+1}}^{u_1}, \cdots, W_{U_{i+x}}^{u_1}\}$ with the anonymous transaction identifiers $Tid' = \{tid_{U_i}, tid_{U_{i+1}}, \cdots, tid_{U_{i+x}}\}$ are retained and decrypted one by one by using the obtained private key SK_{u_1}, ultimately finding the task ciphertext $W_{U_j}^{u_1}$ that matches $t(u_1 \xrightarrow{m} u_2)$.

(3) Subsequently, find the anonymous transaction identifier tid_{U_j} marked with $W_{U_j}^{u_1}$ and tag_{U_j} marked with the key transaction mixer M_k by this identifier. The anonymous transaction identifier tid_{U_j} and tag_{U_j} are decrypted by using the private key SK_S, to obtain the actual transaction amount m corresponding to the target transaction $t(u_1 \xrightarrow{m} u_2)$ and the identity information of the actual transaction initiator U_P and the key transaction mixer M_k.

(4) After obtaining the identity information of the key transaction mixer M_k, S can filter the information table $B_3(Send \quad Receive \quad Message)$ to obtain a task ciphertext $W_{U_j}^{M_k}$ received by M_k and marked with the anonymous transaction identifier tid_{U_j}. S then obtain the private key SK_{M_k} to decrypt the task ciphertext $W_{U_j}^{M_k}$, to obtain the identity information of the actual transaction recipient U_R. Finally, the actual transaction $T(U_P \xrightarrow{m} U_R)$ is output.

2. If the number of $Count\,(W_{Set})$ of the obtained task ciphertexts is equal to 0, it is highly likely that u_1 is the actual transaction initiator and u_2 is the transaction mixer M_1. In this case, S only needs to obtain the identity information of the actual transaction recipient U_R to de-anonymize the transaction.

(1) The supervisor S takes u_1 as the target user to carry out an investigation, filters the information table $B_1(Send \quad Tid)$ to obtain anonymous transaction identifiers $Tid = \{tid_{U_1}^0, tid_{U_1}^1, \cdots, tid_{U_1}^n\}$ corresponding to all anonymous transactions initiated by u_1. Then, S decrypts the set of

anonymous transaction identifiers Tid by using its private key, and performs preliminary filtering to obtain all anonymous transaction identifiers $Tid' = \{tid_{U_1}^i, tid_{U_1}^{i+1}, \cdots, tid_{U_1}^{i+x}\}$ with a same amount m as the target transaction $t(u_1 \xrightarrow{m} u_2)$ and the timestamp tmp before the target transaction occurs, where $x \in N^+$ and $0 \le i \le n - x$.

(2) The supervisor S finds a task ciphertext $W_{u_1}^{u_2}$ received by u_2 and marked with the anonymous transaction identifier of Tid' from the information table $B_3(Send \quad Receive \quad Message)$.

(3) The supervisor S finds $tag_{u_1}^i$ marked with the anonymous transaction identifier $tid_{u_1}^i$ corresponding to $W_{u_1}^{u_2}$ from the information table $B_2(Send \quad Tag)$. And then S decrypts $tag_{u_1}^i$ with its own SK_S to obtain the identity information of M_k.

(4) The supervisor S filters the table $B_3(Send \quad Receive \quad Message)$ to obtain a task ciphertext $W_{u_1}^{M_k}$ received by M_k and marked with the anonymous transaction identifier $tid_{u_1}^i$, and decrypts $W_{u_1}^{M_k}$ by using the key of M_k, to obtain the identity information of the actual transaction recipient U_R. Finally, the actual transaction $T(U_P \xrightarrow{m} U_R)$ is output.

Fund Tracing. The supervisor S can use *fund tracing* to trace a flow of funds of a target user and de-anonymize all anonymous transactions with the target user as an actual transaction initiator. The specific execution process is as follows.

Step 1. When the supervisor S needs to investigate the fund transfer path of a target user U_B, S first filters the table $B_1(Send \quad Tid)$ maintained by MB, to obtain anonymous transaction identifiers $Tid = \{tid_B^0, tid_B^1, \cdots, tid_B^n\}$ corresponding to all anonymous transactions initiated by the target user. And then S decrypts the set of anonymous transaction identifier Tid by using its own SK_S, to obtain a specific amount $m_{Set} = \{m_0, m_1, \cdots, m_n\}$ of each transaction.

Step 2. The supervisor S finds, from the table $B_2(Send \quad Tag)$ maintained by MB, $Tag = \{tag_B^0, tag_B^1, \cdots, tag_B^n\}$ of key transaction mixers marked with the set of anonymous transaction identifier Tid. Subsequently, S decrypts all tags of Tag by using its own SK_S, to obtain the key transaction mixers $M_k = \{M_k^0, M_k^1, \cdots, M_k^n\}$ corresponding to each anonymous transaction.

Step 3. The supervisor S filters $B_3(Send \quad Receive \quad Message)$ to obtain the task ciphertext $W_B^{M_k^i}$ received by each key transaction mixer M_k^i in the set $M_k = \{M_k^0, M_k^1, \cdots, M_k^n\}$ and marked with the anonymous transaction identifier tid_B^i, to finally obtain a task ciphertext set $W_{Set} = \{W_B^{M_k^0}, W_B^{M_k^1}, \cdots, W_B^{M_k^n}\}$.

Step 4. The supervisor S obtains the private key of each key transaction mixer M_k^i from the key center C, and decrypts the task ciphertext one by one to obtain identity information of an actual transaction recipient $U_{Set} = \{U_R^0, U_R^1, \cdots, U_R^n\}$ corresponding to each anonymous transaction.

Step 5. A transaction set $T_{Set} = \{T_0(U_B \xrightarrow{m_0} U_R^0), T_1(U_B \xrightarrow{m_1} U_R^1), \cdots, T_n(U_B \xrightarrow{m_n} U_R^n)\}$ of funds transferred out of the target user U_B is finally output, based on the specific amount $m_{Set} = \{m_0, m_1, \cdots, m_n\}$ of each anonymous transaction initiated by the target user U_B and the identity information of the actual transaction recipient $U_{Set} = \{U_R^0, U_R^1, \cdots, U_R^n\}$ corresponding to each anonymous transaction.

5 Theoretical Analysis

This section mainly analyzes the anonymity and traceability of CTS. Specifically, it analyzes whether CTS can maintain the anonymity of transactions in the face of external attacks, and whether it can still achieve transaction tracing and fund tracing in the presence of malicious transaction originators.

5.1 Anonymity

Theorem 1 (Anonymity). *If the signature algorithm and encryption algorithm used in CTS are based on the elliptic curve discrete logarithm problem, which is computationally hard, and if the generation of the anonymous set is done covertly, with all transaction mixers in M_{Set} assumed to be trusted internal collaborators, then the proposed scheme satisfies transaction anonymity.*

Proof. The concealment of anonymous set M_{Set} generation means that only the transaction initiator U_P knows the transaction mixers contained in the anonymous set M_{Set} and the specific link of each transaction mixer M_i in the anonymous transaction. In addition, this paper assumes that all the transaction mixers in M_{Set} are trusted internal collaborators. That is, they will not actively disclose the information they have about anonymous transactions. Therefore, this paper only considers the influence of external attackers.

A malicious external attacker may launch sybil attacks to forge user identities, to obtain anonymous transaction information. Specifically, the attacker joins M_{Set} through identity disguise, multi-identity, or replay attacks to obtain the anonymous transaction information, and compromises the anonymity of transactions based on this information. In addition to joining M_{Set} to obtain the anonymous transaction information, the attacker may directly attack the message platform MB or the key center C to compromise the anonymity of transactions. Therefore, the attack opponent for the anonymity of transactions is defined as $A_{adv} \in \{A_{syb}, A_{dbc}\}$, where A_{syb} is an adversary launching sybil attacks, and A_{dbc} is an adversary launching database attacks and ciphertext attacks. When the adversaries A_{syb} and A_{dbc} successfully attack any game, A_{adv} can successfully attack transaction anonymity of the present disclosure with a specific probability.

Game 1. Based on anonymous transaction steps of CST, a process in which the adversary A_{syb} simulates sybil attacks is as follows.

Generation Phase. After receiving the anonymous transaction request $Q_P = \{tid_P, Sign_P(tid_P)\}$ broadcast by the transaction initiator U_P, the adversary A_{syb} launches sybil attacks to forge a plurality of user identities $A_{A_{syb}} = \{P_{A_{syb}}^0,$

$P_{A_{syb}}^1, \cdots, P_{A_{syb}}^q\}$, and calls the function $AnsReq(\cdot)$ to generate a plurality of replies $R_{A_{syb}}^i = \{U_{A_{syb}}^i, add_{A_{syb}}^i, Sign_{A_{syb}}^i(tid_p\|U_{A_{syb}}^i\|add_{A_{syb}}^i)\}$ that need to be sent to the transaction originator U_P. Then, it uses the public key PK_{U_P} of U_P to encrypt the reply ciphertexts. The sybil attack may be launched to forge the plurality of user identities through identity disguise, multi-identity, replay attack, or the like. Identity disguise means that the adversary A_{syb} forges digital signatures of other users to pass identity verification of U_P. Multi-identity means that the adversary A_{syb} registers a plurality of user identities on the message platform MB. The replay attack means that the adversary A_{syb} intercepts valid replies sent by the other users and resend the replies to U_P.

Challenge Phase. The adversary A_{syb} sends the generated false replies $R_{A_{syb}}^i$ to the transaction initiator U_P in expectation of joining the anonymous set M_{Set} of U_P.

Guess Phase. It is assumed that ξ_λ is a negligible probability and the quantity of the false replies $R_{A_{syb}}^i$ generated by the adversary A_{syb} is $q(q \geq k)$. According to the anonymous set generation manner, we can calculate the probability of the adversary A_{syb} successfully joins M_{Set} of U_P. If the adversary A_{syb} forges a digital signature of another user through identity disguise, a probability that the signature passes identity verification is extremely low because the signature algorithm used in CTS is ECDSA and the elliptic curve discrete logarithm problem on which the algorithm is based is computationally hard. Moreover, the identity authentication function of MB does not allow the adversary A_{syb} to register a plurality of user identities. If the adversary A_{syb} replays the valid replies sent by the other users, it is almost impossible for the adversary A_{syb} to join the anonymous set by replaying the valid replies from others, as the digital signature of the reply contains the identifier tid of the anonymous transaction corresponding to the reply. It can be learned that a probability that the adversary A_{syb} successfully passes the identity verification of U_P is ξ_λ.

In addition, the probability of A_{syb} successfully joining M_{Set} of U_P is influenced by the privacy requirement k of U_P and the quantity c of users in the network who reply to the anonymous transaction request. That is, when the privacy requirement k of U_P is larger, the quantity c of network users who reply to the anonymous transaction request is smaller. The larger the quantity q of false replies $R_{A_{syb}}^i$ generated by the adversary A_{syb}, the greater the probability of the adversary A_{syb} successfully joining the anonymous set. It can be calculated that a probability that the adversary A_{syb} can forge at least one user identity to successfully join M_{Set} of U_P is

$$Pr[A_{syb} \cap M_{Set} \neq \varnothing] = \begin{cases} (1 - \frac{C_c^k}{C_{q+c}^k}) * \xi_\lambda < \xi_\lambda & c \geq k \\ 1 * \xi_\lambda & c < k \end{cases},$$

and the probability that the adversary A_{syb} can forge k user identities to successfully join M_{Set} of U_P is $Pr[A_{syb} \supset M_{Set}] = \frac{C_c^k}{C_{q+c}^k} * \xi_\lambda < \xi_\lambda$. From this, it can be learned that the probability of the adversary A_{syb} launches the sybil attack to join M_{Set} of U_P can be neglected. Therefore, the adversary A_{syb} cannot obtain the anonymous transaction information by joining the anonymous set.

Game 2. Based on a storage characteristic of the message platform MB and the key center C, a process in which the adversary A_{dbc} simulates the database attack and ciphertext attack is as follows.

Generation Phase. The adversary A_{dbc} first launches the database attack to obtain information stored on the messaging platform MB and the key center C. The database attack of the adversary A_{dbc} is assumed to be maximized. That is, the adversary A_{dbc} can successfully obtain all ciphertext stored on MB and C. After obtaining these ciphertext, the adversary A_{dbc} forges keys needed to decrypt these ciphertext, namely, it forges the private keys of the supervisor S, the message platform MB, and the key center C.

Challenge Phase. The adversary A_{dbc} decrypt the anonymous transaction identifier tid, tag, information tables $\{B_1, B_2, B_3\}$ and key distribution table B_{key} by using the forged private keys SK'_S, SK'_{MB}, and SK'_C, to obtain plaintext corresponding to ciphertext.

Guess Phase. The probability that the adversary A_{dbc} can successfully decrypt and obtain tid, tag, $\{B_1, B_2, B_3\}$ and B_{key}, is largely influenced by the accuracy of private key forgery. Due to the ECC encryption algorithm used in CTS and the difficulty in solving the elliptic curve discrete logarithm problem, the probability of the adversary A_{dbc} successfully forging the user's private key is extremely low. As a result, the chance of success for attacker A_{dbc} to use database attacks and ciphertext attacks to compromise transaction anonymity is greatly low.

Based on the security assumption of CTS in the conclusion, the probability of attackers A_{syb} and A_{dbc} successfully attacking the corresponding game can be ignored. Therefore, this scheme satisfies the anonymity of transactions.

5.2   Traceability

Theorem 2 (Transaction Traceability). *Transaction traceability can be guaranteed when the supervisor S knows the specific transaction amount m, the identity information of the initiator u_1 and the recipient u_2 of the target transaction $t(u_1 \xrightarrow{m} u_2)$.*

Proof. Based on the basic idea of *transaction tracing* in CTS, we should find the anonymous transaction identifier tid_t that matches with the target transaction $t(u_1 \xrightarrow{m} u_2)$ firstly. Then we find tag_t marked with the anonymous transaction identifier and decrypt the tag to obtain the identity information of the key transaction mixer M_k. Thirdly, we get the task ciphertext $W^{M_k}_{u_1}$ with anonymous transaction identifier received by the key transaction mixer M_k through filtering, and then decrypt the task ciphertext to get the identity information of the actual transaction initiator U_P and the transaction receiver U_R as well. Finally, the target transaction can be tracked and we also obtain the information of the actual transaction $T(U_P \xrightarrow{m} U_R)$ is obtained.

According to the transaction tracing process of the *transaction tracing*, the first key step in determining tracing reliability is to find the anonymous transaction identifier tid_t that matches with the target transaction $t(u_1 \xrightarrow{m} u_2)$. It can be learned from the tracing process that the *transaction tracing* can obtain the unique anonymous transaction identifier matching the target transaction $t(u_1 \xrightarrow{m} u_2)$ through three rounds of filtering. The second critical step in determining tracing reliability is to perform decryption to obtain the identity information of the key transaction mixer M_k in tag_t marked with the anonymous transaction identifier tid_t. However, authenticity of the identity information of the key transaction mixer M_k in tag_t is questionable. To secretly transfer illegal funds, the malicious transaction initiator hides identity information of the real key transaction mixer M_k so that the false tag_t is generated finally. If tag_t is false, the identity information of the actual transaction recipient cannot be directly obtained on the premise of obtaining task content marked by tag_t. Consequently, the target transaction cannot be traced. In this case, an alternative is provided by the scheme to settle the malicious behavior that the transaction initiators falsify the identity information of the key transaction mixer M_k. Even knowing that the tag_t is fake, the supervisor can still restore the real transaction by finding all task ciphertext marked with the transaction identifier tid_t corresponding to the false tag_t. Because users cannot forge or tamper with anonymous transaction identifiers.

Theorem 3 (Fund Traceability). *Under the condition that the supervisor S has the identity information of the target user U_B, the scheme in this paper meets the traceability of funds.*

Proof. The basic idea of *fund tracing* in CTS is to find anonymous transactions initiated by the target user U_B and the corresponding anonymous transaction identifiers, and then associate the anonymous transaction identifier with the key transaction mixer M_k corresponding to each transaction, so as to obtain the actual recipient of each transaction and realize the fund tracing of the target user U_B. According to the fund tracing process of *fund tracing*, the authenticity of *tag* is the key to determining the reliability of the tracking. Based on the above analysis, it can be learned that regardless of whether *tag* is true or not, it will not affect the flow of funds for the supervisor S tracing the target user U_B, but will only affect supervision costs.

It can be learned in combination with theorem 2 and theorem 3 that the target transaction and the funds of the target user can be traced when the supervisor S grasps different information. Therefore, CTS has traceability, and tracing results are highly reliable.

6   Experimental Evaluation

In this experiment, the public key cryptographic algorithm SM2 based on elliptic curve issued by the State Cryptography Administration is used to encrypt and sign the transaction information. The algorithm designed in this paper is implemented in Java programming language. And the experimental environment is 11th Gen Intel(R) Core (TM) i5-1135G7 @2.40GHz (8CPUs) 2.4GHz 8192MB RAM and Operating System is Windows10-64bit.

6.1 Transaction Tracing

In this experiment, we set the number of historical transactions N to 1 million. Next, we repeat the execution of the transaction tracing algorithm according to the different privacy requirement k of U_P. Finally, the experiment ultimately derived the average success ratio of executing the transaction tracing algorithm to obtain the real transaction information when U_P provides real tag or false tag, along with the average computation delay and average communication overhead required in this process. The results are shown in Fig. 4.

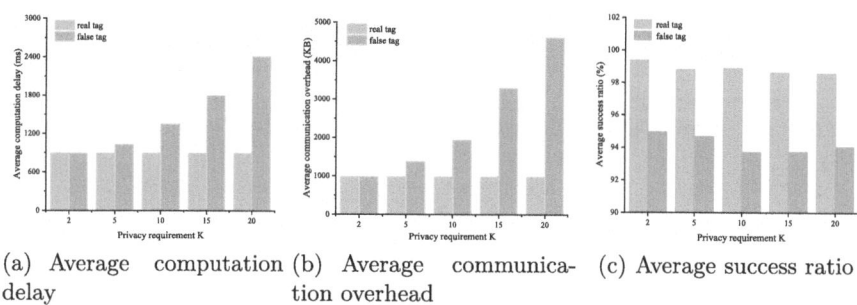

(a) Average computation delay   (b) Average communication overhead   (c) Average success ratio

Fig. 4. Transaction tracing.

When U_P provides real tag, the average computation delay and the average communication overhead required for transaction tracing do not change significantly, regardless of the change of k. However, when U_P provides false tag, both increase as k grows. Here is the reason for this phenomenon. If the tag provided by U_P for marking M_k is authentic, the specific amount of the target transaction can be known through tid and the ciphertext information of M_k, as well as the identity of the actual initiator and receiver of the transaction, which helps to track the target transaction. However, if the tag provided by U_P is false, the identity information of the actual transaction recipient is not available only with the task information owned by the fake M_k. In this case, the only way to correctly determine the identity of the actual transaction recipient is to obtain the ciphertext of each transaction mixer of M_{Set}. That is to say, tracing transactions with false tag requires to decrypt the task ciphertexts of k mixers of M_{Set}. Thus, the average computation delay and the average communication overhead required for transaction tracing will grow with an increase of the privacy requirement k of U_P.

6.2 Impacts of the Number of Historical Transactions on Transaction Tracing

This section analyses the impacts of the number of historical transactions N on the average success ratio of transaction tracing, as well as the required average computation delay and average communication overhead. We first set $k = 10$ and repeat the execution of the transaction tracing algorithm under different values of N and uncertainty in tag authenticity.

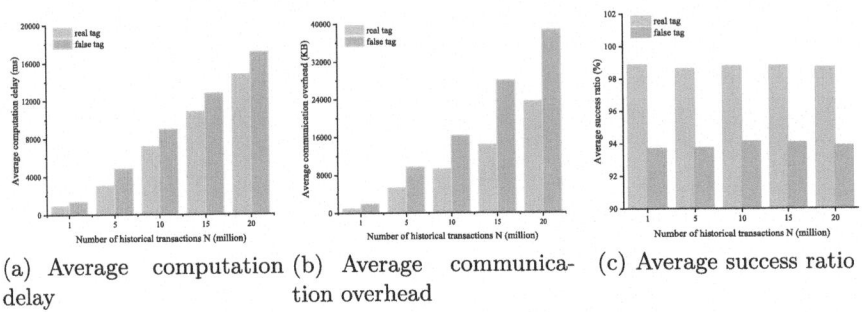

(a) Average computation delay   (b) Average communication overhead   (c) Average success ratio

Fig. 5. Impacts of the number of historical transactions on transaction tracing.

As can be seen from Fig. 5, regardless of the authenticity of tags provided by transaction initiators, the average computation delay and the average communication overhead required for transaction tracing grow as the number of historical transactions increases. However, the success ratio of the transaction tracing algorithm has not been significantly affected. Specifically, as the number of historical transactions increases from 1 million to 20 million, the average computation delay required for completing transaction tracing under real tag increases from 889.139 ms to 14782.386 ms, and the average communication overhead required increases from 971.648KB to 23489.295KB. For false tag, the average computation delay required for completing transaction tracing increases from 1345.514 ms to 17107.351 ms, and the average communication overhead required increases from 1934.712KB to 38631.866KB.

6.3 Fund Tracing

The experiment first sets the number of historical transactions as N to 1 million, and repeats the execution of fund tracing algorithm according to different values of k. Finally, we obtain the average success ratio of fund tracing under real tag or false tag, as well as the average computation delay and the average communication overhead required to do so. The results are shown in Fig. 6.

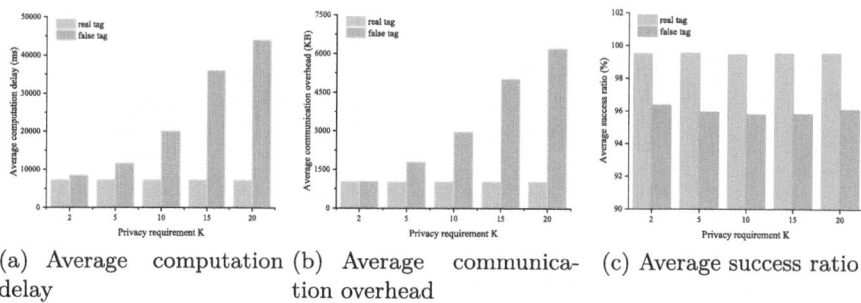

(a) Average computation delay

(b) Average communication overhead

(c) Average success ratio

Fig. 6. Fund tracing.

Through experimental data, we find that when real and false tag are provided by transaction initiators, the average computation delay and the average communication overhead required by the supervisor to complete the fund tracing are relatively limited. When $K = 20$, the average computation delay required to complete the fund tracing under real tag is 7164.697 ms, and the average communication overhead is 1017.147KB. The average computation delay for fund tracing under false tag is 43,958.4992ms and the average communication overhead is 6200.2101KB. Furthermore, the average success ratio of fund tracing is more than 95%.

6.4 Impacts of the Number of Historical Transactions on Fund Tracing

We analyze the impacts of the number of historical transactions N on the average success ratio of fund tracing, as well as the average computation delay and the average communication overhead required. The experiment starts by setting $k = 10$ and repeats the execution of the fund tracing algorithm with different number of historical transactions and uncertain authenticity of *tag*.

As shown in Fig. 7, the average computation delay and the average communication overhead required for fund tracing grow with an increase of historical transactions regardless of the authenticity of tags. As the number of historical transactions increases from 1 million to 20 million, the average computation delay required for completing fund tracing under real tag increases from 7164.684ms to 71323.279ms. Under false tag, the average computation delay required for completing fund tracing increases from 19981.136ms to 189602.644ms. The average communication overhead required for completing fund tracing under real tag increases from 1017.148KB to 11330.759KB, while under false tag, it increases from 2952.481KB to 43008.168KB. On the contrary, the average success ratio of implementing the fund tracing algorithm to find the money flow of the target user has remained stable at 94% and 98.5%.

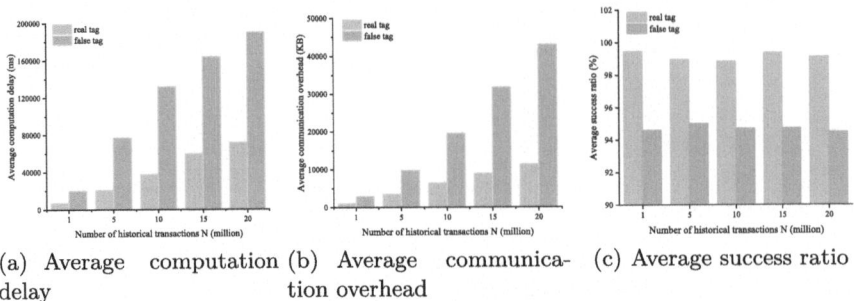

(a) Average computation delay   (b) Average communication overhead   (c) Average success ratio

Fig. 7. Impacts of the number of historical transactions on fund tracing.

7   Conclusion

To solve the problem that strong anonymity makes transactions difficult to trace, this paper proposes a comprehensive scheme for transaction and fund tracing in distributed anonymous transactions. The building blocks of the CTS scheme proposed by us are as follows: the system architecture of a traceable distributed anonymous transaction model, a transaction tracing method and a fund tracing method. Our theoretical analysis and experimental results demonstrated the utility and security of CTS. We will continue to optimize this scheme, focusing on reducing the system overhead required to trace transactions and funds when users submit false tags.

Acknowledgments. This work was sponsored in party by the National Natural Science Foundation of China under Grant 62062017; the Research Fund Program of Guizhou University of Finance and Economics under Grant 2021KYZD06; the Key Laboratory Building Program of Blockchain and Fintech of Department of Education of Guizhou Province under Grant QIAN JIAO JI[2023]014.

References

1. Androulaki, E., Karame, G.O., Roeschlin, M., Scherer, T., Capkun, S.: Evaluating user privacy in bitcoin. In: Sadeghi, A.-R. (ed.) FC 2013. LNCS, vol. 7859, pp. 34–51. Springer, Heidelberg (2013). https://doi.org/10.1007/978-3-642-39884-1_4
2. Herrera-Joancomartí, J.: Research and challenges on bitcoin anonymity. In: Garcia-Alfaro, J., et al. (eds.) DPM/QASA/SETOP -2014. LNCS, vol. 8872, pp. 3–16. Springer, Cham (2015). https://doi.org/10.1007/978-3-319-17016-9_1
3. Shentu, Q.C., Yu, J.P.: Research on anonymization and de-anonymization in the bitcoin system. arXiv preprint arXiv:1510.07782 (2015)
4. Han, X., Yuan, Y., Wang, Y.F.: Security problems on blockchain: the state of the art and future trends. ACTA Automatica Sinica **45**(1), 206–225 (2019)
5. Garman, C., Green, M., Miers, I.: Accountable privacy for decentralized anonymous payments. In: Grossklags, J., Preneel, B. (eds.) FC 2016. LNCS, vol. 9603, pp. 81–98. Springer, Heidelberg (2017). https://doi.org/10.1007/978-3-662-54970-4_5

6. Zhang, J.Y., Wang, Z.Q., Xu, Z.L., Ouyang, Y.F., Yang, T.: A regulatable digital currency model based on blockchain. J. Comput. Res. Dev. **55**(10), 2219–2232 (2018)

7. Wu, Y.B., Fan, H.N., Wang, X.Y., Zou, G.N.: A regulated digital currency. SCIENCE CHINA Inf. Sci. **62**, 1–12 (2019)

8. Lin, C., He, D.B., Huang, X.Y., Khan, M.K., Choo, K.K.R.: DCAP: a secure and efficient decentralized conditional anonymous payment system based on blockchain. IEEE Trans. Inf. Forensics Secur. **15**, 2440–2452 (2020)

9. Wang, Z., Pei, Q., Liui, X., Ma, L., Li, H., Yu, S.: DAPS: a decentralized anonymous payment scheme with supervision. In: Wen, S., Zomaya, A., Yang, L.T. (eds.) ICA3PP 2019. LNCS, vol. 11945, pp. 537–550. Springer, Cham (2020). https://doi.org/10.1007/978-3-030-38961-1_46

10. Li, Y.N., Yang, G.M., Susilo, W., Yu, Y., Au, M.H., Liu, D.X.: Traceable monero: anonymous cryptocurrency with enhanced accountability. IEEE Trans. Dependable Secure Comput. **18**(02), 679–691 (2021)

11. Zhu, Z.Q., Yao, Z.Y., Zhu, W.H., Zhao, H.H., Pan, C.F., Si, X.M.: Anonymous and traceable data trading scheme based on blockchain. J. Appl. Sci. **40**(4), 653–665 (2022)

12. Guo, Y.N., Jiang, W.B., Shuai, Y.E.: Supervisable blockchain anonymous transaction system model. J. Comput. Appl. **42**(9), 2757–2764 (2022)

13. Xue, L., Liu, D.X., Ni, J.B., Lin, X.D., Shen, X.M.: Enabling regulatory compliance and enforcement in decentralized anonymous payment. IEEE Trans. Dependable Secure Comput. **20**(2), 931–943 (2023)

14. Wüst, K., Kostiainen, K., Čapkun, V., Čapkun, S.: PRCash: fast, private and regulated transactions for digital currencies. In: Goldberg, I., Moore, T. (eds.) FC 2019. LNCS, vol. 11598, pp. 158–178. Springer, Cham (2019). https://doi.org/10.1007/978-3-030-32101-7_11

15. Ma, T.J., Xu, H.X., Li, P.L.: Skyeye: a traceable scheme for blockchain. Cryptology ePrint Archive (2020)

16. Chatzigiannis, P., Baldimtsi, F., Chalkias, K.: SoK: auditability and accountability in distributed payment systems. In: Sako, K., Tippenhauer, N.O. (eds.) ACNS 2021. LNCS, vol. 12727, pp. 311–337. Springer, Cham (2021). https://doi.org/10.1007/978-3-030-78375-4_13

17. Chatzigiannis, P., Baldimtsi, F.: MiniLedger: compact-sized anonymous and auditable distributed payments. In: Bertino, E., Shulman, H., Waidner, M. (eds.) ESORICS 2021. LNCS, vol. 12972, pp. 407–429. Springer, Cham (2021). https://doi.org/10.1007/978-3-030-88418-5_20

18. Chen, S., et al.: Tracing method for blockchain transaction execution based on recoding and replay. J. Software **34**(10), 4681–4704 (2023)

19. He, J.J., Chen, Y.L.: Encryption-based auditable privacy-preserving scheme for consortium blockchain. Comput. Eng. **49**(6), 170–179 (2023)

20. Song, J.W., Zhang, D.W., Han, X., Du, Y.: Supervised identity privacy protection scheme in blockchain. J. Software **34**(7), 3292–3312 (2023)

AutoMiner: Reinforcement Learning-Based Mining Attack Simulator

Wei Li[1,5], Lide Xue[2(✉)], Ziyang Han[3], Bingren Chen[1], Xishan Zhang[4,5], and Xuehai Zhou[1]

[1] University of Science and Technology of China, Hefei, China
[2] ShangHai Municipal Big Data Center, Shanghai, China
xldxld@mail.ustc.edu.cn
[3] Anban Tech Co., Shanghai, China
[4] State Key Laboratory of Processor Technologies, Institute of Computing Technology, Chinese Academy of Sciences, Beijing, China
[5] Cambricon Technologies, Beijing, China

Abstract. As blockchain technology rapidly advances, it faces significant security threats from attacks like selfish mining, which exploit consensus algorithm vulnerabilities, undermining system security. Traditional analysis methods, primarily reliant on Markov models, are often utilized to address these complex threats. However, these methods frequently fall short in accurately simulating the multifaceted nature of blockchain attacks, leading to gaps in security measures. To bridge this gap, we introduce AutoMiner, an innovative reinforcement learning-based framework that integrates Miner Monte Carlo Tree Search (MMCTS) with Long Short-Term Memory (LSTM) networks for simulating and detecting potential mining attacks within the Proof of Work (PoW) framework. Our experimental results demonstrate AutoMiner's capability to outperform traditional selfish and honest mining strategies, achieving up to 20% higher profits under specific scenarios. This highlights AutoMiner's potential in enhancing blockchain security by providing a more comprehensive and effective approach to analyzing and mitigating mining attacks.

Keywords: Blockchain · Reinforcement learning · Security analysis

1 Introduction

Blockchain technology has emerged as a significant computational innovation. It provides a decentralized storage solution with Byzantine Fault Tolerance (BFT), ensuring system integrity in the face of faults and attacks [6]. However, maintaining such decentralized systems, particularly those using Proof of Work (PoW), requires substantial storage, computational power, and energy from nodes [1]. To promote the faithful execution of blockchain protocols, public blockchains typically incorporate incentive mechanisms, such as token rewards [5,10,33]. In

© The Author(s), under exclusive license to Springer Nature Singapore Pte Ltd. 2025
T. Zhu et al. (Eds.): ICA3PP 2024, LNCS 15251, pp. 222–241, 2025.
https://doi.org/10.1007/978-981-96-1525-4_12

Bitcoin's blockchain, miners expend computational resources to secure transactions and receive BTC rewards in return. Nevertheless, the PoW mechanism introduces a delay in block confirmation to ensure security, necessitating multiple subsequent blocks for validation. This delay risks block overturning, enabling profitable, malicious forking by dishonest miners, which leads to incentive incompatibility [4].

Building on the discussion of the importance of incentive mechanisms, the issue of incentive incompatibility is underscored by the strategy of selfish mining [11]. This approach not only undermines the integrity of transactions processed by honest nodes but also wastes their computational resources. It challenges the foundational assumption of PoW blockchain security, which posits that malicious actors need majority control over the network's computational power to influence consensus [34]. Despite the various strategies proposed by the research community to address this issue [12,20,25,31], the efficacy of these security solutions remains insufficient in light of several theft incidents in recent years [19,22,37]. This compels developers to seek more robust mechanisms for ensuring incentive compatibility and security.

Exploring security vulnerabilities in blockchain incentive mechanisms is of paramount importance, as it provides critical insights for addressing these weaknesses and informs the design of more robust protocols. However, the predominant methodologies employed for analyzing blockchain attacks—comprising theoretical, simulated, and intuitive approaches—are exceptionally labor-intensive and time-consuming. This challenge is exacerbated by the increasing complexity of blockchain technologies, such as sharding and cross-chain interactions [16,23,24,32,35,36]. While the advent of AI-based agent miners offers a promising direction for automating and accelerating this analysis [7,14,29,30], significant obstacles persist:

Complexity in Modeling: The vast possibilities in miner actions and environments present a significant modeling challenge. Current analyses, constrained by simplicity, often assume that miners work on the newest blocks of either the public main chain or a private one [13,14,30]. This simplification omits a broader spectrum of attack strategies.

Subjective Analysis Focus: Many studies employing AI agents assume specific attack patterns, such as selfish mining [7,13,30]. While this approach is fruitful for detailed study of these attacks, it lacks the generality needed for applicability across varied blockchain protocols.

Markov Process Preference in Modeling: A preference for Markov processes characterizes several AI agent models [2,3,13,17,18,28]. Though beneficial for analytical clarity, this approach diverges from reality, where rational attackers might consider an extensive history of states for optimal decision-making.

Addressing prevalent challenges in blockchain security, particularly within Bitcoin's ecosystem, we introduce AutoMiner. This innovative agent model leverages the novel Miner Monte Carlo Tree Search (MMCTS) and LSTM neural networks to simulate and analyze mining attacks with unprecedented depth and breadth (Fig. 1). AutoMiner's MMCTS algorithm adeptly navigates potential blockchain states, allowing for a wide spectrum of actions, including dynamic allocation of mining power, strategic block broadcasting, or dormancy. Coupled with LSTM networks, AutoMiner benefits from an extended memory horizon, enhancing decision-making with a detailed history of blockchain states. Distinguished by its ability to explore without bias towards predefined patterns and its remarkable efficiency, AutoMiner marks a significant leap forward, operable on conventional computing hardware with substantial rewards optimization.

Our key contributions are as follows:

1. **MMCTS Algorithm:** We unveil the Miner Monte Carlo Tree Search (MMCTS), a pioneering simulation tool that deepens our understanding of mining strategies and their security implications. This algorithm not only broadens the scope of analysis but also enhances the precision of security assessments in blockchain networks.
2. **Neural Network Integration:** By integrating LSTM neural networks, AutoMiner transcends traditional analysis methods, utilizing extensive historical data to predict mining behaviors with unmatched accuracy. This combination of machine learning and blockchain analytics sets a new standard in the field.
3. **Operational Efficiency and Adaptability:** AutoMiner redefines efficiency, enabling complex simulations on conventional hardware while avoiding fixed attack patterns in favor of a dynamic exploration approach. Our strategy extends beyond Bitcoin, offering a robust framework for securing various blockchain architectures against evolving threats.

2   Related Work

2.1   Blockchain Agent Simulation

[15] introduces a cryptocurrency market model with two types of agents—market and miner—within Bitcoin's security infrastructure. [27] extends this by focusing on decentralized exchanges, analyzing the influence of front runners on profits and strategies to counteract price slippage. [8] applies AI agents to simulate cryptocurrency market dynamics, using indicators like the Relative Strength Index (RSI) for strategic hardware management to optimize costs and gains.

2.2   Exploring Blockchain Incentive Manipulations

Our investigation aligns with [7] in examining malicious node strategies, specifically hash power allocation. However, we extend beyond their focus by including

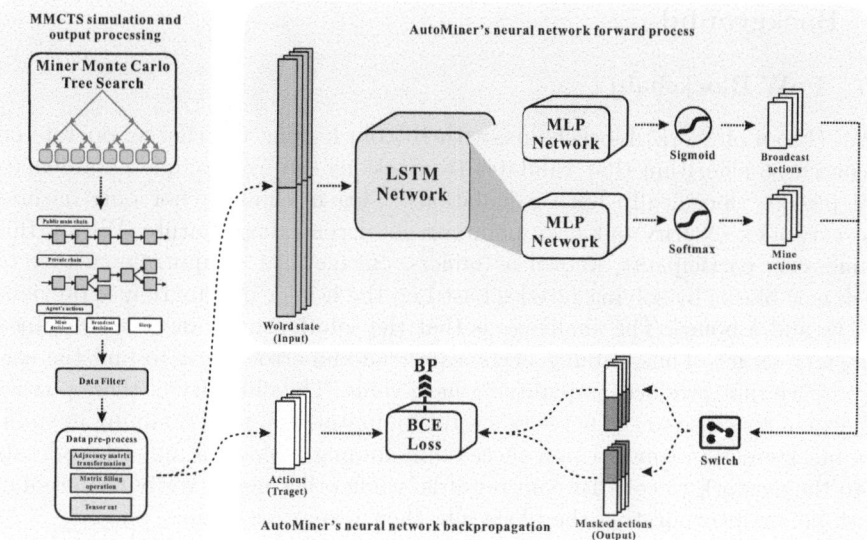

Fig. 1. Illustrative overview of AutoMiner's operation and its strategic decision-making process.

broadcast and sleep actions, unlike their emphasis on statistical analysis which omits selfish mining tactics. [13] targets selfish mining through a Markov model, analyzing the impact of propagation delay on Bitcoin's network, which provides a narrower scope compared to our broader examination of incentive manipulation strategies. [38] optimizes selfish mining thresholds in Ethereum using Probabilistic Termination Optimization (PTO) to mitigate strategic weaknesses. This method complements our approach by focusing on tactical vulnerability reduction. ForkDec [30] presents a simulator for Bitcoin under selfish mining attacks, detailing block specifics and fork data to create a comprehensive dataset for analysis. This diverges from our method by emphasizing data-driven simulation. SquirRL [14] converts consensus protocols into MDPs, applying reinforcement learning for strategy exploration. This contrasts with our method by limiting agent actions and simplifying security analysis but requires more extensive training, highlighting a trade-off between action space specificity and computational efficiency.

2.3   Detecting Blockchain Mining Maneuvers

[22] develops a blockchain simulator enhanced for selfish mining analysis, utilizing machine learning to detect such strategies. This marks a significant step towards identifying and mitigating mining-related security breaches.

Conversely, VDHLA [21] combines learning automata with versatile action sets to counter selfish mining in Bitcoin, showcasing a novel defense mechanism within the blockchain ecosystem.

3   Background

3.1   PoW Blockchain

PoW (Proof of Work) blockchains, with Bitcoin leading the charge, operate on a consensus algorithm that validates transactions through complex mathematical puzzles, specifically hash calculations. This mechanism not only ensures the network's security but also maintains its decentralized nature. Within this framework, participants, known as miners, engage in a computational race to hash new blocks by solving puzzles based on the header information of previous blocks and a nonce. The challenge is that the solution must meet a predefined difficulty target. Thus, mining becomes a trial-and-error quest to find the correct nonce that produces a qualifying hash value. The difficulty of these puzzles adjusts in response to the network's total computational power, aiming to stabilize block creation times. Upon successfully mining a block, a miner broadcasts it to the network to earn Bitcoin rewards, while other nodes verify the proof of work before incorporating the block into their respective chains.

3.2   Fork and Selfish Mining

As illustrated in Fig. 2(a), a blockchain may experience a fork, transitioning from a single-chain structure to a Directed Acyclic Graph (DAG) structure. This phenomenon occurs due to temporary inconsistency when the consensus network fails to achieve global agreement on a block at a certain height. In PoW blockchains like Bitcoin, such scenarios are permissible because the consensus protocol adopts the "longest chain rule." This rule dictates that the chain with the greatest height is considered the unique main chain. Mathematically, this endows the main chain with convergence properties, ensuring that all consensus nodes will eventually agree on a single chain (the public main chain).

While this outlines the intended conduct under Bitcoin's protocols, strategies like "selfish mining" introduce a twist. Here, miners hoard their newly mined blocks, choosing strategic moments to release them. This move aims to replace the public chain with their private one, exploiting the network to gain more rewards than their computational effort warrants. As Fig. 2(b) illustrates, a malicious node might broadcast blocks 4', 5', and 6', superseding blocks 4 and 5 on the main chain and nullifying their transactions. By initially keeping blocks 4', 5', and 6' under wraps, the attacker sidesteps the competition, skewing the reward system in their favor. This tactic not only inflates the attacker's gains disproportionately but also renders the honest miners' efforts on blocks 4 and 5 futile, undermining the blockchain's security infrastructure.

3.3   Monte-Carlo Tree Search (MCTS)

MCTS combines tree structures with the Monte-Carlo evaluation approach, emerging as a heuristic algorithm for navigating decision-making challenges. It relies on massive random sampling to evaluate potential actions, crafting a

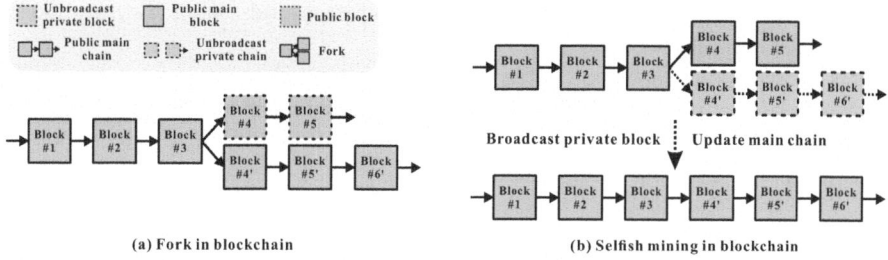

Fig. 2. Schematic diagram of selfish mining case

search tree that guides the decision-making process. At its core, MCTS thrives by expanding this search tree and leveraging insights from Monte-Carlo simulations to identify the most advantageous action. Designed for scenarios laden with complexity and partial information, such as chess, strategic planning, and game theory, MCTS excels [9, 26]. The essence of MCTS unfolds through a series of calculated steps:

1. **Selection:** The journey starts from the current state, selecting a node within the search tree through a calculated strategy. Techniques like the Upper Confidence Bound Applied to Trees (UCT) are particularly effective here, guiding the choice of nodes.
2. **Expansion:** In this step, the selected node is expanded by adding potential next moves and their corresponding child nodes. This expansion relies on domain-specific insights or rules tailored for the task.
3. **Simulation:** With the new nodes in place, the process moves to simulation. Random strategies are employed, navigating through a sequence of unpredictable moves until a terminal state or the simulation limit is reached.
4. **Backpropagation:** Finally, the results of the simulation—whether success or failure—are propagated back to the originating node, updating its statistics such as visit count and accumulated rewards.

Through iterative execution, MCTS methodically broadens the search tree, enhancing its understanding of the actionable value landscape. With each simulation, the algorithm edges closer to the optimal solution, improving the decision-making capabilities of AI agents. Central to this process is the UCT algorithm, which thrives on a crucial balance: navigating the unknown while capitalizing on the known. It leverages the Upper Confidence Bound (UCB) to guide node selection, striking a harmony between exploration and exploitation.

Specifically, UCT assesses each node's UCB value during selection, favoring nodes with the highest UCB for expansion. This value bifurcates into exploration—gauging the untapped potential based on how seldom a node has been explored—and exploitation—evaluating the node's track record of rewards. Nodes less frequented by exploration efforts score higher on the exploration scale, inviting further investigation into their unseen value. Conversely, nodes with a history of higher rewards rank higher for exploitation, indicating they are likely sources of favorable outcomes.

By weaving together exploration and exploitation elements within the UCB equation, UCT adeptly finds its footing in the quest for optimized search amidst incomplete data. For a given non-leaf node n and its children $ch(n) = \{n_i | i = 1, 2, 3, \ldots\}$, the UCB formula is as follows:

$$r_i = v_i + c \cdot \sqrt{\frac{2\ln(\sum_j T_j)}{T_i}} \tag{1}$$

Here, r_i is the evaluation metric, while v_i is the mean of all simulation outcomes for the subtree with node n_i at its base, representing the anticipated reward from node n_i based on current simulations. T_i denotes the number of interactions with node n_i, essentially how often n_i has been selected by the tree's selection mechanics. $\sum T_i$ represents the total interactions with node n, and c is an adjustable constant, finely tuning the balance between the UCT algorithm's exploitation (left term) and exploration (right term) aspects.

3.4 Long Short-Term Memory (LSTM)

Long Short-Term Memory (LSTM) neural networks stand out as a distinct type within the realm of Recurrent Neural Networks (RNNs), crafted for the nuanced task of sequence data modeling and prediction. Their edge over conventional RNNs lies in their adept handling of long-term dependencies.

At the heart of LSTM networks is the concept of a "memory cell", a sophisticated construct engineered to capture and utilize the long-term interdependencies found throughout sequences. As depicted in Fig. 3, this memory cell is fortified with an array of gates: the input gate, forget gate, and output gate. These gates act as meticulous gatekeepers, dynamically managing the information flow—deciding in real-time what to retain, discard, or pass forward.

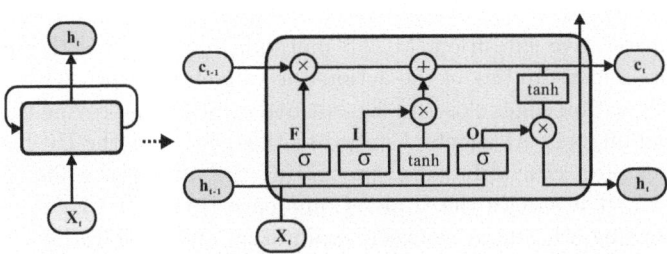

Fig. 3. Schematic structural diagram of the LSTM neural network, where X_t represents the input at time t, h_t represents the hidden state at time t, c_t represents the hidden state at time t, F represents the forget gate, and I represents Input gate (input gate), O represents output gate (output gate).

Within the LSTM framework, the input gate orchestrates the flow of new data, the forget gate decides the fate of existing memories, and the output gate selects the bits of information to be relayed to subsequent layers or earmarked for predictions. Furthermore, LSTMs utilize an internal cell state, which serves as a dedicated reservoir for holding onto information deemed worthy of long-term recall.

LSTMs excel at handling lengthy sequences, adeptly navigating around the pitfalls of gradient vanishing or explosion that plague traditional RNNs. This ability to capture long-term dependencies across sequences makes LSTMs indispensable for a range of tasks, from crafting language models and translating languages to processing auditory signals and forecasting temporal patterns.

4 Problem Modeling and Assumptions

4.1 Bitcoin System Model

The framework operates under the assurance that foundational cryptographic mechanisms (e.g., public and private keys, digital signatures) remain robust. The Public Main Chain PMC_t is defined as $\{B_0^{PMC_t}, B_1^{PMC_t}, B_2^{PMC_t}, \ldots, B_{latest}^{PMC_t}\}$, representing, at any given moment t, the consensus-driven, singular chain recognized by the network's honest nodes. Here, $B_h^{PMC_t}$ denotes the block positioned at height h within the PMC_t chain, with $B_0^{PMC_t}$ marking the network's genesis block. For an arbitrary node i, its local ledger at time t is rendered as $PC_t^i = \{B_0^{i_t}, B_1^{i_t}, B_2^{i_t}, \ldots, B_{latest}^{i_t}\}$, where $B_h^{i_t}$ signifies the block at height h on the PC_t^i chain. For every honest node j, PC_t^j aligns perfectly with PMC_t, while for every malicious node k, PMC_t is a subset of PC_t^k.

4.2 Bitcoin Network Framework

We explore a Bitcoin peer-to-peer network with N nodes, assuming that all nodes are seamlessly interconnected for communication and broadcasting using gossip protocols, where transmission delay is negligible.

4.3 Adversary Agent Model

The narrative posits that the Total Hash Computing Power (THCP) is segmented such that adversary agent A commands a portion of this power, quantified as $p_A \times THCP$. Here, p_A reflects A's share of hash power relative to the entire network's capacity.

Adopting the reinforcement learning perspective for defining agent models, our discussion focuses on framing the adversary agent of AutoMiner through World State and Actions. This approach does not delve into AutoMiner's strategic plans, rewards, or value functions, as AutoMiner transcends conventional reinforcement learning models.

At time t, adversary agent A is characterized by: $World\ State_t^A = (PMC_t, PC_t^A)$, $Actions_t^A \in \{\text{Broadcast}(bclist_t^A), \text{Mine}(minelist_t^A), \text{Sleep}\}$, delineating the three actionable paths—$\text{Broadcast}()$, $\text{Mine}()$, and Sleep—that a node can take, as further explored below:

- $\text{Broadcast}()$: With the parameter $bclist_t^A \subseteq \{B_k^{A_t}|$ for any k such that $B_k^{A_t} \in PC_t^A$ and remains un-broadcasted $\}$, $\text{Broadcast}(bclist_t^A)$ indicates that agent A selects from its private chain PC_t^A a random set of blocks (i.e., $bclist_t^A$) that haven't yet been broadcasted, and sends them out into the network.
- $\text{Mine}()$: The parameter $minelist_t^A = \{(B_k^{A_t}, r_k^{A_t})|$ for any k such that $B_k^{A_t} \in PC_t^A$, and the sum of $r_k^{A_t}$ equals $1\}$, where $B_k^{A_t}$ identifies the block chosen for action, and $r_k^{A_t}$ specifies the portion of computational power allocated. $\text{Mine}(minelist_t^A)$ reveals that agent A selects blocks $B_k^{A_t}$ from its private chain PC_t^A, dedicating a portion of its computational power (calculated as $r_k^{A_t} \times p_A \times THCP$) to mine new blocks on top of $B_k^{A_t}$.
- Sleep: Agent A opts for inaction, standing by without engaging in any activities.

5 AutoMiner Design

5.1 Main Algorithm

AutoMiner's core algorithm unfolds as described here. It begins with the MMCTS algorithm orchestrating simulations to generate a steady stream of data (world states) and corresponding labels (action sets for those states) crucial for neural network training. Following this, the DataFilter function sifts through the raw data to isolate high-reward nuggets for pre-processing via the PreProcess function. This refined data then feeds into training AutoMiner's Neural Network (NN). The culmination of this process involves testing the model in a controlled environment, yielding performance insights. Visual representations of these stages, excluding testing, are captured in Fig. 1, while the testing phase is highlighted in Fig. 7.

Algorithm 1: AutoMiner Main Algorithm

---

while *training samples are insufficient* **do**
    $\lfloor$ Add MMCTS()'s output to $mmcts_{output}$;

$mmcts'_{output}$=PreProcess(DataFilter($mmcts_{output}$));
Use $mmcts'_{output}$ to train $AutoMiner'sNN$;
Test $AutoMiner'sNN$;
return *trained AutoMiner'sNN and test result*

5.2   Algorithm Components

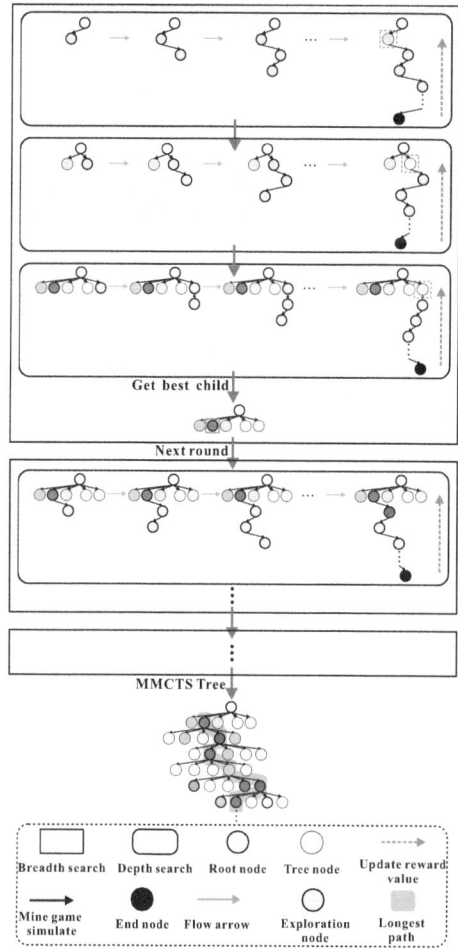

Fig. 4. MMCTS algorithm flow diagram

MMCTS Algorithm: MMCTS is a sophisticated search algorithm tailored for the vast action search space of miner agents, leveraging a tree framework to track nodes' reward values, evolving world states, and their associated actions.

The workflow, depicted in Fig. 4, begins from a current node. Actions are chosen at random to conduct a vertical simulation (a deep search) until the simulation's end, defined by conditions such as reaching a specific depth. Subsequently, the reward values for the offspring of the starting node are recalibrated, capturing the insights gained from the deep search for these child nodes. This cycle of deep searches repeats to explore different "paths" (breadth simulations), accumulating insights across a spectrum of child nodes (indicated by the various shades of blue in Fig. 4). The child node with the highest reward becomes the new focal node, initiating another iteration of this depth-breadth search process.

The parameters and tree node data structure of MMCTS are as follows:

1. MMCTS parameters:
   - Maximum Exploration Width: MAX_EXP_WIDTH
   - Maximum Exploration Depth: MAX_EXP_DEPTH
   - Game Round Limit: $GAME\_ROUND$
   - Miner agent: A
2. Tree node data structures:
   - Value: The average of all simulation rewards obtained from searches with the current node as the root.
   - Info: The world state of this node and the set of actions leading to this node from its parent. For detailed structures of world state and actions, refer to Subsect. 4.3.

The specific details of MMCTS are formalized in Algorithm 2.

Algorithm 2: MMCTS Algorithm

Initialization: $MMCTS\ Tree_0^A$'s $root = \langle\ Value : 0,\ Info : (World\ State_0^A,$
$Actions_0^A)\rangle$, where $World\ State_0^A = (PMC_0, PC_0^A)$ and $Actions_0^A = \{(0, 1)\}$,
Root's children $= \{\}$;

Core algorithm:

Let current node $CurrN = root\ node$;

while $current\ iteration \leq GAME\_ROUND$ **do**

    while $breadth\ search \leq MAX\_EXP\_WIDTH$ **do**

        $World\ State_{new}^A,\ Actions\_list = \texttt{MineGame}(World\ State_{now}^A)$;

        $Child\ Node = \langle Value : 0, Info : (World\ State_{new}^A, Actions\_list)\rangle$;

        $World\ State = World\ State_{new}^A$;

        while $exploration\ depth \leq MAX\_EXP\_DEP\text{-}\ TH$ **do**

            $World\ State_{new}^A,\ Actions\_list = \texttt{MineGame}\ (World\ State)$;

            $World\ State = World\ State_{new}^A$;

        Update the statistical information of the node, i.e., update
        $Child\ Node$'s $Value$;

        $CurrN$ add child $Child\ Node$;

    Select the $Best\ Child$ within $CurrN$'s children based on reward values, and
    set $CurrN = Best\ Child$;

Get the longest path of the search tree;

Output the $Info$ of all nodes on the longest path, and the total rewards of agent A on this path;

Algorithm 3: MineGame Algorithm

Input: $World\ State = World\ State_t^A$
Output: Updated $World\ State$ and $Action\ list$
if random$()\ >\ 0.5$ **then**

> Randomly set $bc_{list} = \{B_k^{A_t} \mid \forall k$ s.t. $B_k^{A_t} \in PC_t^A$, and $B_k^{A_t}$ is un-broadcasted$\}$;
> Broadcast(bc_{list});
> **if** $World\ State_t^A$ changes **then**
>> Update to $World\ State_{(t+1)}^A$;
>> **return** $(World\ State_{(t+1)}^A,\ bc_{list})$;

else

> Randomly set $mine_{list} = \{(B_k^{A_t}, r_k^{A_t}) \mid \forall k$ s.t. $B_k^{A_t} \in PC_t^A$, and $\Sigma(k)r_k^{A_t} = 1\}$;
> Execute Mine$(mine_{list})$, obtain the mining competition result, and update $World\ State = World\ State_{(t+1)}^A$;
> **return** $(World\ State_{(t+1)}^A,\ mine_{list})$;

Data Filter and Pre-process: The data filter meticulously selects valuable data from the plethora of raw data produced by MMCTS simulations, acknowledging that not all data holds value. It employs the simulated reward values as a criterion for this selection process. Based on practical experimentation, AutoMiner adopts a threshold that elevates the expected gains from honest mining by 50%, striking a balance between the neural network's training efficacy and the investment in MMCTS simulations.

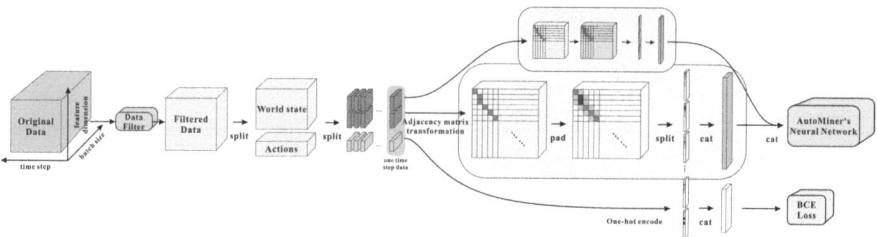

Fig. 5. Data filter and PreProcess processing flow chart

The process of preparing training data, as shown in Fig. 5, begins with breaking down the data into distinct world states and their associated action sets. These world states (PMC and PC) inherently adopt a DAG (Directed Acyclic Graph) configuration. In line with techniques used in graph neural networks, we transform this DAG data into adjacency matrices. For example, the DAG chain illustrated on the left in Fig. 6 is translated into a two-dimensional binary adjacency matrix, capturing all structural nuances of the DAG. In this matrix, a value of 1 at position (i, j) signifies that block i is a descendant of block j,

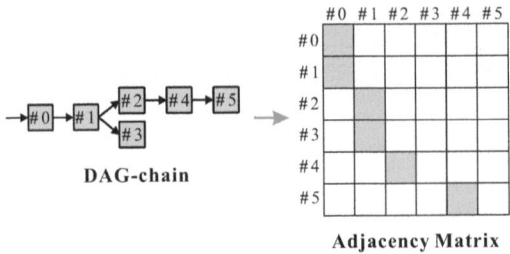

Fig. 6. Transformation from DAG graph to adjacency matrix

while a 0 indicates no such relationship. Given the expansive and sparse nature of the matrix for longer chains, we employ strategic padding to enhance training outcomes. This involves incrementally adjusting the surroundings of 1 s in the matrix (setting the increment at 0.1) and substituting zeroes with the matrix's average value. Following this, we concatenate the adjacency matrix, converting it into an input tensor. For action sets, a straightforward one-hot encoding suffices for broadcasting actions. The specific contents of AutoMiner's data filter and pre-process are formalized in Algorithms 4 and 5.

Algorithm 4: DataFilter algorithm

Input: Original MMCTS's output, $Filter_{Number}$
Output: Filtered MMCTS data
for *each* MMCTS*'s output* **do**
    if *Agent's reward* > $Filter_{Number}$ **then**
        Keep this MMCTS data record;
    else
        Remove this data record;

AutoMiner's NN Train: As depicted in Fig. 1, AutoMiner's neural network architecture comprises a single-layer LSTM (judged to strike an optimal balance between training expense and testing efficacy), two MLP (Multi-Layer Perceptron) networks, and their corresponding sigmoid and softmax activation functions. Key considerations include: 1) Given that AutoMiner's NN often outputs a sparse 0–1 tensor, the loss calculation undergoes a weighted adjustment. This means amplifying the learning weights for non-zero entries and diminishing those for zeros, fostering quicker network convergence. 2) The actions influencing the world state typically oscillate between mining and broadcasting, rendering the label data as tensors peppered with meaningless zeros. These zeros, indicating whether the action is mining or broadcasting, could introduce noise into the model if learned indiscriminately, dampening performance. To circumvent this,

---

Algorithm 5: `PreProcess` algorithm

Input: Filtered `MMCTS` data
Output: Processed `MMCTS` data
foreach *filtered* `MMCTS` *data* **do**
> Split input data according to dimensions of World State, Actions, and time step;
> **foreach each** *time step data* **do**
> > Perform the following operations sequentially on both *PMC* and *PC* in the *World State*: convert to adjacency matrix, fill the matrix, split and Concatenate;
> > One-hot encode the broadcast list in Actions;

return *Processed data*

---

we introduce an action masking strategy during loss computation (as shown in Fig. 1), focusing solely on the loss values for meaningful, non-zero actions and ensuring they benefit from backpropagation. The specific content of the training process is formalized in Algorithm 6.

Algorithm 6: AutoMiner's NN Train

Input: Processed `MMCTS` data, number of epochs, initial states h_t and c_t
Output: AutoMiner's Neural Network (NN)
foreach *epoch* **do**
> $(training\ data, label\ data) = MMCTS_{output}$;
> $output = AutoMiner's\ NN(training\ data)$;
> Create an appropriate weight matrix for *label data*;
> Apply mask to output based on *label data*;
> backward `BCEloss`$((output, label\ data), weight)$;

return *AutoMiner's NN*

AutoMiner's NN Test: Figure 7 demonstrates AutoMiner's NN in a test scenario where it adapts to changes in the world state through actions affecting the simulation, engaging in a cycle of decision and impact until the test concludes. The environment is a virtual blockchain mining field with honest miners, adversaries (e.g., engaging in selfish mining), and AutoMiner competing for rewards indicative of their strategies.

At the heart of the test phase is the Switch module, enabling AutoMiner's NN to choose between broadcasting and mining without backpropagation learning. This unit facilitates strategic decisions, shifting from broadcasting to mining based on action outcomes, without preconceptions of blockchain attacks. The AutoMiner agent is thus defined by the synergy of its NN and the Switch module, emphasizing tactical action within the simulated mining contest. The specific content of the test is formalized in Algorithm 7.

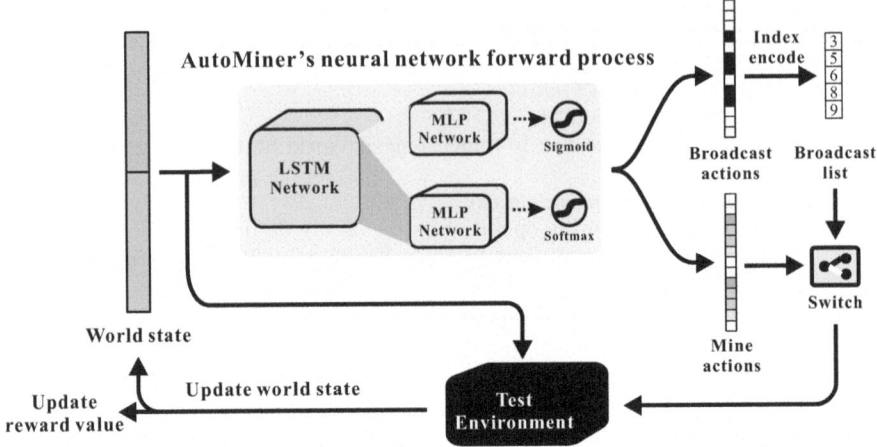

Fig. 7. AutoMiner test flow chart

Algorithm 7: AutoMiner's NN Test

Input: AutoMiner's NN, initial world state, initial h_t and c_t, test rounds
Output: Reward value obtained by Agent A in the test environment
while *test rounds not reached* **do**

$Lstm_{output}, (h_t, c_t) = \text{LSTM}(world\ state, (h_t, c_t))$;
$Tmp\ bc_{actions} = \text{MLPNN}_1\ (Lstm_{output})$;
$bc_{actions} = \text{Sigmoid}(Tmp\ bc_{actions})$;
$Tmp\ mine_{actions} = \text{MLPNN}_2(Lstm_{output})$;
$mine_{actions} = \text{Softmax}(Tmp\ mine_{actions})$;
$Actions = \text{Switch}(bc_{actions}, mine_{actions})$;
$World\ state = \text{Test}(Actions, World\ state)$;

return *the reward value for Agent A from the test environment records*

---

6   Experiments

6.1   Experimental Setup

In this experiment, we selected a PC equipped with an Intel(R) Core(TM) i5-9500 CPU @3.00 GHz as the experimental platform. This environment was chosen to showcase one of AutoMiner's key features: its low resource requirements. The device operates on the Ubuntu 20.04 operating system. On the software side, we built AutoMiner's structure using the PyTorch framework and wrote simulator code to evaluate the effects of honest mining and selfish mining strategies.

6.2   Training Parameters

In the AutoMiner experiment design, we fine-tuned three crucial parameters: Maximum Round Number, Computation Budget, and Game Maximum Rounds, to ensure deep and efficient exploration without overfitting. The Maximum Round Number is capped at 15, guiding the depth of Monte Carlo Tree Search (MCTS) for mining strategy exploration. With a Computation Budget of 100, we permit extensive breadth exploration per game round. Lastly, the Game Maximum Rounds, set at 7, bound each simulation's time frame to sharpen focus and prevent overfitting. These settings collectively streamline the learning and simulation of mining strategies.

In training AutoMiner, we optimized key parameters to enhance neural network learning and simulation accuracy. We set the iterations to 200, allowing for extensive weight adjustments. A batch size of 32 was chosen to balance effective training with memory constraints, preventing issues from oversized batches. The LSTM time step was fixed at 32 to enable the capture of long-term dependencies crucial for blockchain dynamics. With 100 nodes, we aimed for a medium-sized network simulation, facilitating network effects study and experimental control. Local and global mining powers were set to 35 and 65, respectively, creating a competitive environment for AutoMiner to optimize returns under computational limits. These adjustments aimed at optimizing model learning efficiency and simulation precision, targeting optimal training outcomes.

6.3   Experiments Result and Analysis

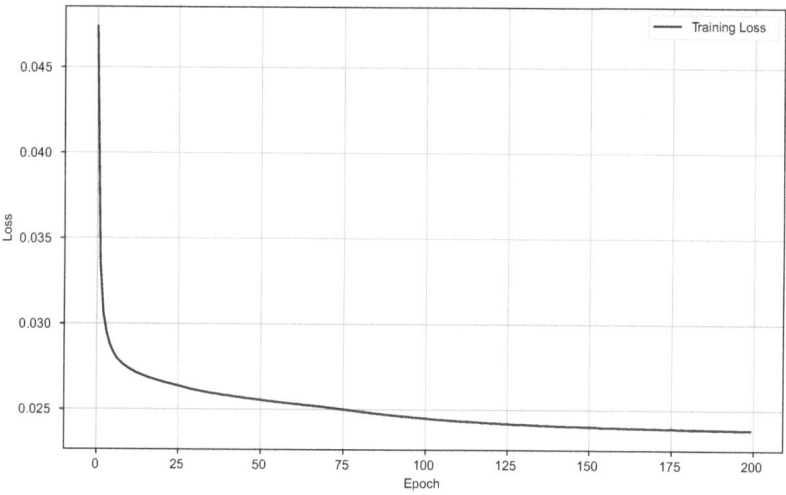

Fig. 8. Training loss of AutoMiner

The Training Process of AutoMiner. Figure 8 shows the loss curve of AutoMiner during the supervised learning training process. In the early stages of training, the total loss decreases rapidly, indicating that the model can quickly learn from the data and improve its predictions. As the number of iterations (epochs) increases, the loss curve stabilizes, gradually approaching a lower loss value. This indicates that the model begins to converge, and the improvements from new learning cycles become minimal. After about 50 epochs, the change in loss value is not significant, showing a relatively flat trend. This suggests that the model may have reached its performance bottleneck under the current configuration and dataset.

Figure 9 displays the reward curve of AutoMiner during the reinforcement learning phase, reflecting the performance improvement of the model with each training epoch. Initially, the reward values fluctuate significantly, but over time, they show an upward trend, suggesting that the model is gradually optimizing its mining strategies. The curve stabilizes and begins to rise noticeably after about 45 epochs, indicating that the model is starting to master effective strategies.

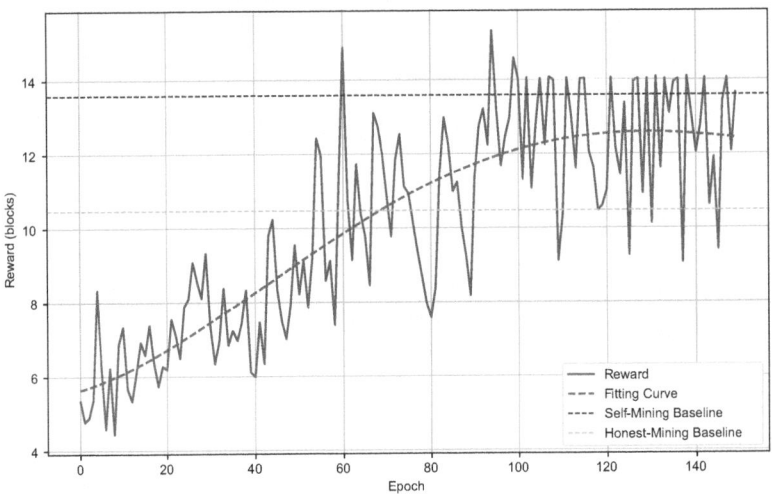

Fig. 9. Comparison between AutoMiner's mining rewards and baseline.

Despite ongoing fluctuations, AutoMiner's performance exceeds the honest mining baseline in most cases and later even surpasses the selfish mining baseline. This demonstrates that AutoMiner, through continuous learning and strategy adjustment, has gained a competitive edge in mining activities, proving the effectiveness and adaptability of its strategies.

Comparison of AutoMiner Earnings. In Table 1, we compare the average and maximum earnings of AutoMiner with those of selfish mining and honest mining across 100 simulated cases. The data shows that AutoMiner has an average earning of 12.23 and a maximum earning of 15.87. In contrast, selfish mining

Table 1. Comparison of Earnings

methods	average	maximum
AutoMiner	12.23	15.87
selfish mining	13.59	14.24
honest mining	10.47	11.39

has a slightly higher average earning of 13.59, but a lower maximum earning of 14.24 compared to AutoMiner. Honest mining performs relatively lower, with average and maximum earnings of 10.47 and 11.39, respectively.

This indicates that while AutoMiner's average earnings are slightly lower than a purely selfish mining strategy, it can achieve higher maximum returns under certain conditions due to its ability to adopt more flexible strategies in response to varying network conditions and incentive mechanisms. For instance, under specific input scenarios, AutoMiner may opt for stubborn mining over selfish mining, resulting in greater potential rewards. Notably, we did not provide AutoMiner with any prior mining strategy hints (such as selfish mining or stubborn mining). This demonstrates that AutoMiner can adaptively adjust its mining strategies based on environmental changes (e.g., chain views, computational power) to maximize returns.

7 Conclusion

In this paper, we present AutoMiner, a novel approach that utilizes Miner Monte Carlo Tree Search in combination with Long Short-Term Memory networks to improve the simulation and optimization of mining strategies in PoW blockchains. This method addresses the limitations of traditional techniques in accurately simulating complex attacks and can be widely applied to identify potential security vulnerabilities in PoW consensus algorithms. Our prototype experiments demonstrate that AutoMiner outperforms existing strategies, achieving profit increases of up to 20%, and, in some cases, providing better returns than selfish mining.

To build on these results, future work will include conducting additional performance and scenario testing to validate AutoMiner's effectiveness and robustness in various blockchain environments. We also plan to expand AutoMiner to support detection and analysis of other types of consensus protocols in blockchain systems, such as Delegated Proof of Stake (DPoS) and Byzantine Fault Tolerance (BFT), further enhancing its adaptability.

References

1. Cryptocurrency market capitalizations (2020). https://coinmarketcap.com/. Accessed 12 May 2024
2. Bai, Q., Xu, Y., Liu, N., Wang, X.: Blockchain mining with multiple selfish miners. IEEE Trans. Inf. Forensics Secur. (2023)
3. Bai, Q., Zhou, X., Wang, X., Xu, Y., Wang, X., Kong, Q.: A deep dive into blockchain selfish mining. In: ICC 2019-2019 IEEE International Conference on Communications (ICC), pp. 1–6. IEEE (2019)
4. Biais, B., Bisiere, C., Bouvard, M., Casamatta, C.: The blockchain folk theorem. Rev. Financ. Stud. **32**(5), 1662–1715 (2019)
5. Böhme, R., Christin, N., Edelman, B., Moore, T.: Bitcoin: economics, technology, and governance. J. Econ. Perspect. **29**(2), 213–238 (2015)
6. Castro, M., Liskov, B., et al.: Practical byzantine fault tolerance. OsDI **99**, 173–186 (1999)
7. Chitra, T., Quaintance, M., Haber, S., Martino, W.: Agent-based simulations of blockchain protocols illustrated via Kadena's chainweb. In: 2019 IEEE European Symposium on Security and Privacy Workshops (EuroS&PW), pp. 386–395. IEEE (2019)
8. Cocco, L., Tonelli, R., Marchesi, M.: An agent based model to analyze the bitcoin mining activity and a comparison with the gold mining industry. Future Internet **11**(1), 8 (2019)
9. Coulom, R.: Efficient selectivity and backup operators in Monte-Carlo Tree Search. In: van den Herik, H.J., Ciancarini, P., Donkers, H.H.L.M.J. (eds.) CG 2006. LNCS, vol. 4630, pp. 72–83. Springer, Heidelberg (2007). https://doi.org/10.1007/978-3-540-75538-8_7
10. Eyal, I., Gencer, A.E., Sirer, E.G., Van Renesse, R.: {Bitcoin-NG}: a scalable blockchain protocol. In: 13th USENIX Symposium on Networked Systems Design and Implementation (NSDI 2016), pp. 45–59 (2016)
11. Eyal, I., Sirer, E.G.: Majority is not enough: bitcoin mining is vulnerable. Commun. ACM **61**(7), 95–102 (2018)
12. Feng, C., Niu, J.: Selfish mining in ethereum. In: 2019 IEEE 39th International Conference on Distributed Computing Systems (ICDCS), pp. 1306–1316. IEEE (2019)
13. Gobel, J., Keeler, H.P., Krzesinski, A.E., Taylor, P.G.: Bitcoin blockchain dynamics: the selfish-mine strategy in the presence of propagation delay. Perform. Eval. **104**, 23–41 (2016)
14. Hou, C., et al.: SquirRL: automating attack analysis on blockchain incentive mechanisms with deep reinforcement learning. arXiv preprint arXiv:1912.01798 (2019)
15. Kaligotla, C., Macal, C.M.: A generalized agent based framework for modeling a blockchain system. In: 2018 Winter Simulation Conference (WSC), pp. 1001–1012. IEEE (2018)
16. Kwon, J., Buchman, E.: Cosmos whitepaper. Netw. Distrib. Ledgers **27**, 1–32 (2019)
17. Li, Q., Chang, Y., Wu, X., Zhang, G.: A new theoretical framework of pyramid Markov processes for blockchain selfish mining. J. Syst. Sci. Syst. Eng. **30**, 667–711 (2021)
18. Li, T., Wang, Z., Yang, G., Cui, Y., Chen, Y., Yu, X.: Semi-selfish mining based on hidden Markov decision process. Int. J. Intell. Syst. **36**(7), 3596–3612 (2021)

19. Madhushanie, N., Vidanagamachchi, S., Arachchilage, N.: BA-flag: a self-prevention mechanism of selfish mining attacks in blockchain technology. Int. J. Inf. Secur., 1–10 (2024)
20. Nayak, K., Kumar, S., Miller, A., Shi, E.: Stubborn mining: generalizing selfish mining and combining with an eclipse attack. In: 2016 IEEE European Symposium on Security and Privacy (EuroS&P), pp. 305–320. IEEE (2016)
21. Nikhalat-Jahromi, A., Saghiri, A.M., Meybodi, M.R.: VDHLA: variable depth hybrid learning automaton and its application to defense against the selfish mining attack in bitcoin. arXiv preprint arXiv:2302.12096 (2023)
22. Peterson, M., Andel, T., Benton, R.: Towards detection of selfish mining using machine learning. In: International Conference on Cyber Warfare and Security, vol. 17, pp. 237–243 (2022)
23. Poon, J., Buterin, V.: Plasma: scalable autonomous smart contracts. White paper, pp. 1–47 (2017)
24. Poon, J., Dryja, T.: The bitcoin lightning network: scalable off-chain instant payments (2016)
25. Sapirshtein, A., Sompolinsky, Y., Zohar, A.: Optimal selfish mining strategies in bitcoin. In: Financial Cryptography and Data Security: 20th International Conference, FC 2016, Christ Church, Barbados, 22–26 February 2016, Revised Selected Papers 20, pp. 515–532. Springer, Cham (2017). https://doi.org/10.1007/978-3-662-54970-4_30
26. Silver, D., et al.: Mastering the game of go without human knowledge. Nature **550**(7676), 354–359 (2017)
27. Struchkov, I., Lukashin, A., Kuznetsov, B., Mikhalev, I., Mandrusova, Z.: Agent-based modeling of blockchain decentralized financial protocols. In: 2021 29th Conference of Open Innovations Association (FRUCT), pp. 337–343. IEEE (2021)
28. Wang, T., Liew, S.C., Zhang, S.: When blockchain meets AI: optimal mining strategy achieved by machine learning. Int. J. Intell. Syst. **36**(5), 2183–2207 (2021)
29. Wang, Y., et al.: A detection method against selfish mining-like attacks based on ensemble deep learning in IoT. IEEE Internet Things J. (2024)
30. Wang, Z., Lv, Q., Lu, Z., Wang, Y., Yue, S.: ForkDec: accurate detection for selfish mining attacks. Secur. Commun. Netw. **2021**, 1–8 (2021)
31. Wang, Z., Liu, J., Wu, Q., Zhang, Y., Yu, H., Zhou, Z.: An analytic evaluation for the impact of uncle blocks by selfish and stubborn mining in an imperfect ethereum network. Comput. Secur. **87**, 101581 (2019)
32. Wood, G.: Polkadot: vision for a heterogeneous multi-chain framework. White Paper **21**(2327), 4662 (2016)
33. Wood, G., et al.: Ethereum: a secure decentralised generalised transaction ledger. Ethereum Project Yellow Paper **151**(2014), 1–32 (2014)
34. Xu, J., Wang, C., Jia, X.: A survey of blockchain consensus protocols. ACM Comput. Surv. **55**(13s), 1–35 (2023)
35. Yu, H., Nikolić, I., Hou, R., Saxena, P.: OHIE: blockchain scaling made simple. In: 2020 IEEE Symposium on Security and Privacy (SP), pp. 90–105. IEEE (2020)
36. Zamani, M., Movahedi, M., Raykova, M.: RapidChain: scaling blockchain via full sharding. In: Proceedings of the 2018 ACM SIGSAC Conference on Computer and Communications Security, pp. 931–948 (2018)
37. Zhou, L., et al.: Sok: decentralized finance (DeFi) attacks. In: 2023 IEEE Symposium on Security and Privacy (SP), pp. 2444–2461. IEEE (2023)
38. Zur, R.B., Eyal, I., Tamar, A.: Efficient MDP analysis for selfish-mining in blockchains. In: Proceedings of the 2nd ACM Conference on Advances in Financial Technologies, pp. 113–131 (2020)

FedSV: A Privacy-Preserving Byzantine-Robust Federated Learning Scheme with Self-validation

Wenhao Jiang[✉], Shaojing Fu[✉], Yuchuan Luo, Lin Liu, and Yongjun Wang

College of Computer, National University of Defense Technology, Changsha, China
{jwh_roy,fushaojing,luoyuchuan09,liulin16,wangyongjun}@nudt.edu.cn

Abstract. In recent years, Federated Learning (FL) has been making significant progress in many areas. Meanwhile, with the deepening of the study in FL, researchers found that FL is vulnerable to some attacks. Malicious users and server can bring different privacy and security threats. To counter these attacks, a multitude of strategies have been proposed. One approach to aggregation based on validation datasets involves using datasets to verify the models submitted by participants. The accuracy of the models is then assessed to determine if they are malicious. However, all aggregation schemes based on validation datasets require clients to upload plaintext. While ensuring the robustness of the aggregation process, these solutions pose a serious privacy threat, making them infeasible in practice. For the first time, we consider client privacy in the aggregation scheme that uses validation datasets, allowing us to maintain both the aggregation performance and privacy protection. We investigate how to efficiently incorporate fully homomorphic encryption (FHE) into algorithms that use validation datasets. Specifically, We find the approximate polynomial to replace the nonlinear operations in neural network so that it can fit the homomorphic operations, and we use clustering algorithm to split user groups based on the validation scores to determine which group is malicious. Therefore, we propose FedSV based on CKKS and server self-validation. The experimental results show that FedSV can significantly improve the robustness of FL and the error of accuracy with the baseline is kept within 5.7%, even if there is only one benign user.

Keywords: Federated Learning · Byzantine Robustness · Privacy Protection · Validation Datasets

1 Introduction

Federated learning (FL) [20], which consists of a central server and multiple clients, interacts model parameter rather than dataset information. This approach allows machine learning to train models without the raw data, reducing the risk of data breaches for clients. FL is an inevitable outcome of the digital

© The Author(s), under exclusive license to Springer Nature Singapore Pte Ltd. 2025
T. Zhu et al. (Eds.): ICA3PP 2024, LNCS 15251, pp. 242–260, 2025.
https://doi.org/10.1007/978-981-96-1525-4_13

age, which aims to solve the problem of data silos and protect privacy. However, with the promotion of its application, security and privacy protection capabilities of FL have been shown to be inadequate [13]. There are two main threats to the existing FL framework:

Malicious Clients: The clients in FL have the characteristics of large number, poor protection, and easy to be broken by the adversary. After the attacker controls one or more clients, it can perform Byzantine attacks, such as data poisoning attack (DP) [16] and model poisoning attack (MP) [19]. They upload malicious models to participate in aggregation, which can easily destroy the global model, so that the global model cannot converge or update in the wrong direction. Arguably, malicious client is the biggest vulnerability point in FL [11].

Malicious Server: Uploading model parameters by the client will result in the leakage of the raw data. The server can extrapolate the original information through gradient inversion attack [7] or infer the identities of the participants through member inference attacks [29]. In the latest research, preference profiling attack (PPA) [35] can profile the private preferences of a local user.

In this article, we consider a scenario that the central server has a small clean dataset (0.2% of training dataset). For example, three hospitals A, B, and C hope to shorten the diagnosis time of common diseases by generating a disease prediction model, but the pathology data of patients is private and cannot be shared. If we adopt the typical FL method that encrypting the parameters and sending them to the central server for training, it is necessary to add a participant, that is, the central server D. This increases authentication and communication overhead, as well as designing robust aggregation algorithms to combat Byzantine attacks, sybil Attack [32], and more. The idea based on validation dataset is to use a certain institution, such as A, as the central server to collect the encrypted models from B and C and use its own dataset to verify their effect and determine whether they are malicious updates.

There have been some methods based on validation dataset to resist attacks from malicious clients [4,18,22,24]. These aggregation algorithms verify the models uploaded by participants, and determine whether the update is malicious according to the prediction accuracy of the client's model. However, these approaches do not consider the possibility of malicious server obtaining privacy data of participants. In general, measures to protect data privacy lead to data being invisible, and algorithms that resist Byzantine attacks almost always rely on the characteristics of the data. In the scenario of this article, if clients upload encrypted models, even if there is a dataset on the server, it cannot be used for verification. To solve this problem, we adopted fully homomorphic encryption (FHE) technology that allows addition and multiplication operations on ciphertext. We applied polynomial approximation to the nonlinear operations in neural network training to adapt the operations of FHE.

Our Work: In this work, we propose a novel Byzantine-robust and privacy-preserving FL method called FedSV. We integrate client's privacy protection into the aggregation scheme based on validation datasets for the first time, which enhances Byzantine fault tolerance while protecting privacy. Specifically, we identify an approximate polynomial to replace the nonlinear operations within the neural network, enabling it to be compatible with homomorphic operations. Additionally, we employ a clustering algorithm to categorize user groups according to their validation scores, which helps us identify the malicious groups. The experimental results indicate that FedSV maintains consistency with the baseline accuracy, with an error within 5%, regardless of whether it is under the independent and identically distributed (IID) setting with DP and MP attacks or under the non independent identically distributednon (non-IID) setting with MP attacks. At the same time, FedSV can protect privacy and significantly improve the robustness of FL even if there is only one benign user.

2   Background and Related Work

2.1   Privacy-Preserving Byzantine-Robust Federated Learning

Federated learning's data interaction mechanism ensures clients' original dataset privacy and security, but risks of data leakage exist in other aspects. For instance, gradient inversion schemes [7] infer user dataset information from gradients. Additionally, member inference attacks [29] identify which client uploads specific data. Several attacks on Federated Learning experiments intensify the need to prioritize user privacy and security. There are some classic mainstream solutions proposed to resolve these privacy and security issues. Multi-party secure computing (MPC) was introduced by Bonawitz et al. [3]. This approach enables the aggregation server to perform an aggregation operation while keeping the gradient value unknown. However, the data overhead for communication between client and server is high. Sav et al. [27] proposed Multi-Party Homomorphic Encryption Technology (MHE) that encrypts uploaded client data and allows the server to perform a ciphertext aggregation operation, ensuring privacy and security. Mo et al. [21] introduced the trusted execution environment method (TEE). TEE has low efficiency in aggregation operations due to limited resources in the execution environment. Differential privacy schemes were proposed in literature [28]. Although it enhances security, it negatively impacts the final training model efficacy. These schemes enhance Federated Learning system security to a certain extent, but they have shortcomings and neglect the presence of malicious clients.

In order to defend against poisoning attacks [16,19] and inference attacks [7] [29] simultaneously discussed in our previous chapter, Researchers have devised a lot of Byzantine-robust and privacy-preserving FL algorithms. So et al. [1] proposed BREA based on integrated random quantization, verifiable outlier detection, and secure model aggregation methods, while ensuring Byzantine resilience, privacy, and convergence. SEAR put forward by Zhao et al. [17] relying on a trusted execution environment, Intel SGX. SEAR protects the privacy of clients' models while enabling Byzantine resilience. DisBezant proposed by Ma et al. [15] adopting a trustworthiness based mechanism in non-IID and two-party security calculation. Hu et al. [34]proposed a method that utilizes compressive sensing (CS) as a lightweight encryption to protect data privacy, while maintaining Byzantine robustness using cosine similarity to update the encryption (compression) model, reducing the computational and communication costs significantly.

The ideas behind these methods to defence Byzantine attacks can be summarized as follows: The server can analyze the differences between updates via complex calculations based on the different features of all collected models, and distinguish malicious updates from benign updates.

2.2   Self-validation

To defend against malicious clients, another idea is to use validation dataset to verify the models uploaded by participants, and determine whether the models are malicious according to their prediction accuracy, requiring the server to have a dataset similar to the benign participants. Wang et al. [24] designed a basic method: The server calculates the prediction accuracy of each client's uploaded gradient on the validation dataset. If the accuracy is lower than the set threshold, it is classified as malicious and excluded. Then, the server calculates the average value of the filtered gradients as global model gradient. On this basis, Tan et al. [22] added a deep learning model to the server, which can guide the server to select benign participants for aggregation in the next iteration based on the historical behavior of the clients and the results of this round of validation, reducing training costs. The Zeno aggregation algorithm proposed by Xie et al. [33] uses the validation dataset to calculate the random decrease score for each user, and then takes the average updated model of the top n-m participants with the highest score as the aggregation result. The scheme proposed by Cao et al. [10] takes the training results of the server itself as the trust root, and takes the similarity of model updates between the server and participants as the assessment criteria. For each participant model w, Fang et al. [12] used the validation data set to calculate the validation error rates of the global model aggregated w and non-aggregated w respectively, and took the difference between the two as the participant's score. After removing the first m participants with the highest scores, they adopt other aggregation algorithms to aggregate the remaining participants.

These validation based schemes against Byzantine attacks requiring clients to upload data in plaintext, otherwise, aggregate calculation cannot be completed, and the protection of user data privacy is ignored. Table 1 shows a comparison of FL schemes based on validation datasets.

Table 1. Comparison of FL schemes based on validation datasets.

Defence Scheme	Resisting DP Attack	Resisting MP Attack	Protecting Privacy	Other Setting
Wang [24]	✓	✓	✗	Setting Threshold
Tan [22]	✓	✓	✗	Adding a Model
Xie [33]	✓	✓	✗	Knowing adversaries' number
Cao [10]	✓	✓	✗	–
Fang [12]	✓	✓	✗	Knowing adversaries' number
FedSV, this work	✓	✓	✓	–

3   Preliminaries

3.1   Federated Learning

In a traditional federated learning system, there are several clients and a central server. Each client has part of the data set D_i, $i \in [1, 2, ..., n]$, that is, the total data set can be expressed as $D=\{D_1, D_2, ..., D_n\}$. The user's goal is to get a better trained model w_i without exposing his/her privacy. In a certain iteration round t, each client obtains the global model w^t from the central server and then updates the model gradients with the local data set. The training purpose can be described as the objective function shown in Eq. 1.

$$F(x, w, y) = \min_{w} E_{(x,y) \sim D} L(x, w; y) \tag{1}$$

where x and y are training data and labels respectively, w is the model parameters, and $L(x, w; y)$ is the loss function. The client C_i sends the local model w_i to the central server, which aggregates the models of all participants through Eq. 2 to get the global model:

$$w^t = \sum_{i=1}^{n} \zeta_i w_i \tag{2}$$

where $\zeta_i = \frac{|D_i|}{|D|}$, $\sum_{i=1}^{n} \zeta_i = 1$ and $|D|$ means the number of the dataset D's samples. The central server sends the global model w^t to all clients, and each client updates the local model $w_i^{t+1} = w^t - \eta \times g_i^t$, where η is the learning rate of local training and g_i^t is the gradient of w^t under D_i.

3.2   A Fully Homomorphic Encryption Scheme: CKKS

The CKKS (Cheon-Kim-Kim-Song) homomorphic encryption [5] is a fully homomorphic encryption scheme [30] that allows arbitrary computations on ciphertexts, yielding correct results upon decryption. It consists of the following 7 parts: **Setup, Key Generation, Encryption (Enc), Decryption (Dec),**

Addition, Multiplication, Rescaling. Homomorphic encryption consists of five parts:

- $(pk, sk) \leftarrow$ KenGen : Select a cyclotomic polynomial $\Phi_N(X)$ of degree N, a large modulus q, and a small modulus Δ. Randomly sample a secret key $s \in \mathbb{Z}_q$ from an appropriate distribution. Compute $a = \Delta \cdot s \mod q$ and $b = \Phi_N(a)$. The public key $pk = (a, b)$. The secret key $sk = s$.
- $\text{ct} \leftarrow \text{Enc}(pk, m)$: For plaintext $m \in \mathbb{C}$, sample two noises $e_0, e_1 \in \mathbb{C}^N$ from a discrete Gaussian distribution. Compute $c_0 = \langle a, m \rangle + e_0 \mod q$ and $c_1 = \langle b, m \rangle + e_1 \mod q$, where $\langle \cdot, \cdot \rangle$ denotes the inner product. The ciphertext is $\text{ct} = (c_0, c_1)$.
- $m \leftarrow \text{Dec}(sk, \text{ct})$: Compute $\beta = \Delta \cdot s \mod q$. For each coefficient, compute $m = \frac{c_0 - c_1 \cdot \beta}{2} \mod q$.
- $\text{ct}_{\text{sum}} \leftarrow \text{Addition}(\text{ct}_1, \text{ct}_2)$: Given two ciphertexts $\text{ct}_1 = (c_{0,1}, c_{1,1})$ and $\text{ct}_2 = (c_{0,2}, c_{1,2})$, the homomorphic addition is performed as follows: Compute $c_{0,\text{sum}} = (c_{0,1} + c_{0,2}) \mod q$. Compute $c_{1,\text{sum}} = (c_{1,1} + c_{1,2}) \mod q$. The resulting ciphertext is $(c_{0,\text{sum}}, c_{1,\text{sum}})$.
- $\text{ct}_{\text{prod}} \leftarrow \text{Multiplication}(\text{ct}_1, \text{ct}_2)$: Given two ciphertexts $\text{ct}_1 = (c_{0,1}, c_{1,1})$ and $\text{ct}_2 = (c_{0,2}, c_{1,2})$, the homomorphic multiplication is performed as follows: Compute $c_{0,\text{prod}} = (c_{0,1} \cdot c_{0,2}) \mod q$. Compute $c_{1,\text{prod}} = (c_{1,1} \cdot c_{1,2}) \mod q$. The resulting ciphertext is $\text{ct}_{\text{prod}} = (c_{0,\text{prod}}, c_{1,\text{prod}})$.

The CKKS scheme supports homomorphic addition and multiplication, enabling encrypted data to be processed without decryption. The rescaling operation is also a key feature that manages noise growth, maintaining the balance between operational depth and decryption error.

3.3 Neural Networks

Neural networks abstract human brain neural networks from the perspective of information processing, establish some simple model, and form different networks according to different connection methods. Two kinds of neural networks are used in this paper: MLP and AlexNet. We need to know their internal structure to figure out how to make predictions after encrypt them. The simplest M-P neuron can be expressed as:

$$y = f(\sum_{i=1}^{n} w_i \cdot x_i + b) \tag{3}$$

where w_i means the weight of neuron, x_i is input and b is bias. y is output for this neuron and input for the next neuron. The activation function $f(\cdot)$ is used to map the function values to $[0, 1]$ or $[-1, 1]$.

MLP [26] has an input layer, some hidden layers, and an output layer. They are all consist of the simplest units of neuron. As long as we can modify the non-linear activation function $f(\cdot)$ to be a polynomial of addition and multiplication, we can perform homomorphism operations on the encrypted network.

AlexNet [14] consists of 5 convolution layers and 3 fully connected layers, including 3 pooling layers. Max pooling finds the maximum value in the matrix, which is unfriendly to homomorphism operations. It is difficult to transfer the max pooling, so we choose the average pooling as a substitute, which has been proven that this substitution has little impact on model accuracy [31].

4   Problem Setup

4.1   Threat Model

FedSV aims to build a FL system that protects privacy and can resist Byzantine attacks, so it is assumed that both the client and server are malicious but not collude. We assume that KGC (Key Generation Center) is an honest and trustworthy third party. For S (Server), we set it to be honest and curious, meaning that it will execute the protocol correctly but infer the raw data by observing the models uploaded by C (Client). This is consistent with the assumption in the literature [7,25]. For the clients, attackers have the following knowledge about FL systems: local training data and local models, loss functions, and learning rates [2,6]. We consider full-knowledge setting to show that our method can defend against strong attacks.

4.2   System Model

FedSV contains three entities, which are KGC, S and C. They mainly serve the following functions:

- KGC : KGC is responsible for generating and distributing keys, and goes offline after completing the task.
- S : S verifies the encryption models of the clients and distinguishes malicious users, ensuring that only benign user models are aggregated. S attempts to recover raw data from the models uploaded by the clients and steals privacy.
- C : C has local datasets and trains local models. Benign clients want to benefit from the global model and get a better performing model, while malicious clients try to destroy or control the global model.

4.3   Defense Goals

Our goal is to design a FL method that can protect privacy and resist Byzantine attacks without sacrificing fidelity and excessive efficiency. In particular, we use FedAvg [20] under no attacks as the baseline for discussing fidelity. We set the novel scheme FLTrust [10] proposed by Cao et al. as the baseline for discussing robustness because it does not require any other settings compared to other solutions, that is, our method should be robust to malicious clients and similar to FLTrust, be secure to malicious server, and be as accurate as FedAvg under no attacks. Specifically, we need to achieve the following objectives:

- **Security**. The method should protect privacy of clients and prevent malicious server from stealing user's raw data.

- **Robustness**. The method should distinguish most malicious clients and eliminate their negative impacts. We compare FLTrust with it.
- **Fidelity**. The method should not reduce the prediction accuracy under normal circumstances. We compare FedAvg with it.

5   Our FedSV

5.1   Overview

FedSV balances privacy preserving and Byzantine robustness. Specifically, clients encrypt and upload their models. Server implements prediction of image labels on encrypted neural network models to obtain scores for each client. Then the server performs binary clustering based on the scores and aggregates on benign user groups to update global model (c.f. Fig. 1).

5.2   Details of FedSV

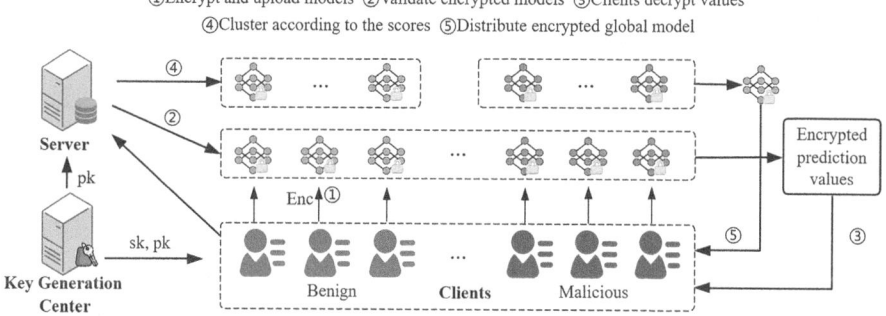

Fig. 1. System flow chart of FedSV.

FedSV includes four steps: Key distribution and system initialization, Local training and uploading of encrypted models, Server self-validation process, and Global model distribution. We use FHE functions introduced in Sect. 3.2. We now describe the above process in detail.

- **Key distribution and system initialization:** This stage involves key generation and distribution, as well as setting global parameters. KGC chooses a security parameter λ, generates pk and sk using KenGen function, and distribute pk and sk to all clients C. S Generates a randomly initialized model w and sets the learning rate η, the number of local training rounds E and batch size B. Then send these parameters to C.

Algorithm 1: FedSV

Input: C_i with local datasets D_i, $i = 1, 2, ..., n$, public key pk and private key sk; S with dataset D_s; global training epochs T; local training epochs E; local learning rate η; batch size B.

Output: Global encrypted model $[[w]]^{t+1}$.

1   $w \leftarrow random\ initialization$ //Server side;

2 **for** t *from* 1 *to* T **do**

3   //The server sends the current round of encrypted $[[w]]^t$ to clients;

4   //Client side;

5   **for** i *from* 1 *to* n **do**

6        $w^t \leftarrow Dec([[w]]^t, sk)$;

7        $w_i^0 \leftarrow w^t$;

8   **for** e *from* 1 *to* E **do**

9   Randomly sample a batch D_B from D_i;

10           $w_i^e = w_i^{e-1} - \eta \nabla loss(D_B; w_i^{e-1})$;

11   **end**

12        $[[w]]_i^{t+1} \leftarrow Enc(w_i^E, sk)$;

13   Send $[[w]]_i^{t+1}$ to S;

14   **end**

15   //Server side;

16   **for** i *from* 1 *to* n **do**

17        $[[Value_i]] \leftarrow Validate([[w]]_i^{t+1}, D_s)$;

18   Send $[[Value_i]]$ to C_i;

19        $Value_i \leftarrow Dec([[Value_i]], sk)$ //Client side;

20   Send $Value_i$ to S;

21        $Score_i \leftarrow$ Evaluate the accuracy of $Value_i$ //Server side;

22   **end**

23   **Score** $\leftarrow Score_{1,2,...,n}$;

24     $Benign\ C_i\ group \leftarrow max(k\text{-}means(\textbf{Score}, clusters = 2))$;

25     $[[\mathbf{w_b}]]_i^{t+1} \leftarrow [[w]]_i^{t+1}$ of Benign C_i group, $i \in benign\ numbers$;

26     $[[w]]^{t+1} \leftarrow FedAvg([[\mathbf{w_b}]]_i^{t+1})$

27 **end**

- **Local training and uploading of encrypted models:** In the t-th round, C_i uses local dataset D_i to train on w^t, obtaining w^{t+1}. After encrypting w^{t+1}, C_i uploads $[[w]]_i^{t+1} = \text{Enc}(w^{t+1}, pk)$ to S. We mark malicious models as $[[w_m]]_i^{t+1}$ and benign models as $[[w_b]]_i^{t+1}$.

- **Server self-validation process:** For each $[[w]]_i^{t+1}$, S scrambles and records the sample order of the small clean validation dataset on itself to validate the encrypted model and gains a group of encrypted predicted values. We will discuss this process in detail separately in Sect. 5.3. S sends the encrypted values back to the C_i for decryption, and C_i upload their decrypted values to S. We will discuss the security of this process detailly in Sect. 5.4. Then, S evaluates C_i according to their values, gaining a $Score_i$ for each C_i. Using the k-means clustering algorithm, S divides the scores into two categories.

The group with a higher average value of $Score_i$ is labeled as the benign user group $[[\mathbf{w}_b]]_i^{t+1}$. S aggregates them with FedAvg using homomorphic operation and obtains the global model $[[w]]^{t+1}$ of next epoch.

- **Global model distribution:** S sends $[[w]]^{t+1}$ to each C_i. C_i decrypts it with sk, getting $w^{t+1} = Dec([[w]]^{t+1}, sk)$. Then jump to the second step until E rounds training completed.

5.3 Method to Validate Encrypted Model

The prediction process of neural network can be seen as a combination of a series of linear and nonlinear operations. FHE does not support nonlinear operations, so the key to solving the problem is to approximate nonlinear operations as polynomial consisting only of addition and multiplication operations.

In the experiment, we use MLP and Alexnet model, which require approximation of the activation function ReLU. We need to first convert it from a discontinuous piecewise function to a continuous function. Fortunately, in 2017, Ramachandran et al. discovered the Swish activation function [23] that can always replace ReLU (c.f. Fig. 2(a)). ReLU can be expressed as:

$$ReLU(x) = \begin{cases} x, & x \geq 0 \\ 0, & x < 0 \end{cases} \tag{4}$$

And Swish can be expressed as:

$$Swish(x) = \frac{x}{1 + e^{-x}} \tag{5}$$

We define Taylor-n as the n-th expansion of the Swish function. We obtain a series of approximate polynomials by expanding the Taylor expansion of the Swish function at $x = 0$: Taylor-2(x)$= \frac{x}{2} + \frac{x^2}{4}$, Taylor-4(x)$= \frac{x}{2} + \frac{x^2}{4} - \frac{x^4}{48}$, Taylor-6(x)$= \frac{x}{2} + \frac{x^2}{4} - \frac{x^4}{48} + \frac{x^6}{480}$, (c.f. Fig. 2(b)).

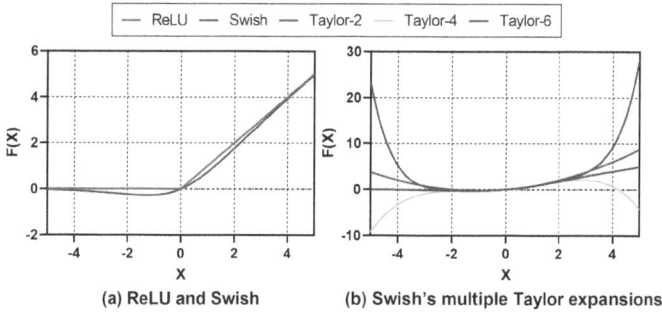

(a) ReLU and Swish   (b) Swish's multiple Taylor expansions

Fig. 2. ReLU, Swish and it's Taylor expansions.

We demonstrate in the experiments that Taylor-2 has higher accuracy and lower time cost compared to other expansions. So we choose Taylor-2 instead of the ReLU function.

5.4 Security of Uploading Decrypted Values

In the server self-validation process, S sends the encrypted values back to the C_i for decryption, and C_i upload their decrypted values to S. We will now introduce the security of this step.

For benign users, S only obtains the predicted label values of their models on the dataset on S, with no additional information. The model remains in ciphertext form, and relying solely on the labels, S cannot perform feature or attribute inference attacks on the client.

For malicious users, we consider DP attacks and MP attacks separately. They are unaware of the label values of the dataset on S, and S shuffles the sample order each time, so they do not know how their uploaded models perform. When an adversary performs a DP attack, he/she intentionally flip certain labels to other labels, which leads to them being unable to distinguish whether a value is its original value or a flipped value after decrypting the predicted label values. For example, if the adversary's model flips "4" to "6," then he/she cannot determine whether the decrypted predicted label value of "6" was originally "4" or "6." Therefore, from S's perspective, an adversary performing a DP attack has a very low accuracy rate on some labels, thereby detecting this type of attack. When an adversary performs an MP attack, he/she expect to disrupt the global model by aggregating random noise models through S. From S's perspective, the models of adversaries performing an MP attack have a consistently low accuracy rate on every label, which are easy to distinguish from benign users.

Meanwhile, adversaries don't have a chance to achieve high accuracy through speculation. Assuming dataset on S has L labels. The probability of malicious user guessing n items correctly is $\frac{1}{L^n}$. In our experiment, the accuracy corresponding to MNIST and Cifar-10 ultimately stabilize at 0.9 and 0.7. For these two datasets, $L = 10$. If we set 100 items on S, the probability of malicious user satisfying accuracy is $\frac{1}{10^{0.9 \times 100}} = 10^{-90}$ and $\frac{1}{10^{0.7 \times 100}} = 10^{-70}$, which is negligible. Because the number of clients and training rounds generally do not reach the level of 10^{70}, even if malicious users control all clients and participate in training every round, they cannot meet the accuracy requirements.

6 Experiments

6.1 Experimental Setup

Datasets: Assuming there are M classes in a dataset and a dataset has N items. We use the following two datasets:

- **MNIST:** MNIST is a 10-class digital image classification dataset consisting of 60000 training samples and 10000 testing samples. For this dataset, $M = 10$, $N = 60000$.
- **Cifar-10:** Cifar-10 is a is a color image classification dataset classification dataset consisting of 50000 training samples and 10000 testing samples. For this dataset, $M = 10$, $N = 50000$.

FL Settings: Assuming we have 10 users, all of whom participate in each round of training. We conduct experiments under independent and identically distributed (IID) and non-independent and identically distributed (non-IID) settings. For the former, we divide the training dataset into 10 groups, with an equal number of items in each group, and the number of items for each label in each group is $\frac{N}{10 \times M}$. Then, these 10 groups are assigned to 10 clients. Each user has $\frac{N}{10}$ data blocks from the training samples, with $\frac{N}{10 \times M}$ for each label. For the latter, we assign each user a portion of the dataset corresponding to $\frac{M}{5}$ to $\frac{4M}{5}$ labels, with the number of samples for each label randomly ranging from $\frac{N}{10 \times M}$ to $\frac{N}{M}$. On our server, there are 100 test samples, with 10 samples for each label.

Poisoning Attacks: We evaluate the Byzantine robustness of algorithms by assessing the prediction accuracy under data poisoning (DP) and model poisoning (MP). For DP, we use the classic label flipping attack, flipping label l to $M - l - 1$. We flip half labels: $l = 0, 1, 2, 3, 4$, reversing the labels to 5,6,7,8,9. For MP, we use random gradient attack, directly generating random noise as model gradient without training. We conducted experiments on 0, 3, 6, and 9 malicious users out of 10 users.

Baseline Settings: To compare the effectiveness of algorithms under no attack, we use FedAvg [20] as the benchmark for comparison. However, in the scenario described in this article, there is a small amount of dataset on the server, which is different from the settings on the FedAvg. So we set FLTrust as the benchmark, which takes the training results of the server itself as the trust root, and takes the similarity of model updates between the server and participants as the assessment criteria. As expected, FedSV should have comparable accuracy to FLTrust in any attack setting.

6.2   Experimental Results

Taylor-2 of Swish Can Replace ReLU: In order to meet the requirements of FHE operations, we must transform the nonlinear operations in neural networks into polynomial. The AlexNet model and MLP model we use both adopt the activation function ReLU, which is similar to Swish. We conduct comparative experiments on the second-order, fourth-order, and sixth-order Swish Taylor expansions. We used Taylor expansions of different orders as activation functions to train 50 rounds on the MNIST and Cifar-10 and found that the accuracy of Taylor-2 is higher than others (c.f. Fig. 3).

At the same time, it can be intuitively assumed that Taylor-2 consumes the least amount of time compared to Taylor-4 and Taylor-6, as it lacks some higher-order computations. From these results, it can be determined that in the different Taylor expansions of Swish, we choose Taylor-2 instead of the Swish function.

Next, we compare the training effects of ReLU, Swish, and Taylor-2 as activation functions for model training. We conduct experiments on FedAvg and FedSV

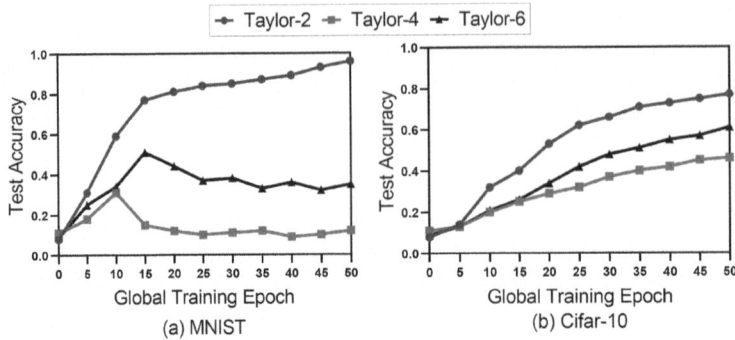

Fig. 3. Comparison of prediction accuracy for different Swish Taylor expansions.

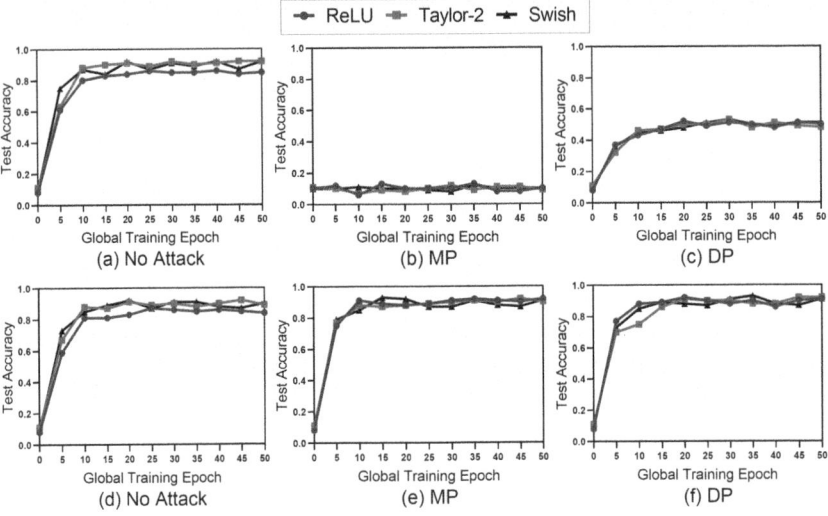

Fig. 4. The top three images generated by FedAvg algorithm, and the bottom three generated by FedSV algorithm. Training with MLP model on MNIST, using ReLU, Swish, and Taylor-2 as activation functions, and comparing prediction accuracy under no attack, MP, and DP with 90% malicious clients.

respectively. We set the number of malicious users to 9 under no attack, MP and DP separately (c.f. Fig. 4). In any algorithm, Taylor-2 substitution exhibits similarity and even surpasses ReLU in some cases. In Fig. 4(a), replacing ReLU with Taylor-2 increases accuracy by 6.6%. This is because ReLU causes the gradient to disappear, while Taylor-2 flexibly maintains a small amount of negative numbers. All results show that replacing ReLU with Taylor-2 as the activation function has positive impact on accuracy.

Finally, we test the time cost of predicting 100 MNIST samples with different model settings and take the average of five experimental results (c.f. Table 2).

Table 2. Time cost of predicting 100 MNIST samples using MLP model with different settings.

Activation Function	Encryption Scheme	Time Cost (s)
ReLU	–	0.00249
Swish	–	0.00254
Taylor-2	–	0.00235
Taylor-2	Gentry scheme [8]	20.38613
Taylor-2	GHS [9]	18.29487
Taylor-2	CKKS [5]	0.22186

In this table, Taylor-2 is 5.6% faster than ReLU. Based on the above experiments, we can conclude that, Taylor-2 of Swish can replace ReLU as activation function. At the same time, we compare the time cost of different encryption schemes on Taylor-2. Gentry Scheme [8] and GHS [9] only support the encryption of a single parameter, which means that the unit of matrix operations in the forward propagation process is a ciphertext parameter. In contrast, CKKS supports vector encryption, with the unit of operation being a vector, resulting in an order of magnitude improvement. Compared to the unencrypted scenario, although the CKKS scheme increases the time cost by 1000 times, the time cost of approximately 0.2 s per user is acceptable due to the low base time cost, the small number of prediction samples, and the encryption of vectors instead of individual parameters.

Byzantine Robustness of FedSV: We compare the performance of FedAvg, FLTrust, and FedSV aggregation algorithms in different attack scenarios containing $0/3/6/9$ malicious users on MNIST and Cifar-10 datasets. We set m representing the number of malicious users. For example, DP-m represents m malicious users under the DP attack. When $m = 0$, which means no attack, the test accuracy of these three algorithms is significantly similar (c.f. Fig. 6).

In Fig. 5 and Fig. 7, the test accuracy of FedAvg under MP attacks stabilizes at 0.1, which is equivalent to guess randomly among 10 tags. This indicates that MP attacks have a significant impact on the global model, even if there are only a few malicious users. But under DP attacks, especially label flipping attack, FedAvg can tolerate a little number of adversaries, because most benign users eliminate the influence of malicious models that have the similar small data as other clients in average aggregation (c.f. Fig. 5(a)). However, as the number of malicious users passes half, the impact on aggregation increases significantly. From Table 3, when 9 out of 10 users are malicious, prediction accuracy of FedAvg reduces to close to 0.5 under DP attack.

For FLTrust and FedSV, neither DP nor MP can pass the validation of the dataset on the server, and they are not allowed to participate in aggregation. The number of benign users and the model quality have a significant impact

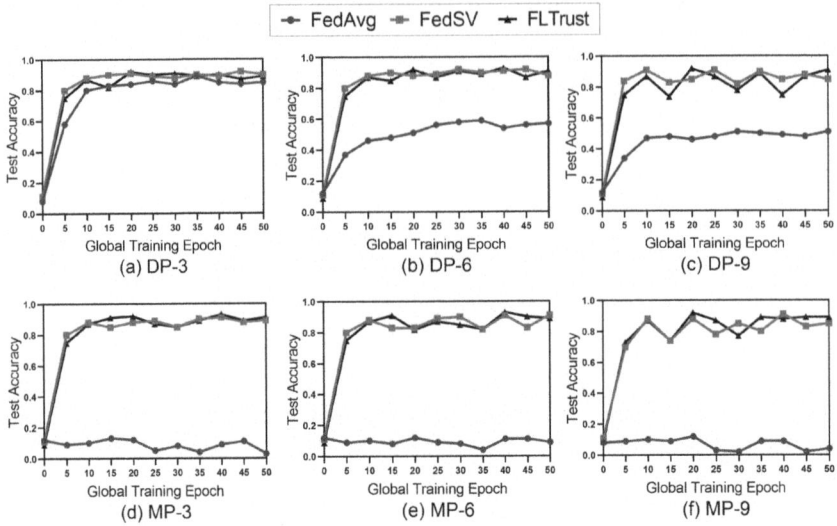

Fig. 5. The top three images generated under DP attack, and the bottom three generated under MP attack. Training with MLP model on MNIST.

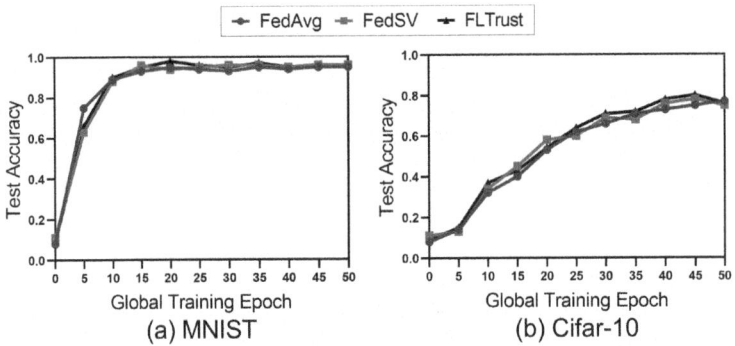

Fig. 6. Training on MNIST and Cifar-10 under no attack.

Table 3. Average Prediction accuracy of last 3 training rounds using different method with 0/3/6/9 malicious clients adopting DP and MP on 2 datasets

Datasets	AGR	No attack	MP-3	MP-6	MP-9	DP-3	DP-6	DP-9
MNIST	FedAvg	0.9247	0.1001	0.1000	0.1001	0.8970	0.5662	0.5130
(MLP)	FLTrust	0.9136	0.9125	0.9108	0.8864	0.9108	0.9024	0.8907
	FedSV	0.9325	0.9130	0.9057	0.8970	0.9101	0.9018	0.8897
Cifar-10	FedAvg	0.7324	0.1001	0.0989	0.1000	0.4873	0.3624	0.3042
(AlexNet)	FLTrust	0.7126	0.7532	0.7287	0.5003	0.6994	0.6471	0.5940
	FedSV	0.7298	0.7569	0.7346	0.5104	0.7132	0.6328	0.6087

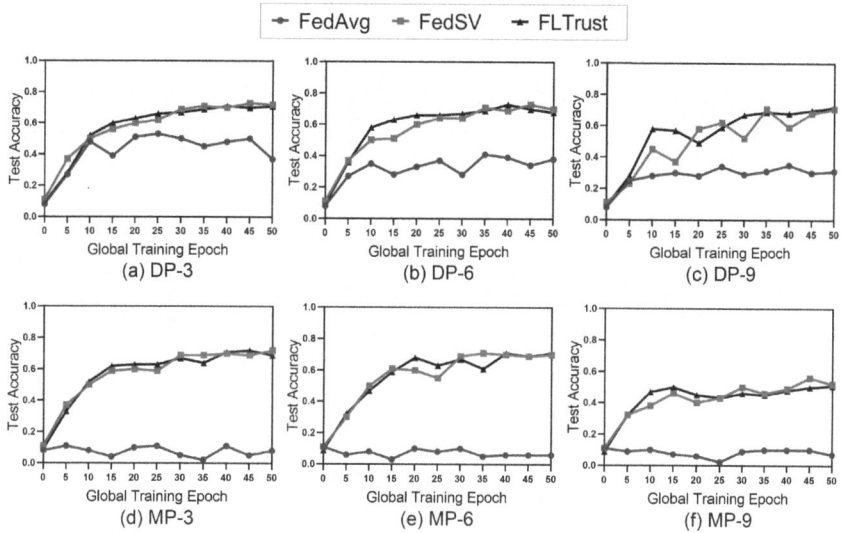

Fig. 7. The top three images generated under DP attack, and the bottom three generated under MP attack. Training with AlexNet model on Cifar-10.

on accuracy. When 9 out of 10 users are malicious, FedSV is equal to FLTrust, finding the benign model which gets the highest score and training on it. This results in a wide range of oscillations in accuracy because the global model depends entirely on one model. We calculate the average prediction accuracy of last 5 training rounds using 3 methods with 0/3/6/9 malicious clients under DP and MP attacks on 2 datasets (c.f. Table 3).

Impact of Non-IID Setting: When each client's dataset does not contain all labels and the distributions vary, even in the absence of attacks, the accuracy of the FedAvg algorithm is reduced by about 30% compared to the IID setting, and FLTrust and FedSV also correspondingly decrease by around 30% (c.f. Fig. 8(a)).

As previously set, adversaries can flip half of labels in DP attack. We find that S can detect MP attack but has difficulty resisting DP attack when the dataset held by the client are incomplete. This is because when predicting benign users' models with feature data corresponding to labels that benign users do not have, it is easy to incorrectly predict them as some known local labels, which is very similar to the behavior of DP attack and indistinguishable for S. Therefore, for FedSV, S is unable to identify benign user groups, resulting in low accuracy similar to the attacked FedAvg. As for MP attack, the adversary's model produces incorrect labels for every sample, which is easily distinguishable from benign users.

By comparing the results, we found that: FedSV doesn't reduce the prediction accuracy under no attack, and it can distinguish most malicious clients and eliminate their negative impacts. Meanwhile, all model parameters are encrypted for

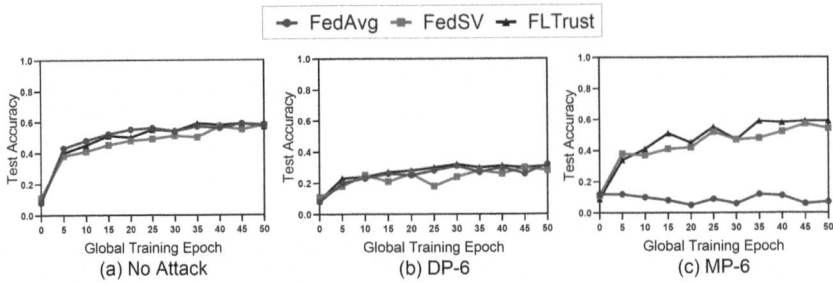

Fig. 8. Training with MLP model on MNIST dataset under no attack, DP-6 and MP-6 attack in non-IID scenarios.

the server and it cannot implement member inference attacks or model inversion attacks. Our FedSV achieves the security, robustness and fidelity goal.

7   Conclusion

We propose and evaluate a novel method of FL called FedSV, which aims to protect privacy and improve Byzantine robustness. The difference between FedSV and existing methods based on validation datasets is that in FedSV's scenario, we consider the privacy of clients. We use a FHE method, CKKS to protect privacy and achieve prediction of images on encrypted neural networks. Our experiments show that FedSV has strong Byzantine robustness, even if there is only one benign client. Further future work could include: 1)dealing with the non-IID datasets under DP attack by collaborating with other aggregation schemes that support non-IID settings and 2)designing lightweight and efficient FedSV.

Acknowledgements. This work is supported by National Nature Science Foundation of China (No. 62102429, No. 62102422, No. 62102430, No. 62072466), Natural Science Foundation of Hunan Province, China (Grant No. 2023JJ30640).

References

1. So, J., Guler, B., Avestimehr, A.S.: Byzantine-resilient secure federated learning. IEEE J. Sel. Areas Commun. **39**, 2168–2181 (2021)
2. Bagdasaryan, E., Veit, A., Hua, Y., Estrin, D., Shmatikov, V.: How to backdoor federated learning. In: International Conference on Artificial Intelligence and Statistics, pp. 2938–2948. PMLR (2020)
3. Bonawitz, K., Ivanov, V., Kreuter, B., Marcedone, A., McMahan, H.B.: Practical secure aggregation for federated learning on user-held data. CoRR abs/1611.04482   (2016).   http://dblp.uni-trier.de/db/journals/corr/corr1611. html#BonawitzIKMMPRS16
4. Chen, Z., Tian, P., Liao, W., Yu, W.: Zero knowledge clustering based adversarial mitigation in heterogeneous federated learning. IEEE Trans. Network Sci. Eng. **8**, 1070–1083 (2021)

5. Cheon, J.H., Kim, A., Kim, M., Song, Y.: Homomorphic encryption for arithmetic of approximate numbers. In: Takagi, T., Peyrin, T. (eds.) ASIACRYPT 2017. LNCS, vol. 10624, pp. 409–437. Springer, Cham (2017). https://doi.org/10.1007/978-3-319-70694-8_15

6. Fang, M., Cao, X., Jia, J., Gong, N.: Local model poisoning attacks to byzantine-robust federated learning. In: 29th USENIX Security Symposium (USENIX Security 2020), pp. 1605–1622 (2020)

7. Geiping, J., Bauermeister, H., Dröge, H., Moeller, M.: Inverting gradients - how easy is it to break privacy in federated learning? In: Proceedings of the 34th International Conference on Neural Information Processing Systems, NIPS 2020, pp. 105–121. Curran Associates Inc., Red Hook, NY, USA (2020)

8. Gentry, C.: Fully homomorphic encryption using ideal lattices. In: Proceedings of the Forty-First Annual ACM Symposium on Theory of Computing, pp. 169–178 (2009)

9. Gentry, C., Halevi, S., Smart, N.P.: Fully homomorphic encryption with polylog overhead. In: Pointcheval, D., Johansson, T. (eds.) EUROCRYPT 2012. LNCS, vol. 7237, pp. 465–482. Springer, Heidelberg (2012). https://doi.org/10.1007/978-3-642-29011-4_28

10. Cao, X., Fang, M., Liu, J., Gong, N.Z.: FLTrust: byzantine-robust federated learning via trust bootstrapping (2022)

11. Jere, M.S., Farnan, T., Koushanfar, F.: A taxonomy of attacks on federated learning. IEEE Secur. Priv. 19(2), 20–28 (2020)

12. Jia, M.F.C.: Local model poisoning attacks to byzantine-robust federated learning. In: SEC 2020: Proceedings of the 29th USENIX Conference on Security Symposium (2020)

13. Kairouz, P., McMahan, H.B., Avent, B.: Advances and open problems in federated learning. Found. Trends® Mach. Learn., pp. 1–210 (2021)

14. Krizhevsky, A., Sutskever, I., Hinton, G.E.: ImageNet classification with deep convolutional neural networks. Adv. Neural. Inf. Process. Syst. 25, 2031–2051 (2012)

15. Ma, X., Jiang, Q., Shojafar, M., Alazab, M., Kumar, S., Kumari, S.: DisBezant: secure and robust federated learning against byzantine attack in IoT-enabled MTS. IEEE Trans. Intell. Transp. Syst. 24, 2492–2502 (2023)

16. Li, H., Ditzler, G.: Targeted data poisoning attacks against continual learning neural networks. In: 2022 International Joint Conference on Neural Networks (IJCNN), pp. 312–331 (2022)

17. Zhao, L., Jiang, J., Feng, B., Wang, Q., Shen, C., Li, Q.: SEAR: secure and efficient aggregation for byzantine-robust federated learning. IEEE Trans. Dependable Secure Comput., 1631–1654 (2021)

18. Lu, S., Li, R., Chen, X., Ma, Y.: Defense against local model poisoning attacks to byzantine-robust federated learning. Front. Comput. Sci., 163–185 (2022)

19. Xia, G., Chen, J., Yu, C., Ma, J.: Poisoning attacks in federated learning: a survey. IEEE Access, 10708–10722 (2023)

20. McMahan, B., Moore, E., Ramage, D., Hampson, S., y Arcas, B.A.: Communication-efficient learning of deep networks from decentralized data. In: Artificial Intelligence and Statistics, pp. 1273–1282. PMLR (2017)

21. Mo, F., Haddadi, H., Katevas, K., Marin, E., Perino, D., Kourtellis, N.: PPFL: privacy-preserving federated learning with trusted execution environments. In: Proceedings of the 19th Annual International Conference on Mobile Systems, Applications, and Services, MobiSys 2021, pp. 94–108. Association for Computing Machinery, New York, NY, USA (2021). https://doi.org/10.1145/3458864.3466628

22. Tan, J., Liang, Y.-C., Luong, N.C., Niyato, D.: Toward smart security enhancement of federated learning networks. IEEE Network, 340–347 (2021)

23. Ramachandran, P., Zoph, B., Le, Q.V.: Searching for activation functions. arXiv preprint arXiv:1710.05941 (2017)

24. Wang, Y., Zhu, T., Chang, W., Shen, S., Ren, W.: Model poisoning defense on federated learning: a validation based approach. In: Kutyłowski, M., Zhang, J., Chen, C. (eds.) Network and System Security, NSS 2020. LNCS, vol. 12570, pp. 161–182. Springer, Cham (2020). https://doi.org/10.1007/978-3-030-65745-1_12

25. Shokri, R., Marco Stronati, V.S.: Membership inference attacks against machine learning models. Statistics, 265–279 (2016)

26. Rumelhart, D.E., Hinton, G.E., Williams, R.J.: Learning representations by back-propagating errors. Nature **323**(6088), 533–536 (1986)

27. Sav, S., et al.: POSEIDON: privacy-preserving federated neural network learning. CoRR abs/2009.00349 (2020). https://arxiv.org/abs/2009.00349

28. Shokri, R., Shmatikov, V.: Privacy-preserving deep learning (conference paper). In: Proceedings of the ACM Conference on Computer and Communications Security, pp. 1310–1321 (2015)

29. Truex, S., Liu, L., Gursoy, M.E., Yu, L., Wei, W.: Demystifying membership inference attacks in machine learning as a service. IEEE Trans. Serv. Comput., 435–452 (2019)

30. van Dijk, M., Gentry, C., Halevi, S., Vaikuntanathan, V.: Fully homomorphic encryption over the integers. In: Gilbert, H. (ed.) EUROCRYPT 2010. LNCS, vol. 6110, pp. 24–43. Springer, Heidelberg (2010). https://doi.org/10.1007/978-3-642-13190-5_2

31. Gilad-Bachrach, R., Dowlin, N., Laine, K., Lauter, K., Naehrig, M., Wernsing, J.: CryptoNets: applying neural networks to encrypted data with high throughput and accuracy. In: International Conference on Machine Learning, pp. 201–223 (2016)

32. Xiao, X., Tang, Z., Li, C., Xiao, B., Li, K.: SCA: Sybil-based collusion attacks of IIoT data poisoning in federated learning. IEEE Trans. Ind. Inform. **19**, 2608–2618 (2023)

33. Xie, C., Koyejo, S., Gupta, I.: Zeno: distributed stochastic gradient descent with suspicion-based fault-tolerance. In: International Conference on Machine Learning, pp. 6893–6901. PMLR (2019)

34. Hu, G., Li, H., Fan, W., Zhang, Y.: Efficient byzantine-robust and privacy-preserving federated learning on compressive domain. IEEE Internet Things J. **11**, 7116–7127 (2024)

35. Zhou, C., et al.: PPA: preference profiling attack against federated learning. arXiv preprint arXiv:2202.04856 (2022)

A Dual-Defense Self-balancing Framework Against Bilateral Model Attacks in Federated Learning

Xiang Wu[1], Aiting Yao[1], Shantanu Pal[2], Frank Jiang[2],
Xuejun Li[1(✉)], Jia Xu[1], Chengzu Dong[3], Xuefei Chen[1],
Xiuyi Zhang[1], and Xiao Liu[2]

[1] Anhui Provincial International Joint Research Center for Advanced Technology in Medical Imaging, School of Computer Science and Technology, Anhui University, Anhui, China
{xw,aitingyao,cxf,xiuyiz}@stu.ahu.edu.cn, {xjli,xujia}@ahu.edu.cn
[2] Deakin University, Melbourne, Australia
{shantanu.pal,frank.jiang,xiao.liu}@deakin.edu.au
[3] School of Data Science, Lingnan University, Hong Kong, China

Abstract. With the rapid expansion of Artificial Intelligence (AI) services, smart devices generate a large amount of user data at the edge network, which urgently needs to be protected while effectively extracting information. Federated learning (FL) is an important technology for handling dispersed data and strict privacy requirements in this context. However, the security threats caused by model inversion attacks and poisoning attacks can affect the mutual trust between the client and server. Yet, for these two types of attacks, the existing defense mechanisms are contradictory in terms of whether the model parameters are publicly disclosed. In addition, the data distribution of the clients is imbalanced which will increase the bias of model, reducing its practicality. To address this issue, this study proposes a dual defense self-balanced federated learning (DDSFL) framework, aiming to introduce a novel lightweight defense mechanism during the model parameter aggregation stage, combating these two types of attacks simultaneously by applying differential privacy and adjusting learning rates. In addition, this method also integrates a middleware-based reordering algorithm to enhance the robustness of the framework. Experimental results show that DDSFL effectively improves the ability to resist imbalanced data, forged data, and malicious behavior, significantly enhancing the generalization performance and security of the FL system.

Keywords: Federated Learning · Edge Network · Model Inversion Attacks · Model Poisoning Attacks · Differential Privacy

1 Introduction

Federated learning (FL) is a machine learning framework designed for edge computing scenarios [1–4], supporting multi-party collaborative development and

© The Author(s), under exclusive license to Springer Nature Singapore Pte Ltd. 2025
T. Zhu et al. (Eds.): ICA3PP 2024, LNCS 15251, pp. 261–270, 2025.
https://doi.org/10.1007/978-981-96-1525-4_14

model sharing while maintaining data decentralization. It protects privacy by training locally on edge devices, reduces the risk of data leakage, and minimizes data transmission between devices and central servers, making it suitable for resource-constrained environments. FL enhances model performance, ensures data privacy, and offers improved security, resource efficiency, and collaborative learning opportunities.

This paper discusses the challenges facing FL in real-world environments, including training data faces non-identically distributed (non-IID) scenarios that lead to decreased prediction accuracy [5], as well as Model Poisoning Attacks (MPA) and Model Inversion Attacks (MIA) [6–9]. MPA compromises the security of the global model by uploading forged model updates, while MIA reconstructs participants' private data by analyzing model updates. The existence of dual-model attacks further complicates the issue. Therefore, it is necessary to strengthen defenses against these attacks to ensure the security and credibility of the FL environment.

This paper proposes the Dual-Defense Self-Balancing Federated Learning (DDSFL) framework, which addresses the non-IID problem and enhances model security. It tackles the non-IID issue using Z-score data augmentation and multi-client rescheduling algorithms, and combats data leakage and model poisoning attacks with differential privacy masking [11] and robust learning rates [12]. The framework protects data privacy and model security through partial weight uploads and dual masking, and experimental validation demonstrates its effectiveness, significantly reducing the success rate of dual-model attacks.

The contributions of this paper are summarized as follows:

- We propose DDSFL framework that enables secure sharing and joint training of multi-party data, resulting in a highly accurate, secure, and reliable model, providing better assurance for data sharing and research.
- Our proposed dual-defense algorithm successfully maintains data privacy and model security through the upload of partial weights and the adoption of dual masking between the terminals and the aggregator. To our knowledge, this is one of the pioneering works that considers both model inversion attacks and model poisoning attacks in FL and develops effective solutions without causing conflicts.
- The effectiveness of the framework is verified by experiments, which shows that the method can fully resist model inversion attacks, and the success rate of model poisoning attacks is reduced to a safe value.

2   Related Work

In this section, we will review the relevant work on federated learning of imbalanced data, defense against poisoning attacks of the model, and model inversion attacks.

2.1   Imbalanced Data Federated Learning

Imbalanced data federated learning is a classic problem in the field of data science [13]. The main approaches to address this issue are sampling-based learning and ensemble learning. Undersampling methods involve sampling the dataset to obtain a balanced subset, which is relatively easy to implement. However, this method requires a large dataset, while the local databases of FL clients are typically small. Chawla et al. [14] proposed an oversampling method called SMOTE, which generates synthetic minority class samples to balance the dataset. Han et al. [15] improved SMOTE by considering the data distribution of the minority class. However, the aforementioned methods are not applicable to FL since the client's data is distributed and private. Some ensemble methods, such as AdaBoost [16] and Xgboost [17], can learn from misclassifications to reduce bias. However, these machine learning algorithms are sensitive to noise and outliers, which are common in distributed datasets.

2.2   Against Model Attacks

Robust training of outliers is a classic research topic in distributed learning, and median-based gradient descent algorithms are often considered an effective method to solve this problem. Specifically, algorithms are used to defend against Byzantine attacks with arbitrary outliers [18]. To avoid backdoor attacks, such as model poisoning attacks against FL, McMahan et al. [19] showed that the use of norm pruning can effectively alleviate attacks without sacrificing overall performance. Li et al. [20] also proposed a robust FL runtime framework, in which the aggregator uses a variational autoencoder model for anomaly detection. The above work did not consider the issue of weight privacy leakage. Or to detect suspicious privacy attacks using the unique capabilities of the client [21]. In response to concerns about model inversion attacks, Cui et al. [23] reconciling utility and privacy in FL via User-configurable privacy defense.

In summary, global imbalances can lead to a loss of accuracy for models trained through FL. In addition, when the two models attack at the same time, the existing methods can not effectively prevent; The central focus of this paper is to design a collaborative training framework to realize the security sharing and joint training of multi-party data, so as to provide better protection for data sharing and research.

3   Proposed Framework

This section describes our proposed Dual-Defense Self-Balancing Federated Learning (DDSFL) framework. The framework design is inspired by Te-Chuan et al. [24], where the authors applied a mask at each end device. We adopt this double-layer mask application strategy and extend it to a dual packet protection version to cope with bilateral model attacks. In this section, we will introduce the detailed design of DDSFL framework. The framework can be divided into two layers: the model training layer, the model aggregation layer, Subsect. 3.1 and Subsect. 3.2 will provide detailed descriptions of these two layers.

3.1   Model Training Layer

To address accuracy issues in FL, balancing training data across clients is key. Redistributing data among clients directly can raise privacy concerns and increase communication costs, while sequential updates by each client can be inefficient. We address this by adding an intermediary layer between the FL server and clients.

The model training process involves initialization, rebalancing, and training phases. Initially, the FL server collects local data distribution information from participating devices to initialize the neural network model's weights and optimizer. For rebalancing, we employ Z-score-based data augmentation and downsampling to correct global data imbalances, with specific thresholds (t_d and t_a) for identifying majority and minority classes for downsampling or augmentation, respectively [10]. The intermediary layer manages client updates, adjusts training frequency, and forwards models to the FL server.

3.2   Model Aggregation Layer

The model aggregation stage is the core of the framework, which is used to provide defense functions for bilateral model attacks. After each round of training, the client adds noise to a portion of the model weights and uploads them to the server, aims to obscure the specific contributions of individual data points, thereby offering a shield of privacy. However, the introduction of noise to secure varying levels of privacy protection may inadvertently compromise data accuracy, potentially diminishing its effectiveness and usability. By strategically adding noise to a select subset of the weights instead of the entire set, we manage to significantly alleviate the burden associated with model weight transmission and minimize the costs tied to the generation of Gaussian noise. Upon receiving these weighted inputs, the server performs a second-level defense. It adjusts the learning rates of different dimensions based on certain conditions, as shown in Eq. 1.

$$\eta_{\theta,i} = \begin{cases} \eta & \left| \sum_{k \in S_t} sgn(\Delta_{t,i}^k) \right| \geq \theta \\ -\eta & otherwise \end{cases} \tag{1}$$

Concretely, we introduce a hyperparameter called learning threshold θ at the server-side. For every dimension where the sum of signs of updates is less than θ, the learning rate is multiplied by -1 [12]. The aggregated weights are then sent back to the clients for the next round of training. We adopt this approach because we believe that adversaries must overcome the influence of honest devices in order to embed backdoors into the model. Specifically, in our scenario, adversaries attempt to map base class instances with Trojan patterns to the target class (adversarial mapping), while honest devices attempt to map them to the base class (honest mapping). By assigning different learning rates to honest agents and adversaries based on their different task objectives, we aim to achieve effective defense against such attacks.

4 Experiment

In this section, we will deploy and evaluate our proposed DDSFL framework to explore its practicality and performance. By comparing it with traditional FL methods (such as FedAvg) and other baseline approaches, we will demonstrate the effectiveness of the framework in defending against model inversion attacks and model poisoning attacks, as well as its ability to maintain accuracy in scenarios with imbalanced data distribution. To thoroughly assess the DDSFL framework, our experiments will be divided into qualitative and quantitative analyses.

4.1 Evaluation Setup

We evaluate our system on two well-known benchmark datasets, namely CIFAR-10 [26] and MNIST [27], for image classification tasks. We implement model inversion attacks, which utilize GANs to generate synthetic images using only the model weights. This work is targeted at collaborative training scenarios such as FL, which aligns with our objectives. For model poisoning attacks, we replicate a recent model poisoning attack called Distributed Backdoor Attack (DBA), which is one of the latest model poisoning attacks that operates in a distributed manner. In DBA, attackers insert triggers for inversion attacks or targeted model poisoning attacks into normal data to poison the inferences made by the global model.

To validate accuracy in the presence of imbalance, we set up three types of imbalances: *local imbalance*, where each client has a different degree of class imbalance; *size imbalance*, where each client has the same class distribution but different data volumes, and *global imbalance*, where the class distribution of the overall dataset is imbalanced.

4.2 System Accuracy Verification

Against Bilateral Model Attacks. To effectively evaluate defenses against model inversion attacks, we implemented a model inversion attack method inspired by the approach outlined in literature, utilizing Generative Adversarial Networks (GAN) to generate synthetic images purely from the information encapsulated within model weights [9]. To detect and evaluate model inversion attacks, we have adopted the qualitative assessment method proposed in literature [9]. This method, involves analyzing the visual recognizability and fidelity of the generated images to gauge the success of inversion attempts. When the generated images become difficult for humans to recognize, we consider the model inversion attack to have been effectively "defeated" [9].

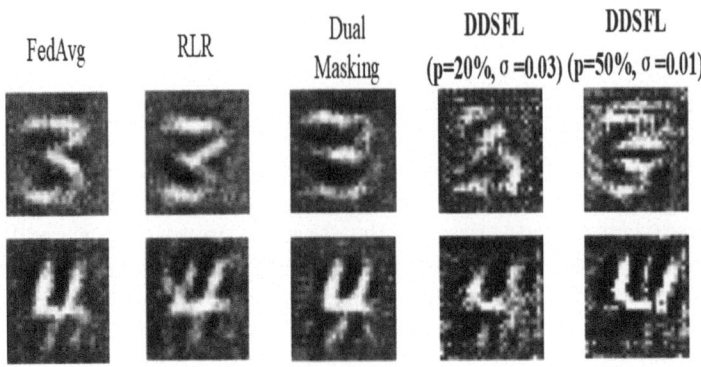

Fig. 1. Defense results of MIA.

Our investigation extends to the performance evaluation of five distinct aggregation methods: (i) FedAvg, (ii) DualMasking (masking ratio = 50%) [5], (iii) RLR [12], (iv) DDSFL ($p = 20\%, \sigma = 0.03$), and (v) DDSFL ($p = 50\%, \sigma = 0.01$). These methods were selected based on their potential relevance and varied approaches to securing model weights against inversion. As shown in Fig. 1. Among these, the DDSFL method distinguished itself by rendering the images less recognizable, thereby indicating a higher degree of protection. The Dual-Masking method, despite its strategy of nullifying a portion of the weight updates in each upload round, revealed that with an increase in training iterations, GANs could more adeptly train on the overall image model weights, suggesting that MIA might still achieve substantial results. It should be noted that although the RLR method performs well in defending against model poisoning attacks, it is not capable of defending against model inversion attacks. These results demonstrate the strong feasibility and effectiveness of our algorithms in defending against model inversion attacks.

To effectively evaluate the defense against model poisoning attacks, we designed an experiment in which we simulated FL across 40 clients for 200 rounds, with 4 of the clients being malicious. After receiving and integrating these updates, we used validation data to measure two key performance metrics of the model: validation accuracy and backdoor accuracy. Validation accuracy was calculated based on the provided validation data, while backdoor accuracy was determined using data that included all base category instances from the original validation data and instances that had been modified with the trojan horse pattern and relabeled as belonging to the target category.

Notably, within the DDSFL framework, we adapted the value of θ based on the experimental dataset: setting θ to 4 for experiments on the MNIST dataset

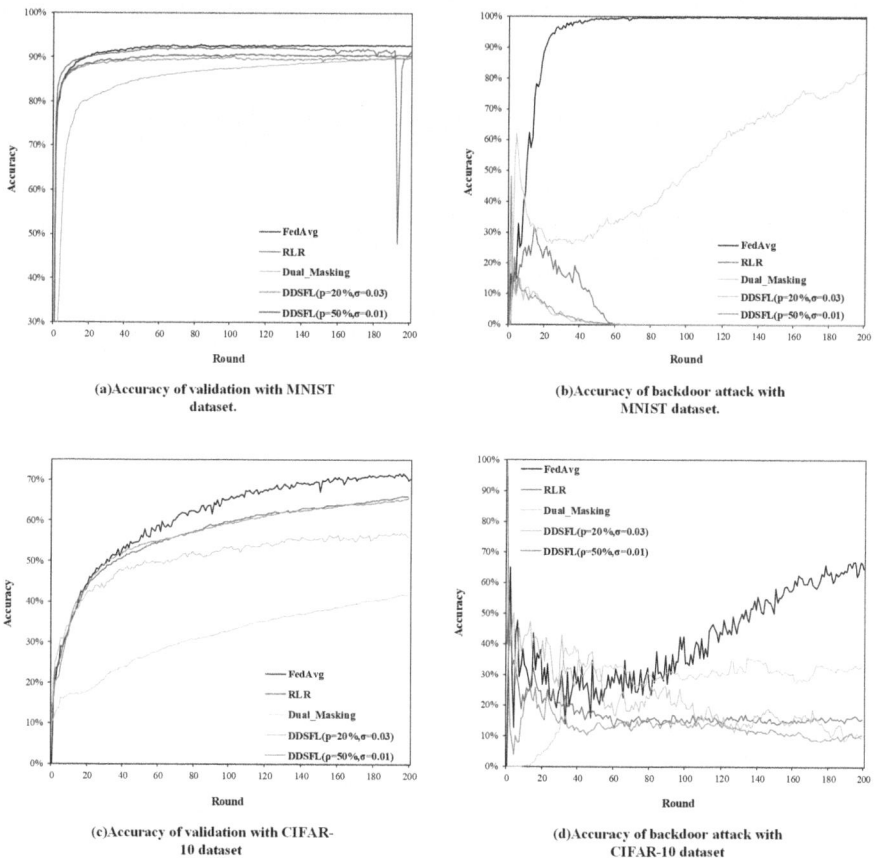

Fig. 2. Defense results of MPA with MNIST dataset and CIFAR-10 dataset

and to 8 for experiments on the CIFAR-10 dataset. Finally, we compared our system with other FL systems to assess the performance of our defense mechanism against model poisoning attacks.

The experimental results show that this method significantly reduces the effectiveness of trojan backdoor attacks. In addition, by adding the first defense layer, not only model inversion attacks were prevented, but also the defense capability against model poisoning attacks was improved. As shown in Fig. 2, after adding the first defense layer, the accuracy of backdoor attack dropped faster than other methods. The experimental results show that DDSFL defends attacks faster with very low accuracy drops. In addition, DDSFL has the ability to defend against model reverse attacks, which RLR cannot achieve. In particular, one reason why DDSFL outperformed RLR in the CIFAR-10 may be that the addition of noise amplifies the impact gap between honest and malicious devices, widening the difference in learning rates, and better preventing learning of the poisoning task (Table 1).

Table 1. Final backdoor, validation and base class accuracies for different aggregations in MNIST and CIFAR-10 dataset

Aggregation	p	σ	MNIST		CIFAR-10	
			Backdoor (%)	Validation (%)	Backdoor (%)	Validation (%)
FedAvg	0	0	99.6	92.6	64.6	70.3
RLR	0	0	0	91.3	15.3	63.1
Dual Masking	0	0	82.2	89.8	33.2	41.7
DDSFL	20 %	0.03	0	89.9	9.6	55.8
DDSFL	50 %	0.01	0	90.3	8.3	66

Accuracy Verification. We choose the vanilla FL algorithm FedAvg as the baseline [1], which has been applied to Google keyboard for improving query suggestions [25]. We validate the effects of three different imbalanced distributions on the MNIST (Table 2).

Table 2. Imbalance Test Accuracy

Imbalance Type	Baseline Test Accuracy	**DDSFL Test Accuracy**
Size Imbalance	81.74%	89.05%
Global Imbalance	87.04%	89.02%
Local Imbalance	88.42%	89.46%

The self-balance model outperformed the baseline model in all three conditions. There are 7.31% improvement on the size imbalance situation, 1.98% on local imbalance, 1.04% on global imbalance with the same experiment setup. From the results, it can be seen that the self balancing method has played an important role in solving the problems of scale balance and global balance. The main goal of the rescheduling strategy is to achieve local equilibrium, which cannot alleviate global imbalances. Therefore, by combining augmentation and downsampling methods, accuracy can be significantly improved.

5   Conclusion

This research is built upon the existing foundation in the field of FL. It aims to address the imbalance in client data by using an intermediary and implementing dual-defense mechanisms to protect client data privacy while safeguarding the server from client-side attacks. Experimental results demonstrate that the proposed DDSFL framework can improve the validation accuracy of imbalanced datasets, mitigate the negative impact of model poisoning attacks, and prevent model inversion attacks. Furthermore, DDSFL still exhibits a certain degree of accuracy reduction in local and complex tasks. We aim to address these limitations in the future.

Acknowledgement. This work was supported by the National Natural Science Foundation of China Project (No. 62372004).

References

1. McMahan, B., Moore, E., Ramage, D., Hampson, S., y Arcas, B.A.: Communication-efficient learning of deep networks from decentralized data. In: Artificial Intelligence and Statistics, pp. 1273–1282. PMLR (2017)
2. Hard, A., et al.: Federated learning for mobile keyboard prediction. arXiv preprint arXiv:1811.03604 (2018)
3. Bonawitz, K., et al.: Towards federated learning at scale: system design. Proc. Mach. Learn. Syst. 1, 374–388 (2019)
4. Yang, Q., Liu, Y., Chen, T., Tong, Y.: Federated machine learning: concept and applications. ACM Trans. Intell. Syst. Technol. (TIST) **10**(2), 1–19 (2019)
5. Zhao, Y., Li, M., Lai, L., Suda, N., Civin, D., Chandra, V.: Federated learning with non-iid data. arXiv preprint arXiv:1806.00582 (2018)
6. Bhagoji, A.N., Chakraborty, S., Mittal, P., Calo, S.: Analyzing federated learning through an adversarial lens. In: International Conference on Machine Learning, pp. 634–643. PMLR (2019)
7. Bagdasaryan, E., Veit, A., Hua, Y., Estrin, D., Shmatikov, V.: How to backdoor federated learning. In: International Conference on Artificial Intelligence and Statistics, pp. 2938–2948. PMLR (2020)
8. He, Z., Zhang, T., Lee, R.B.: Model inversion attacks against collaborative inference. In: Proceedings of the 35th Annual Computer Security Applications Conference, pp. 148–162 (2019)
9. Hitaj, B., Ateniese, G., Perez-Cruz, F.: Deep models under the GAN: information leakage from collaborative deep learning. In: Proceedings of the 2017 ACM SIGSAC Conference on Computer and Communications Security, pp. 603–618 (2017)
10. Duan, M., Liu, D., Chen, X., Liu, R., Tan, Y., Liang, L.: Self-balancing federated learning with global imbalanced data in mobile systems. IEEE Trans. Parallel Distrib. Syst. **32**(1), 59–71 (2020)
11. Sun, L., Qian, J., Chen, X.: LDP-FL: practical private aggregation in federated learning with local differential privacy. arXiv preprint arXiv:2007.15789 (2020)
12. Ozdayi, M.S., Kantarcioglu, M., Gel, Y.R.: Defending against backdoors in federated learning with robust learning rate. In: Proceedings of the AAAI Conference on Artificial Intelligence, vol. 35, pp. 9268–9276 (2021)
13. He, H., Garcia, E.A.: Learning from imbalanced data. IEEE Trans. Knowl. Data Eng. **21**(9), 1263–1284 (2009)
14. Chawla, N.V., Bowyer, K.W., Hall, L.O., Kegelmeyer, W.P.: SMOTE: synthetic minority over-sampling technique. J. Artif. Intell. Res. **16**, 321–357 (2002)
15. Han, H., Wang, W.-Y., Mao, B.-H.: Borderline-SMOTE: a new over-sampling method in imbalanced data sets learning. In: Huang, D.-S., Zhang, X.-P., Huang, G.-B. (eds.) ICIC 2005. LNCS, vol. 3644, pp. 878–887. Springer, Heidelberg (2005). https://doi.org/10.1007/11538059_91
16. Rätsch, G., Onoda, T., Müller, K.R.: Soft margins for AdaBoost. Mach. Learn. **42**, 287–320 (2001)

17. Chen, T., Guestrin, C.: XGBoost: a scalable tree boosting system. In: Proceedings of the 22nd ACM SIGKDD International Conference on Knowledge Discovery and Data Mining, pp. 785–794 (2016)

18. Yin, D., Chen, Y., Kannan, R., Bartlett, P.: Byzantine-robust distributed learning: towards optimal statistical rates. In: International Conference on Machine Learning, pp. 5650–5659. PMLR (2018)

19. Sun, Z., Kairouz, P., Suresh, A.T., McMahan, H.B.: Can you really backdoor federated learning? arXiv preprint arXiv:1911.07963 (2019)

20. Li, S., Cheng, Y., Wang, W., Liu, Y., Chen, T.: Learning to detect malicious clients for robust federated learning. arXiv preprint arXiv:2002.00211 (2020)

21. Ma, M., Zhang, Y., Arachchige, P.C.M., Zhang, L.Y., Chhetri, M.B., Bai, G.: LoDen: making every client in federated learning a defender against the poisoning membership inference attacks. In: Proceedings of the 2023 ACM Asia Conference on Computer and Communications Security, pp. 122–135 (2023)

22. Bonawitz, K., et al.: Practical secure aggregation for privacy-preserving machine learning. In: Proceedings of the 2017 ACM SIGSAC Conference on Computer and Communications Security, pp. 1175–1191 (2017)

23. Cui, Y., Meerza, S.I.A., Li, Z., Liu, L., Zhang, J., Liu, J.: RecUP-FL: reconciling utility and privacy in federated learning via user-configurable privacy defense. In: Proceedings of the 2023 ACM Asia Conference on Computer and Communications Security, pp. 80–94 (2023)

24. Chiu, T.C., Lin, W.C., Pang, A.C., Cheng, L.C.: Dual-masking framework against two-sided model attacks in federated learning. In: 2021 IEEE Global Communications Conference (GLOBECOM), pp. 1–6. IEEE (2021)

25. Yang, T., et al.: Applied federated learning: improving google keyboard query suggestions. arXiv preprint arXiv:1812.02903 (2018)

26. Krizhevsky, A., Hinton, G.: Learning multiple layers of features from tiny images. In: Handbook of Systemic Autoimmune Diseases, vol. 1, no. 4 (2009)

27. Xiao, H., Rasul, K., Vollgraf, R.: Fashion-MNIST: a novel image dataset for benchmarking machine learning algorithms (2017)

PrivARM: Privacy-Preserving Association Rule Mining in the Cloud

Yuxin Zhang[1] , Hui Han[1(✉)] , Guangliang Sun[1], Wei Wu[2] ,
and Lin Liu[1(✉)]

[1] National University of Defense Technology, Changsha 410073, China
{zhangyuxin.cn,liulin16}@nudt.edu.cn, yufengliu163@163.com
[2] Information Engineering University, Zhengzhou 450052, China

Abstract. The fusion of big data and cloud computing has established cloud services as the prime choice for data mining, making it a notable increase in individuals and organizations turning to cloud providers for data mining and storage needs. However, it raises significant concerns regarding data privacy. As a result, data owners commonly encrypt their data before sending it to the cloud, leading to complexities in balancing data privacy, result accuracy, and mining efficiency. To tackle these challenges, we introduce a set of privacy-preserving computational modules in a dual-cloud setup. Expanding on these modules, we present Apriori-based frequent itemset mining (FIM) and association rule mining (ARM) schemes, along with querying schemes, collectively designated as PrivARM. PrivARM caters to both cloud-defined and user-defined thresholds, ensuring strong privacy, result accuracy verification, and users' offline. Theoretical analysis confirms the semi-honest security, while experiments in real transaction databases show practical feasibility. Moreover, comparative studies highlight the schemes' advantages, including low computational overhead and minimal data transfer, achieving up to a $4\times$ speedup.

Keywords: Data security and privacy · Cloud computing · Association rule mining · Secret sharing

1 Introduction

The advent of IoT and cloud computing has ushered in a data-rich digital age. Data mining, is crucial for deriving insights across fields like market basket analysis [16], education [5], healthcare [14], and social media [15], faces challenges for many organizations and individuals due to expertise gaps and resource limitations. Consequently, outsourcing data mining tasks to cloud service providers offers a solution, leveraging their superior storage, computing power, accessibility, and cost-effectiveness to enhance task accuracy and efficiency.

Despite these advantages, data leakage in clouds presents serious risks, including financial losses [25], and raises privacy concerns over user data collection and

© The Author(s), under exclusive license to Springer Nature Singapore Pte Ltd. 2025
T. Zhu et al. (Eds.): ICA3PP 2024, LNCS 15251, pp. 271–290, 2025.
https://doi.org/10.1007/978-981-96-1525-4_15

breaches. Privacy-preserving Data Mining (PPDM) addresses these issues by developing analytical models without access to precise individual data, balancing privacy and data utility. PPDM methods have evolved into randomization-based methods, like random perturbation and data anonymization, which focus on efficiency but may reduce mining accuracy, and cryptography-based methods, like homomorphic encryption to maintain accuracy despite higher computational demands.

Beyond safeguarding users' outsourced data [26], data mining on cloud platforms also involves protecting query data, processing results, and mining models from external threats. Balancing efficient privacy protection with effective data mining represents a fundamental principle for future advancements. In this work, we propose PrivARM, a novel approach using secret sharing to design secure computation modules for privacy-preserving analysis of frequent itemsets and association rules. The main contributions of this paper are as follows.

- We proposed a series of secure computation protocols including secure comparison, secure frequent 1-itemset mining, and secure support computation. Based on them, we designed a secure frequent itemsets and association rules mining scheme.
- We designed two kinds of secure frequent itemsets and association rules query schemes, both supporting user-defined and cloud-defined threshold queries. In these schemes, the query users can be offline during the query process.
- We verified the security of PrivARM under the semi-honest model and tested its performance on real-world databases. Compared to the state-of-the-art work [21], our scheme achieves 4× speed up.

This paper investigates frequent itemset and association rule mining and querying technologies, focusing on privacy in cloud environments and proposes PrivARM. It addresses the technical challenges of securing data in outsourced mining scenarios, meeting data protection needs while contributing to the theoretical and practical advancement of data mining technology in cloud computing.

2   Related Work

Agrawal [2,3] introduced the concept of association rule mining (ARM) for analyzing *basket* data, which has been proved beneficial in sales planning, coupon design, and item placement [17,27,31]. However, many organizations face challenges in independently mining rules due to expertise gaps and resource limitations. Outsourcing data mining tasks to the cloud has become a viable option, prompting research in privacy-preserving association rule mining (PPARM). [4,11,28,34,35]

Various studies have explored PPARM in different contexts. For instance, [32] employed a federated learning framework for IOT applications, achieving faster training with fewer scans of different datasets. The STHE algorithm [18], based on homomorphic encryption (HE) as in [12,24], analyzed datasets like adult income, bank marketing and lung cancer. Combining techniques, [22] used FP-Growth and Genetic Algorithms (GA) to handle association rules and protect sensitive data, while [20] leveraged one-hot encoding and LU decomposition [19] for distributed data mining tasks.

In medical research, PPARM has been used to analyze correlations between patient data and disease outbreaks, making the detection and treatment of disease critical to saving more lives. For example, the FYS algorithm [10] summarised association rules in decentralized environments, examining the relationship between heart disease and dietary habits. Nikunj et.al [12] utilized an efficient ElGamal algorithm for distributed association rule mining. The proposed method can detect risk factors for breast cancer and heart disease, and discussed applications in COVID-19. Pang et.al [24] proposed a homomorphic encryption system which supports multiple cloud users with different public keys.

Despite these advancements, existing methods often struggle to balance privacy, accuracy, and efficiency. Our work addresses these challenges by proposing a novel privacy-preserving association rule mining scheme under a dual-cloud architecture, enabling multiple parties to securely upload and analyze data. This approach leverages secret sharing to enhance privacy and computation.

3   Preliminaries

In this section, we briefly introduce the definitions of FIM and ARM as well as the mining algorithm Apriori and ASS.

3.1   Frequent Itemset and Association Rule

Denote $I = \{i_1, i_2, ..., i_k\}$ as a set of items, where k is the total number of items and i_m ($1 \leq m \leq k$) represents all possible subitems. For example, in shopping basket analysis, i_m may be beer, milk, etc. A transanction t is a set of items where $t \subseteq I$. Denote the set of transactions as $D = \{t_1, t_2, ..., t_n\}$, then D is a database containing n transactions. The frequency and association relationship of transactions in a database can be measured by *support* and *confidence* respectively. Assume we have two itemsets, X and Y where $X \subseteq I$, $Y \subseteq I$ and $X \cap Y = \varnothing$, then the support of X, denoted as $supp(X)$, refers to the number of transactions containing X in the database D. An expression like $X \Rightarrow Y$ is called an association rule and the confidence of $X \Rightarrow Y$ refers to the number of transactions containing $X \cup Y$ in the database D. The support and confidence level with respect to the itemsets X and Y can be computed as:

$$supp(X \Rightarrow Y) = P(X \cup Y) = supp(X \cup Y) \tag{1}$$

$$conf(X \Rightarrow Y) = \frac{P(X \cup Y)}{P(X)} = \frac{supp(X \cup Y)}{supp(X)} \tag{2}$$

Denote the minimum threshold of support as $supp_{min}$ and the minimum threshold of confidence as $conf_{min}$, then X is a frequent item if $supp(X) \geq supp_{min}$, and rule $X \Rightarrow Y$ is a strong association rule if $supp(X \Rightarrow Y) \geq supp_{min}$ and $conf(X \Rightarrow Y) \geq conf_{min}$.

Table 1. An example of the shopping basket and the inner product of vectors

	Beer	Diaper	Bread	Butter	
t_1	1	1	0	1	$q_{\{Beer\}} = (1, 0, 0, 0)$
t_2	1	0	1	0	$q_{\{Bread, Butter\}} = (0, 0, 1, 1)$
t_3	1	1	1	1	$q_{\{Beer, Diaper\}} = (1, 1, 0, 0)$
t_4	1	1	0	0	$q_{\{Diaper, Beer, Butter\}} = (0, 1, 1, 1)$

t_i contains $\{Beer\} \Leftrightarrow t_i \cdot q_{\{Beer\}} = 1$
t_i contains $\{Bread, Butter\} \Leftrightarrow t_i \cdot q_{\{Bread, Butter\}} = 2$
t_i contains $\{Beer, Diaper\} \Leftrightarrow t_i \cdot q_{\{Beer, Diaper\}} = 2$
t_i contains $\{Diaper, Beer, Butter\} \Leftrightarrow t_i \cdot q_{\{Diaper, Beer, Butter\}} = 3$

In this paper, we represent transactions by binary vectors, transforming a database with n transactions and k items into a $n \times k$ matrix. For any transaction $t_j (1 \leq j \leq n)$, if it contains item i_m, then $e_{j,m} = 1$; otherwise, $e_{j,m} = 0$ (as shown in Table 1). By computing the inner products of these vectors, we can determine the support of specific itemsets, thereby identifying frequent itemsets and associated rules.

3.2 Apriori

The Apriori algorithm is a classic method for ARM, employing an iterative, layer-by-layer search approach. Algorithm 1 gives the pseudocode of apriori, where t represents a transaction and *apriori_gen* is a function that generates candidates. Detailed information about the Apriori algorithm is available in [3].

Algorithm 1. Apriori

Input: D, $supp_{min}$, $conf_{min}$;
Output: Frequent itemsets L, association rules R
1: $L = \{\}$;
2: $L_1 = $ find_frequent_1-itemsets(D);
3: **for** $\{k = 2; L_{k-1} \mathrel{!=} NULL; k++\}$ **do**:
4:     $C_k = $ apriori_gen(L_{k-1});
5:     **for** t in D **do**:
6:         $C_t = $ subset(C_k, t);
7:         **for** c in C_t **do**:
8:             c.count ++;
9:         **end for**
10:    **end for**
11:     $L_k = \{c$ in $C_k \mid c.count \geq supp_{min}\}$
12:     $L = L \cup L_k$
13:     return L;
14: **end for**
15: $R = $ association_rules_gen(L, $conf_{min}$);
16: return R;

3.3 Additive Secret Sharing

Additive Secret Sharing (ASS), also referred to as arithmetic sharing [9], is a technique where a l-bit secret x is split into parts such that $\langle x \rangle^A + \langle x \rangle^B \equiv x \pmod{2^l}$, where $\langle x \rangle^A$, $\langle x \rangle^B \in \mathbb{Z}_{2^l}$ and held by two servers, A and B, respectively. Each party receives a share that reveals no information about the original secret. To reconstruct the secret, the shares are combined. This method ensures data privacy by allowing secure computation without revealing individual data points. ASS is simple and computationally efficient, making it suitable for secure computation in privacy-preserving data mining.

4 Problem Statement

In this section, we briefly describe our system model, threat model and design objectives, as well as challenges.

4.1 Challenges

As discussed above, the difficulty in implementing FIM and ARM in the cloud is to simultaneously balance data privacy, result accuracy, and mining efficiency. The result accuracy is the most fundamental goal, which can be verified by comparing the mining results on the ciphertext dataset with those in plaintext. To the best of our knowledge, current researches either apply a series of complex cryptographic techniques to protect privacy and ensure correctness, while reducing the mining efficiency greatly; or achieve the desired efficiency, but fail to protect personal data privacy in all aspects, and even fail to ensure the result accuracy. Therefore, it is challenging to achieve all three goals efficiently in cloud outsourcing data mining scenarios. As such, our goal is to achieve all three security goals without using heavy cryptographic techniques, i.e., to obtain the best compromise between privacy and efficiency while ensuring correctness.

4.2 System Model and Threat Model

This paper adopts the dual-cloud model, as in [21,29,33]. In this approach, two non-colluding cloud servers work together to process and mine the encrypted data uploaded by data owners. The assumption that the parties don't collude is reasonable since they both need to ensure their model security. The model involves a total of four participants, the Cloud Service Provider (CSP), the Evaluation Service Provider (ESP), the Data Owner (DO), and the Query User (QU). The functions and responsibilities of these four components are as follows. Figure 1 shows the system model.

- *Cloud Service Provider*: As a key entity in mining and querying, CSP receives encrypted data from DOs and encrypted query vectors from QUs. In collaboration with ESP, CSP performs operations like addition, multiplication, and comparison to fulfill the privacy-preserving mining tasks.

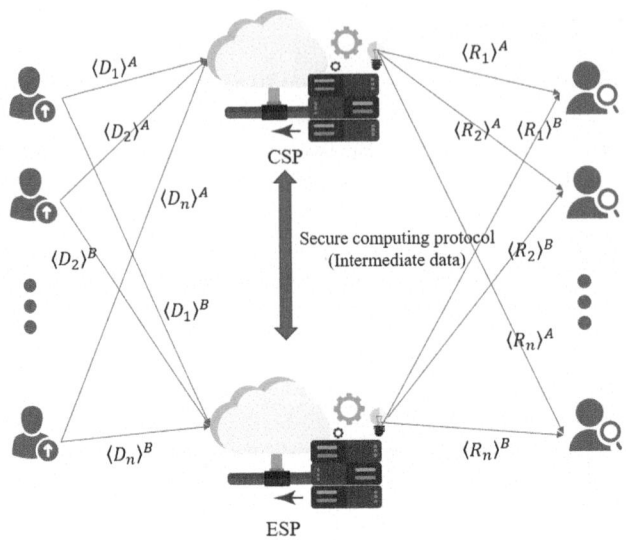

Fig. 1. System model of the dual-cloud architecture

- *Evaluation Service Provider*: The ESP, like CSP, receives encrypted data from DOs and QUs. The ESP's functions are largely similar to those of CSP, with minor variations in the computation process.
- *Data Owner*: The DO, which could be an organization which gathers considerable private data from its users. To safeguard privacy, DO divides the data into two parts by ASS before uploading it to CSP and ESP.
- *Query User*: The QU may possess its own transaction database uploaded for mining. To obtain mining results, QU typically uploads shared query vectors and receives encrypted results in return.

In privacy-preserving contexts, the semi-honest model is often considered, where an attacker adheres to the protocol but also tries to retain the intermediate computational states. In this paper, we operate under the assumption that all participants are semi-honest and no collusion between CSP and ESP. Additionally, we consider the presence of an external attacker, denoted as $\mathcal{A}$, who seeks to access outsourced data, query data, and query results outside the confines of the protocol. $\mathcal{A}$ is presumed to have the capability to:

- monitor the communication channel between the two cloud servers.
- masquerade as any one of the cloud platforms, but not both at a time.

If $\mathcal{A}$ can impersonate both cloud servers simultaneously, it contravenes the non-collusion assumption. This restriction aligns with conventional practices in adversary models, as noted in [1, 24].

4.3 Design Goals

Based on the system model and threat model described above, we propose the design goal of this paper below.

1. Data Privacy: There are three aspects for data privacy: user outsourcing data, user query data, and mining results. For the first two, users may not want their sensitive data to be leaked. For the last one, according to [13], through model inversion attacks, attackers can obtain information from the data used to train models.
2. Result Accuracy: Mining results should reflect the true patterns and trends as accurately as possible. So it is necessary to ensure that the ciphertext results are the same as those of plaintext.
3. Mining Efficiency: In the process of data mining, efficiency should be maximized to reduce the consumption of computational resources and time. As far as possible, the mining efficiency on ciphertext should be close to the mining efficiency on plaintext.

In summary, in the ideal scenario of this model, the cloud server remains unaware of the user's query data and result, yet can return accurate mining outcomes. Conversely, the user remains oblivious to the cloud's underlying data mining model, with knowledge limited only to the query result.

5 Privacy-Preserving Frequent Itemset Mining and Association Rule Mining

To address existing research gaps, this section introduces the mining solution of PrivARM. Additionally, to maximize the utilization of cloud computational resources, the system permits DOs to go offline. Key notations and definitions employed are summarized in Table 2.

5.1 Overview

DO splits its own database D into two parts $\langle D \rangle^A$ and $\langle D \rangle^B$ using ASS, which are sent to CSP and ESP respectively. Moreover, DO is required to represent the transactions as boolean vectors where the presence of an item in a transaction is marked as 1, and the absence as 0. Therefore, for any position t, if $t = 1$, there is $\langle t \rangle^A + \langle t \rangle^B \equiv 1(\mathrm{mod}\ 2^l)$, otherwise $\langle t \rangle^A + \langle t \rangle^B \equiv 0(\mathrm{mod}\ 2^l)$. When a DO uploads the database, it also shares $supp_{min}$ as $\langle supp_{min} \rangle^A$ and $\langle supp_{min} \rangle^B$, and $conf_{min}$ as $\langle conf_{min} \rangle^A$ and $\langle conf_{min} \rangle^B$, along with sharing of the number of mined items num_q, to enable computation of support using vector products and Beaver Triples [6] for SFIM and SARM. CSP and ESP then collaboratively process the data to identify all the frequent itemsets and strong association rules, ensuring the results remain encrypted and secure.

Table 2. Notations and definitions

Notations	Definitions
$supp(X)$	The support of X
$conf(X \Rightarrow Y)$	The confidence of $X \Rightarrow Y$
$supp_{min}$	The minimum support
$conf_{min}$	The minimum confidence
$\langle x \rangle^A (\langle x \rangle^B)$	CSP's (ESP's) share of secret x
$\langle \alpha \rangle / \langle \beta \rangle$	The share of $conf_{min} = \alpha/\beta$
$\mathbb{Z}_{2^l}$	The ring of modulo 2^l, i.e., $\{0, 1, ..., 2^l - 2, 2^l - 1\}$
SC	Secure Comparison algorithm
SF1M	Secure Frequent 1-itemset Mining algorithm
SSC	Secure Support Computation algorithm
SFIM	Secure Frequent Itemset Mining algorithm
SCC	Secure Confidence Comparison algorithm
SARM	Secure Association Rule Mining algorithm
U-SFIQ	User-defined Threshold Secure Frequent Itemset Query
C-SFIQ	Cloud-defined Threshold Secure Frequent Itemset Query
U-SARQ	User-defined Threshold Secure Association Rule Query
C-SARQ	Cloud-defined Threshold Secure Association Rule Query

5.2 Secure Frequent Itemset Mining

SFIM is an algorithm for mining frequent itemsets. It is mainly comprised of two major algorithms, SF1M and SSC. These two are used to mine frequent 1-itemsets and compute support respectively. When the support of an itemset is greater than a specific threshold, it can be determined as frequent.

Secure Comparison (SC). Secure Comparison returns a greater-than result between two numbers without revealing the specific value. In the scenario where CSP and ESP hold parts of two numbers x and y by ASS, comparing these numbers securely without disclosing individual shares is a challenge. The *Secure Comparison* (SC) algorithm addresses this by allowing the secure determination of whether $x > y$ or $x \leq y$, with both numbers being less than 2^l. In SC, CSP holds $\langle x \rangle^A$ and $\langle y \rangle^A$, while ESP holds $\langle x \rangle^B$ and $\langle y \rangle^B$. SC processes these inputs and outputs t: $t = 1$ signifies $x > y$, and otherwise, $x \leq y$. This secure comparison ensures that the comparative result is obtained without revealing x or y to the other party.

In SC, the Millionaire Comparison Protocol (MCP) [8] is required. Concrete steps are detailed in Algorithm 2. It can be seen that ESP at Step 3 has $C + E = (\langle x \rangle^A - \langle y \rangle^A + r) + (\langle x \rangle^B - \langle y \rangle^B) = x - y + r$ and takes it as an input, CSP owns r and uses it as another input, if $x > y$, then there is $x - y + r > r$, returning 1; otherwise returning 0. It is important to note that the result is a boolean value

---

Algorithm 2. Secure Comparison (SC)

Input: $\langle x \rangle$, $\langle y \rangle$;
Output: t $(\langle t \rangle^A, \langle t \rangle^B)$, $t = x > y$
 1: CSP: generates r;
 2: CSP: computes $C = \langle x \rangle^A - \langle y \rangle^A + r$;
 3: ESP: receives C, computes $E = C + \langle x \rangle^B - \langle y \rangle^B$;
 4: CSP and ESP: $\langle t \rangle = $ MCP(r, E);

---

split and randomly assigned to the two parties, ensuring privacy since neither knows the other's result.

Secure Frequent 1-Itemset Mining (SF1M). The SF1M algorithm is pivotal for generating candidate 1-itemsets and identifying frequent 1-itemsets in the Apriori's first step. It uses column vector data $\langle \alpha \rangle$ in the shared database between CSP and ESP, where $\langle \alpha_i \rangle$ indicates the shared value in the i-th row. Both parties aggregate the support values for each item across $\langle D \rangle$'s columns, with CSP and ESP calculating $\sum \langle \alpha_i \rangle^A$ and $\sum \langle \alpha_i \rangle^B$ respectively. An item is considered frequent, with $t = 1$, if its support count meets the minimum support threshold; otherwise, it's labeled infrequent. Details are in Algorithm 3.

Algorithm 3. Secure Frequent 1-itemset Mining (SF1M)

Input: $\langle D \rangle$, $\langle supp_{min} \rangle$;
Output: All frequent 1-itemsets;
 1: CSP: generates r, calculates $C = \sum \langle \alpha_i \rangle^A - \langle supp_{min} \rangle^A + r$;
 2: ESP: calculates $E = \sum \langle \alpha_i \rangle^B - \langle supp_{min} \rangle^B$;
 3: CSP and ESP: $t = $SC(r, C+E).

The SC assesses whether the support count $\langle \alpha_i \rangle$ meets $supp_{min}$ to identify frequent itemsets. This scheme mandates collaborative mining between both parties, necessitating shared knowledge of frequent itemset outcomes. Therefore, the final step involves both parties exchanging their SC results.

Proof. We can see that:

$$C + E = \left(\sum \langle \alpha_i \rangle^A - \langle supp_{min} \rangle^A + r\right) + \left(\sum \langle \alpha_i \rangle^B - \langle supp_{min} \rangle^B\right)$$
$$= \sum \langle \alpha_i \rangle^A + \sum \langle \alpha_i \rangle^B - supp_{min} + r \qquad (3)$$
$$= \sum \langle \alpha_i \rangle - supp_{min} + r$$

Secure Support Computation (SSC). When a vector $G = \{g_1, g_2, ..., g_m\}$ performs an inner product with $D = \{d_1, d_2, ..., d_n\}$ in plaintext, resulting in $\lambda = k$ if all k queried items in D are 1, it indicates the presence of these items in a transaction. CSP and ESP have shared values $\langle D \rangle$ and $\langle G \rangle$, with which they do Secure Inner Product(SIP) [21]. Afterwards, SC is used to compare the shared $\langle \lambda \rangle$ and the shared count $\langle k \rangle$, distinguishing between the cases $\lambda \leq k$, effectively separating $\lambda < k$ from $\lambda = k$. Since the results of SC are boolean values, to compare the support counts of the entire database with $supp_{min}$, they also need to be converted to arithmetic shared values, hence B2A [8] is executed.

Algorithm 4. Secure Support Computation (SSC)

Input: $\langle G \rangle$, $\langle D \rangle$, $\langle \lambda \rangle$ and $\langle k \rangle$;
Output: All frequent k-itemsets ($k \geq 2$);
1: **for** $g[j]$ in $\langle G \rangle$ **do**
2:     **for** $d[i]$ in $\langle D \rangle$ **do**
3:         $\langle \lambda[i] \rangle = \text{SIP}(g[j], d[i])$
4:     **end for**
5:     $\langle t[j] \rangle = \text{B2A}(\text{SC}(\langle \lambda \rangle, \langle k \rangle))$
6:     $\langle t[j] \rangle = \text{SC}(t[j], supp_{min})$
7: **end for**

It is important to note that in this context, SC compares the entire λ array after all SIP operations in a round, consisting of n values, rather than just a single value $\lambda[i]$. This approach eliminates the need for SC in loops, accelerating the computation process.

Algorithm 5. Secure Frequent Itemset Mining (SFIM)

Input: $\langle D \rangle$, $\langle G \rangle$;
Output: All frequent itemsets;
1: CSP and ESP: run SF1M;
2: CSP and ESP: run $apri\_gen$ to generate all candidate 2-itemsets;
3: CSP and ESP: run SSC to get all frequent 2-itemsets;
4: CSP and ESP: repeat Step1-3 until all frequent itemsets are found.

Secure Frequent Itemset Mining (SFIM). Upon the DO's upload of the transaction database share, SFIM initiates. It starts with SF1M for all frequent 1-itemsets, followed by using $apri\_gen$ to create candidate binomial sets, where the $apri\_gen$ algorithm, integral to Apriori for generating candidate itemsets, sees its efficiency largely influenced by the count of frequent itemsets from the preceding layer. Afterwards, SSC processes these candidates to find frequent binomial sets. This process extends to mining frequent 3-itemsets and beyond until no further frequent itemsets are discovered. The steps of SFIM are detailed in Algorithm 5.

5.3   Secure Association Rule Mining

SARM, aimed at mining strong association rules in a dual-cloud setup, is primarily facilitated by Secure Confidence Comparison (SCC). It typically follows SFIM's completion, underscoring its importance in ensuring that for association rules to be deemed strong, the associated rules must initially be recognized as the frequent ones.

Secure Confidence Comparison (SCC). In SFIM, support counts replace support frequencies to avoid the complexities of decimal numbers. In SARM, however, confidence is naturally expressed in frequency values. To manage this problem, DOs convert confidence decimals into fractions, sharing the integer numerator and denominator separately. Thus, if plaintext confidence is α/β, CSP and ESP receive $\langle\alpha\rangle$ and $\langle\beta\rangle$, respectively.

To assess if a rule's confidence meets the minimum threshold, one must evaluate if $\frac{supp(X,Y)}{supp(X)} > \frac{\alpha}{\beta}$. Directly computing division in ciphertext is complex and significantly increases overhead. However, by applying arithmetic rules, the comparison can be simplified as Eq. 4 to avoid direct division, facilitating a more efficient evaluation.

$$supp(X,Y) \cdot \beta > supp(X) \cdot \alpha \tag{4}$$

CSP and ESP each have $\langle supp(X,Y)\rangle$, $\langle supp(X)\rangle$ and $\langle\alpha\rangle$, $\langle\beta\rangle$. They first collaborated to operate Beaver Triple [6] to get the two formulas of $\langle supp(X,Y) \cdot \beta\rangle$ and $\langle supp(X) \cdot \alpha\rangle$, and then the comparison result is obtained by SC. If it returns $t = 1$, it means that the confidence of the rule meets the minimum confidence, then the rule is a strong association rule; otherwise, it is weak.

Algorithm 6. Secure Association Rule Mining (SARM)

Input: $\langle supp(X,Y)\rangle$, $\langle supp(X)\rangle$, $\langle\alpha\rangle$, $\langle\beta\rangle$ and all frequent itemsets;
Output: All association rules;
1: **for** s_item in $frequentSet$ **do**
2:     **for** x_item in $subSet$ **do**
3:         **if** $SCC(s\_item, x\_item) == 1$ **then**
4:             $Rule+ = (s\_item, x\_item)$
5:         **end if**
6:     **end for**
7: **end for**

Secure Association Rule Mining (SARM). Algorithm 6 outlines SARM. Following the mining of all frequent itemsets, the two participants identify all non-empty true subsets for each frequent itemset. Subsequently, they compose all possible association rules in pairs from these subsets. For each candidate association rule, SCC is employed to assess the relationship between the rule's confidence level and $conf_{min}$. This assessment determines whether the rule is considered a strong association or not.

6   Privacy-Preserving Frequent Itemset Query and Association Rule Query

Beyond DOs, there exists another user group: the cloud service user. These users engage as Query Users (QUs), submitting query vectors instead of databases to inquire about the frequency and associations of transaction data. For query schemes, it's crucial to safeguard not only the cloud mining models but also both query vectors and query results, aiming to protect the confidentiality of the query process and its outcomes. Therefore, this section introduces the querying solution of PrivARM.

6.1   Overview

Similar to DOs, QUs upload boolean transaction queries. Based on these data, the cloud servers collaborate to execute the query algorithm and then return. The design scheme of the query offers QU two options when uploading:

- uploads only the shared query vectors.
- uploads the query as well as user-defined thresholds of $supp_{min}$ and $conf_{min}$.

It provides QU with the flexibility to choose between querying from the mining results and conducting a customized query with given thresholds.

6.2   Secure Frequent Itemset Query

User-Defined Threshold Secure Frequent Itemset Query (U-SFIQ). In U-SFIQ, QUs specify the minimum support threshold as $supp_{min}$. In addition, for each query vector q, QU provides num_q for reference, indicating the count of 1 s. CSP and ESP perform SSC on their transaction databases to get the support $supp(\langle q \rangle)$ of the itemsets. Subsequently, with $\langle supp_{min} \rangle$, CSP and ESP execute SC to identify all vectors that where $supp(q) > supp_{min}$. It can be seen that in U-SFIQ, a return result of 1 indicates that the itemset is frequent; otherwise, it is not.

Cloud-Defined Threshold Secure Frequent Itemset Query (C-SFIQ). Steps 1-3 of C-SFIQ are identical to U-SFIQ. Starting from step 4, it binds a random number to each result to hide the cloud's mining results from QU. Details are in Algorithm 7.

6.3   Secure Association Rule Query

Query vectors for SARQ are somewhat different from SFIQ. For an association rule $X \Rightarrow Y$, SARQ requires not only the query vector for X but also for $X \cup Y$, with both vectors presented as a pair (q_{i1}, q_{i2}) for the query. The core concept is to first check the frequency of each vector separately. If either of them is not frequent, it implies that the rule must not be strongly associated, allowing the query process for that combination to be halted. However, if both vectors meet the frequency threshhold, only then does the query process proceed.

Algorithm 7. Cloud-defined Threshold Secure Frequent Itemset Query (C-SFIQ)

Input: $Q = \{q_1, q_2, ..., q_n\}$;
Output: t'
1: **for** q_i in Q **do**
2:     $\langle supp(q_i)\rangle = \text{SSC}(q_i)$;
3:     $\langle t_i \rangle = \text{SC}(\langle supp(\langle q_i \rangle)\rangle, \langle supp_{min}\rangle)$;
4:     select ε_i, $\langle t_i'\rangle \leftarrow \varepsilon_i \cdot \langle t_i \rangle$, i.e., $\langle t_i'\rangle^A \leftarrow \varepsilon_i \cdot \langle t_i \rangle^A$ and $\langle t_i'\rangle^B \leftarrow \varepsilon_i \cdot \langle t_i \rangle^B$;
5: **end for**
6: CSP and ESP: put $\{\langle t_1'\rangle, \langle t_2'\rangle, ..., \langle t_n'\rangle\}$ in a certain randomized order, send it back to QU.

User-Defined Threshold Secure Association Rule Query (U-SARQ).
In U-SARQ, besides Q and the corresponding num_q, QU still needs to upload $supp_{min}$ and $conf_{min}$ based on ASS, where $conf_{min} = \alpha/\beta$. Details of U-SARQ are in Algorithm 8.

Algorithm 8. User-defined Threshold Secure Association Rule Query (U-SARQ)

Input: $Q = \{(q_{11}, q_{12}), (q_{21}, q_{22}), ..., (q_{n1}, q_{n2})\}$, $supp_{min}$, $conf_{min}$;
Output: t'
1: **for** (q_{i1}, q_{i2}) in Q **do**
2:     $\langle supp(q_{i1})\rangle = \text{SSC}(q_{i1})$;
3:     $\langle t_{i1}\rangle = \text{SC}(\langle supp(\langle q_{i1}\rangle)\rangle, \langle supp_{min}\rangle)$;
4:     **if** $t_{i1} == 1$ **then** repeat Step 2-3 for q_{i2};
5:     **else** return 0;
6:     **end if**
7:     $\langle t_i \rangle = \text{SCC}((\langle supp(\langle q_{i1}\rangle)\rangle, \langle supp(\langle q_{i2}\rangle)\rangle), conf_{min})$
8: **end for**

Cloud-Defined Threshold Secure Association Rule Query (C-SARQ).
The process and idea of the C-SARQ is basically the same as that of U-SARQ but at the end a random number binding is applied to the results to hide the mining results as detailed in C-SFIQ.

7   Security Analysis

This section is devoted to an in-depth examination of the security of PrivARM. Prior to this, it is necessary to give the definition of security under the semi-honest model [23].

Definition 1. *Security in the Semi-honest Model: Under the semi-honest model, we assume that the input and output of a participant, designated as P_j in the protocol π, are represented by I_j and O_j, respectively. And the execution view is $\prod_i(\pi)$. In this case, the protocol π is considered secure if the distribution of the simulated view $\prod_i^S(\pi)$ obtained from I_j and O_j is indistinguishable from $\prod_i(\pi)$ [23].*

In addition to the security definition, we still need to give a basic lemma here to lead all the proofs.

Lemma 1. *If all the subprotocols are perfectly simulatable, a protocol is perfectly simulatable [7].*

7.1   Security of SFIM and SARM

Theorem 1. *SC is secure under a semi-honest model.*

Proof. For plaintext x, y, their shared copies for CSP and ESP are $\langle x \rangle^A$, $\langle y \rangle^A$ and $\langle x \rangle^B$, $\langle y \rangle^B$ respectively. In SC, the execution views of CSP and ESP are $\prod_{CSP}(r, E)$ and $\prod_{CSP}(r, C)$, where $E = \langle x \rangle^B - \langle y \rangle^B$, $C = \langle x \rangle^A - \langle y \rangle^A + r$. Initially, $\langle x \rangle^A$, $\langle y \rangle^A$ are randomly chosen from $\mathbb{Z}_{2^l}$, and r is generated randomly by CSP. Then, through addition and subtraction, the result C is obtained, remaining random. Furthermore, $\langle x \rangle^B, \langle y \rangle^B$ are defined such that $\langle x \rangle^B = x - \langle x \rangle^A$, $\langle y \rangle^B = y - \langle y \rangle^A$, with x, y being random values of 1 or 0 for CSP and ESP, making the result E random as well. Assuming the simulation views for both are $\prod_{CSP}^S(r', E')$ and $\prod_{ESP}^S(r', C')$ respectively, and given that SC is semantically secure, it follows that (r, E) and (r', E'), (r, C) and (r', C') are both computationally indistinguishable from each other. This means $\prod_{CSP}(SC)$ and $\prod_{CSP}^S(SC)$, $\prod_{ESP}(SC)$ and $\prod_{CSP}^S(SC)$ are computationally indistinguishable. In summary, according to Lemma 1, SC is secure under the semi-honest model.

Theorem 2. *SF1M is secure under a semi-honest model.*

Proof. In SF1M, the execution views of CSP and ESP are $\prod_{CSP}(r, E)$ and $\prod_{CSP}(r, C)$, respectively, where $E = \langle x \rangle^B - \langle y \rangle^B$, $C = \langle x \rangle^A - \langle y \rangle^A + r$. The proof is similar to the security analysis of SC in Theorem 1 and it can be proved that SF1M is secure under the semi-honest model.

Theorem 3. *SSC is secure under a semi-honest model.*

Proof. The computational modules in SSC include SC, SIP, B2A and addition, of which the proof of SC is given above. And the proof of the others are in [8,21,30] respectively. According to 1, it can be shown that SSC is secure under the semi-honest model.

Theorem 4. *SFIM and SARM is secure under a semi-honest model.*

Proof. The computational modules for SFIM include SF1M and SSC, the proof of which are given in Theorem 2 and 3. The same applies to SARM. According to 1, it can be proved that the SFIM and SARM are secure under the semi-honest model.

7.2   Security of SFIQ and SARQ

Theorem 5. *SFIQ and SARQ are secure under a semi-honest model.*

Proof. The computational modules for SFIQ include SC and SSC, the proof of which is given in Theorem 1, Theorem 3 separately. The same applies to SARQ. According to Lemma 1, it can be proved that the SFIQ and SARQ are secure under the semi-honest model.

8   Performance Analysis

This section focuses on the performance analysis of PrivARM, including both theoretical analysis and empirical testing on real datasets.

8.1   Theoretical Performance Analysis

Assuming DO's database comprises m transactions and n items, formatted as transaction vectors. On it, a theoretical analysis is conducted to assess both computation and communication overheads.

Table 3. Theoretical Computation and Communication Overhead of Algorithms

Algorithm	Computation Overhead	Communication Overhead
SC	$O(n)$	l
SF1M	$O(n)$	l
SSC	$O(mn)$	$4mnl + 3l$
SFIM	$O(n + mn \cdot 2^n)$	$l + (4mnl + 3l) \cdot 2^n$
SARM	$O(2^n)$	-
SFIQ	$O(mn)$	$4mnl + 3l$
SARQ	$O(2mn)$	$2 \cdot (4mnl + 3l)$

1) Computation Overhead

Table 3 lists the theoretical computation overheads in the middle column. For the security computing modules, given their foundation on ASS, they uniformly exhibit a theoretical overhead of $O(n)$. In fact, SF1M processed up to n items; for a single itemset, SSC executes SIP [21] operation across the whole transaction database, leading to a theoretical overhead of $O(mn)$. SFIM, applied to the whole transaction database, faces its worst-case scenario when all itemsets are frequent, necessitating $2^n - n$ SSC operations plus 1 SF1M operation. In scenarios where all rules are associated, SARM's computation overhead can peak at nearly 2^n. For a query vector, SFIQ's complexity is $O(mn)$ as it involves running SSC for the database; SARQ, however, typically deals with combined query vectors, thus doubling the complexity to $O(2mn)$.

2) Communication Overhead

Table 3 lists the theoretical communication overheads in the rightmost. Here, since all of the secure computing modules primarily require the transmission of random numbers within $\mathbb{Z}_{2^l}$, they basically incur low communication overhead while SIP demands more randomness with $4l$ overhead. SF1M requires just a single interaction per term operation; for an itemset, SSC performs SIP operations across the database, alongside SC. The analysis of other algorithmic overheads aligns with that discussed above.

8.2   Experimental Analysis

Environment and Datasets. This subsection details experiments on three well-known datasets: *chess*, *mushroom*, and *connect*, comparing results with those reported in [21]. The main code was in C++ for single-threaded execution, with Python for key data processing. Both cloud servers and the client were simulated on Windows 11 with an AMD Ryzen 7 4800H processor.

Results. Notably, the experimental results in this paper matched the plaintext ones, verifying the algorithm's accuracy. Figure 2a and Fig. 2b show user overheads and mining results, respectively.

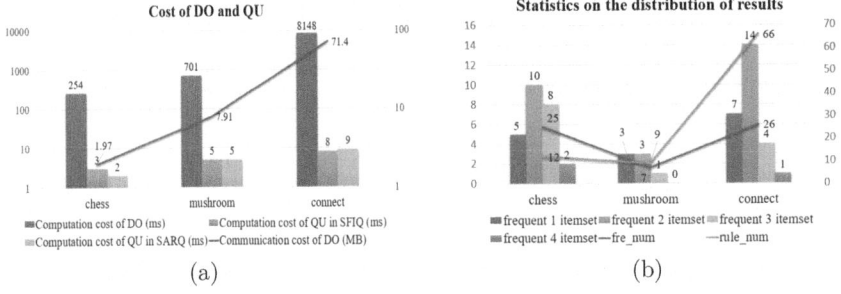

Fig. 2. (a) Communication and computation cost of DO and QU; (b) Statistics on the distribution of SFIM and SARM results

The *chess* dataset, with 3,196 transactions and up to 75 items per record, is analyzed using $supp_{min} = 3120$ and $conf_{min} = 0.99$ in SFIM, following [21]. This process identified 25 frequent itemsets across various item numbers, with a computation time of 254 ms and a communication load of 3,938 KB from DO's data duplication. The method significantly outperformed [21]'s 72.1-minute runtime with an average of 17.7 min, and communication overheads were 131 MB for CSP and 160 MB for ESP. Additionally, 12 association rules were found in 3.29 s, with CSP and ESP incurring 0.4 MB and 2 MB in communication costs, respectively. For SFIQ and SARQ, QU's database sharing took about 3ms for 10 queries, with CSP and ESP processing times around 0.83 and 0.75 min per query, and communication overheads of 5.9 MB and 5 MB for CSP, and 7.2 MB and 6.8 MB for ESP, respectively.

The *mushroom* dataset, an 8,124 × 119 matrix, was analyzed by $supp_{min} = 7800$ and $conf_{min} = 0.99$ and identified 7 frequent itemsets across single, binomial, and trinomial sets similar to [21]. The DO's computation and communication overhead were 701ms and 7,91 3KB*2 respectively. The process averaged 13.1 min, significantly quicker than 86.3 min in [21], with communication costs at 93MB for CSP and 91 MB for ESP. Additionally, 9 association rules were found in 0.27 s, with minimal communication overheads. For SFIQ, 10 queries averaged 34 min and 457 MB (233MB + 224 MB), while SARQ took 23.1 min and 228MB (116 MB + 112 MB), with QU's data sharing taking about 5ms for both.

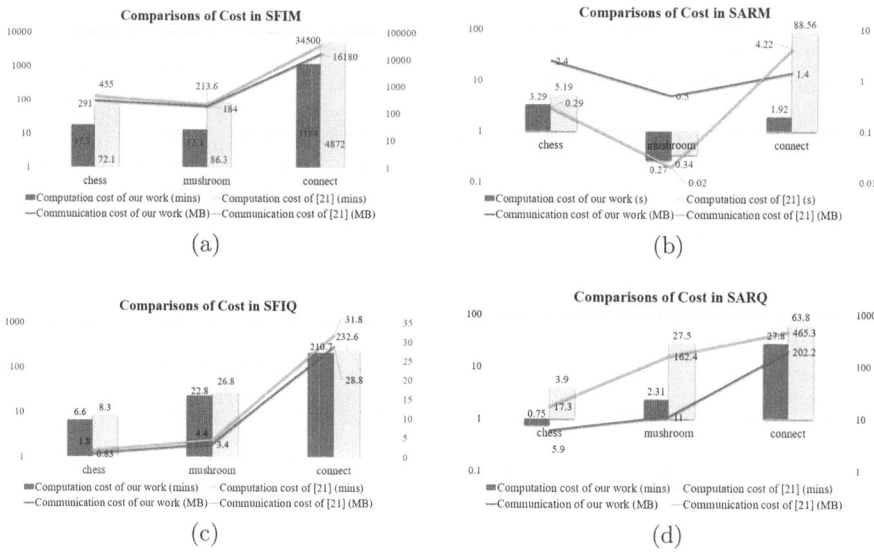

Fig. 3. Comparisons of Communication and Computation Cost in SFIM (a), SARM (b), SFIQ (c), SARQ (d) between Our Work and [21]

The *connect* dataset, a 67,557× 129 matrix in which we set $supp_{min} = 66950$ and $conf_{min} = 0.991$ following [21]. This led to 26 frequent itemsets in 18.9h instead of 81.2 h in [21], including a mix of 1-itemsets to 4-itemsets, and with a total communication load of 16180 MB (8397 MB + 7783 MB). The computation and communication overheads were 8148 ms and 71.35 MB*2 of DO, respectively. This large dataset size resulted in 66 association rules in 1.92 s with communication load of 1.4 MB in total. For SFIQ and SARQ, the overheads increased notably, with runtimes of 28.8 and 27.8 min and communication costs of 421.5 MB and 404.4 MB, respectively. However, QU's data-sharing time remained low at approximately 8.5 ms for both methods.

In Fig. 3, our comparative analyses show that our protocol significantly reduces communication and computation overhead, achieving up to a 4× speedup. User overhead remains low across all datasets, with primary user involvement being offline. Despite the *mushroom* dataset's larger size compared to *chess*, the count of mined frequent itemsets and association rules doesn't correlate linearly with size due to the *mushroom*'s sparser composition. Although SARQ theoretically should have double the running overhead of SFIQ, it is only about 1.5 times higher in practice. This efficiency is due to SARQ's preliminary frequency check, which stops further processing if either set is not frequent, saving time and resources.

9   Conclusions and Future Work

In this paper, we propose PrivARM which concludes a series of secure computing modules including SC, SF1M, SSC and SCC, based on which we design SFIM and SARM, respectively. In addition, corresponding SFIQ and SARQ are designed for cloud-defined and user-defined thresholds, which support users' offline operations. Finally, the security theory analysis is carried out under the semi-honest model and the performance is tested on real datasets. It is proven that PrivARM guarantees data privacy, result accuracy and mining efficiency.

In future work, we will focus on improving computational efficiency for association rule analysis on large encrypted databases without compromising privacy and security standards. Additionally, we will consider the "malicious adversary" model to protect sensitive information in hostile environments.

Acknowledgement. This work is supported by National Natural Science Foundation of China (Grant No.62102430, 62102447), Science Research Plan Program by NUDT (Grant No.Z K22-50).

References

1. Adnan, M., Kalra, S., Cresswell, J.C., Taylor, G.W., Tizhoosh, H.R.: Federated learning and differential privacy for medical image analysis. Sci. Rep. **12**(1), 1953 (2022)
2. Agrawal, R., Imieliński, T., Swami, A.: Mining association rules between sets of items in large databases. In: Proceedings of the 1993 ACM SIGMOD international conference on Management of data, pp. 207–216 (1993)
3. Agrawal, R., Srikant, R., et al.: Fast algorithms for mining association rules. In: Proceedings of the 20th International Conference Very Large Data Bases, VLDB. vol. 1215, pp. 487–499. Santiago (1994)
4. Aljehani, S.S., Alotaibi, Y.A.: Preserving privacy in association rule mining using metaheuristic-based algorithms: A systematic literature review. IEEE Access (2024)
5. Batool, S., Rashid, J., Nisar, M.W., Kim, J., Kwon, H.Y., Hussain, A.: Educational data mining to predict students' academic performance: A survey study. Educ. Inf. Technol. **28**(1), 905–971 (2023)

6. Beaver, D.: Efficient multiparty protocols using circuit randomization. In: Advances in Cryptology—CRYPTO'91: Proceedings 11, pp. 420–432. Springer (1992). https://doi.org/10.1007/3-540-46766-1_34

7. Bogdanov, D., Laur, S., Willemson, J.: Sharemind: a framework for fast privacy-preserving computations. In: Jajodia, S., Lopez, J. (eds.) ESORICS 2008. LNCS, vol. 5283, pp. 192–206. Springer, Heidelberg (2008). https://doi.org/10.1007/978-3-540-88313-5_13

8. Chandran, N., Gupta, D., Rastogi, A., Sharma, R., Tripathi, S.: Ezpc: Programmable and efficient secure two-party computation for machine learning. In: 2019 IEEE European Symposium on Security and Privacy (EuroS&P), pp. 496–511. IEEE (2019)

9. Demmler, D., Schneider, T., Zohner, M.: Aby-a framework for efficient mixed-protocol secure two-party computation. In: NDSS (2015)

10. Dhinakaran, D., Joe Prathap, P.M.: Ensuring privacy of data and mined results of data possessor in collaborative ARM. In: Ranganathan, G., Bestak, R., Palanisamy, R., Rocha, Á. (eds.) Pervasive Computing and Social Networking: Proceedings of ICPCSN 2021, pp. 431–444. Springer Nature Singapore, Singapore (2022). https://doi.org/10.1007/978-981-16-5640-8_34

11. Dhinakaran, D., Prathap, P.: Preserving data confidentiality in association rule mining using data share allocator algorithm. arXiv preprint arXiv:2304.14605 (2023)

12. Domadiya, N., Rao, U.P.: Elgamal homomorphic encryption-based privacy preserving association rule mining on horizontally partitioned healthcare data. J. Inst. Eng. (India): Series B **103**(3), 817–830 (2022)

13. Fredrikson, M., Jha, S., Ristenpart, T.: Model inversion attacks that exploit confidence information and basic countermeasures. In: Proceedings of the 22nd ACM SIGSAC Conference on Computer and Communications Security, pp. 1322–1333 (2015)

14. Guo, C., Chen, J.: Big data analytics in healthcare. In: Knowledge technology and systems: Toward establishing knowledge systems science, pp. 27–70. Springer (2023). https://doi.org/10.1007/978-981-99-1075-5_2

15. Hagemann, L., Abramova, O.: Sentiment, we-talk and engagement on social media: Insights from twitter data mining on the us presidential elections 2020. Internet Res. **33**(6), 2058–2085 (2023)

16. Hou, R., Ye, X., Zaki, H.B.O., Omar, N.A.B.: Marketing decision support system based on data mining technology. Appl. Sci. **13**(7), 4315 (2023)

17. Kaur, M., Kang, S.: Market basket analysis: identify the changing trends of market data using association rule mining. Proc. Comput. Sci. **85**, 78–85 (2016)

18. Kumar, G.S., Premalatha, K., Maheshwari, G.U., Kanna, P.R.: No more privacy concern: a privacy-chain based homomorphic encryption scheme and statistical method for privacy preservation of user's private and sensitive data. Expert Syst. Appl. **234**, 121071 (2023)

19. Liu, J., Liang, Y., Ansari, N.: Spark-based large-scale matrix inversion for big data processing. IEEE Access **4**, 2166–2176 (2016)

20. Liu, J., Tian, Y., Zhou, Y., Xiao, Y., Ansari, N.: Privacy preserving distributed data mining based on secure multi-party computation. Comput. Commun. **153**, 208–216 (2020)

21. Liu, L., et al.: Towards strong privacy protection for association rule mining and query in the cloud. IEEE Transactions on Cloud Computing (2023)

22. Menaga, D., Saravanan, S.: Ga-pparm: constraint-based objective function and genetic algorithm for privacy preserved association rule mining. Evol. Intel. 15, 1487–1498 (2022) (2024)
23. Oded, G.: Foundations of cryptography: Volume 2, basic applications (2009)
24. Pang, H., Wang, B.: Privacy-preserving association rule mining using homomorphic encryption in a multikey environment. IEEE Syst. J. **15**(2), 3131–3141 (2020)
25. Prajapati, P., Shah, P.: A review on secure data deduplication: Cloud storage security issue. J. King Saud Univ.-Comput. Inform. Sci. **34**(7), 3996–4007 (2022)
26. Rong, H., Liu, J., Wu, W., Hao, J., Wang, H., Xian, M.: Toward fault-tolerant and secure frequent itemset mining outsourcing in hybrid cloud environment. Comput. Secur. **98**, 101969 (2020)
27. Santoso, M.H.: Application of association rule method using apriori algorithm to find sales patterns case study of indomaret tanjung anom. Brilliance: Res. Artif. Intell. **1**(2), 54–66 (2021)
28. Saygin, Y., Verykios, V.S., Elmagarmid, A.K.: Privacy preserving association rule mining. In: Proceedings Twelfth International Workshop on Research Issues in Data Engineering: Engineering E-Commerce/E-Business Systems RIDE-2EC 2002, pp. 151–158. IEEE (2002)
29. Servan-Schreiber, S., Langowski, S., Devadas, S.: Private approximate nearest neighbor search with sublinear communication. In: 2022 IEEE Symposium on Security and Privacy (SP), pp. 911–929. IEEE (2022)
30. Shamir, A.: How to share a secret. Commun. ACM **22**(11), 612–613 (1979)
31. Ünvan, Y.A.: Market basket analysis with association rules. Commun. Stat.-Theor. Methods **50**(7), 1615–1628 (2021)
32. Wu, J.M.T., Teng, Q., Huda, S., Chen, Y.C., Chen, C.M.: A privacy frequent itemsets mining framework for collaboration in iot using federated learning. ACM Trans. Sensor Netw. **19**(2), 1–15 (2023)
33. Wu, W., Liu, J., Wang, H., Hao, J., Xian, M.: Secure and efficient outsourced k-means clustering using fully homomorphic encryption with ciphertext packing technique. IEEE Trans. Knowl. Data Eng. **33**(10), 3424–3437 (2020)
34. Zehtabchi, S., Daneshpour, N., Safkhani, M.: A new method for privacy preserving association rule mining using homomorphic encryption with a secure communication protocol. Wireless Netw. **29**(3), 1197–1212 (2023)
35. Zhang, L., Wang, W., Zhang, Y.: Privacy preserving association rule mining: taxonomy, techniques, and metrics. IEEE Access **7**, 45032–45047 (2019)

Outliers are Real: Detecting VLM-Generated Images via One-Class Classification

Baoping Liu[1], Bo Liu[1(✉)] ⓘ, and Ming Ding[2] ⓘ

[1] Australian Artificial Intelligence Institute, School of Computer Science, University
of Technology Sydney, Ultimo, Australia
`Baoping.Liu@student.uts.edu.au, Bo.liu@uts.edu.au`
[2] Data61, CSIRO, Eveleigh, Australia
`Ming.Ding@data61.csiro.au`

Abstract. Recent generative artificial intelligence (AI) advancements have enabled high-quality images with cross-modal controllability, e.g., text-to-image generation. Vision-language models (VLMs) play an indispensable role in such cross-modal generation tasks by aligning the representations of conceptions in different modalities, which inspires us to detect VLM-generated images with traces and fingerprints of VLMs. Therefore, in this paper, we propose a one-class classification (OOC) framework, namely **O**utliers are **Real** (OaReal) to recognize VLM-generated images. In sight of the rarity of VLMs due to the high cost of training, we regard VLM-generated images as normal samples and explore their distributions in CLIP latent space. During the testing phase, samples far from the explored distribution (outliers) are regarded as real images, while samples in the explored distribution are classified as VLM-generated samples. Compared with binary classifiers, the OOC design of our OaReal significantly relieves training from learning complex patterns with diverse real and fake images. Furthermore, we propose a **hard** sample **a**ware **co**ntrastive **l**oss (Harsacol) that takes edited samples into consideration and improves the inclusiveness of the explored space for both VLM-synthetic and VLM-edited samples. We conducted comprehensive experiments to test the performance of our OaReal. The results suggested the superiority of our methods in terms of both effectiveness and efficiency.

Keywords: vision-language model · anomaly detection · image forensics · contrastive loss

1 Introduction

Recent years have seen generative AI gaining the capacity to generate high-quality text, images, videos and audio. With the recent research effort on cross-modal alignment, generative AI has reached a new era of cross-modal generation, a remarkable stage where information from one modality, such as text, can

© The Author(s), under exclusive license to Springer Nature Singapore Pte Ltd. 2025
T. Zhu et al. (Eds.): ICA3PP 2024, LNCS 15251, pp. 291–308, 2025.
https://doi.org/10.1007/978-981-96-1525-4_16

be understood to create content in an entirely different modality, like images. One example of cross-modal generation is text-driven image generation (text-to-image generation), where sophisticated models can synthesize or edit detailed and contextually relevant images based solely on textual descriptions. These advancements have been largely driven by vision-language models (**VLMs**) such as Contrastive Language–ImagePre-training (**CLIP**) [29], which integrate visual and textual data to understand and generate content. VLMs greatly lower the barrier between text and images and enable generative models such as generative Adversarial Networks (GANs) [13] and Diffusion Models (DMs) [15] to generate images with text prompts. Table 1 shows some prominent generative models that adopt VLMs to achieve text-to-image generation. It can be seen that in both image synthesizing and editing tasks, CLIP is the most adopted VLM, while other VLMs, such as VisualGPT, are also adopted by some works according to the generation tasks (Fig. 1).

Fig. 1. Examples of VLM-generated images, including VLM-synthesized (left) and VLM-edited (right) images.

At the same time as generative AI evolves, the forensics of AI-generated images has also attracted much research attention to protect people from deceptive information. Recent generators combining powerful generative units (e.g. DMs) and cross-modal translators (e.g. VLM) can generate realistic images indistinguishable from human eyes, which calls for more powerful and specialized forensic tools. While many existing detectors only focus on the fake content and identify various artifacts such as abnormal spatial textures [1,35,50], inconsistent sequential patterns [23,49] and features capture by frequency analysis [16,17,43] and powerful networks (e.g. Transformer) [9,24,41]. Better detec-

Table 1. Text-to-image generators and the adopted vision-language models.

Model	Task	VLM
DALL-E 2 [30]	Synthesizing	CLIP
Stable diffusion [34]	Synthesizing	CLIP
VQGAN-CLIP [8]	Synthesizing/editing	CLIP
Imagen [37]	Synthesizing	CLIP
GLIDE [27]	Synthesizing/editing	CLIP
Imagic [18]	Editing	CLIP
Latent Diffusion Model [33]	Synthesizing	CLIP
Visual chatgpt [46]	Synthesizing/editing	VisualGPT

tion performance has also been achieved when some work looks further into generation pipelines [11,40,44,47]. However, most Deepfake detectors encounter generalization challenges that limit their performance in unseen datasets. The generalization challenge occurs mainly for two reasons: (1) From the generation perspective, fake images generated by different approaches naturally follow different data distributions. (2) From the detection perspective, binary classifiers need to learn from large-scale datasets containing diverse real images and fake images from various generators so that they can possibly learn some discriminative features applicable to unseen datasets and generators.

Motivated by the popularity of text-to-image generation and the crucial role of VLMs in bridging text and image modalities, we focus on traces and fingerprints left by VLMs and propose a generalizable classifier (OaReal) for real and VLM-generated images. Regardless of diverse downstream generation pipelines, there are few VLMs to choose from due to the need for massive data and computation to train a VLM. For example, the training of CLIP took 400 million text-image pairs as input and up to 592 V100 GPU working for 18 days. Although some works propose subsequent VLMs by fine-tuning CLIP, the prohibitive cost to train a backbone VLM naturally leads to the rarity of VLMs. The rarity of VMLs makes it possible for detectors to simply capture the traces and fingerprints of VLMs with much less cost but achieve great generalization to various seen and unseen VLM-driven generators. Therefore, our OaReal regards the detection of VLM-generated images as a one-class classification (OOC) task, where only normal samples are required. In detail, we regard the **VLM-generated images** as **normal samples** to learn the distribution during the training phase; the learned feature indicates the fingerprints shared by VLM-generated images. The test samples are passed through the same feature extraction pipeline during the inference phase. The extracted features are then compared with the learned distribution to determine anomaly scores. Outliers with higher anomaly scores are classified as real images. By adopting the OOC to recognize VLM-generated images, OaReal can be trained without analyzing real images, which is costly due to the real-world diversity. By focusing on the

VLM fingerprint rather than diverse downstream pipelines, the proposed OaReal can generalize well to unseen VLM-generated images (the generation side). Since the OaReal is free of real images during training, it relieves the difficulties in pattern analysis during training and can further improve the generalization (the detection side). In addition, we propose the **hard sample aware contrastive loss** (Harsacol) to improve the inclusiveness of the learned feature space so that our detector can recognize both synthetic and edited samples.

In summary, we propose an effective OOC framework in this paper to detect VLM-generated images. The contributions of this paper are:

- We propose a framework, namely **Outliers** are **Real** (OaReal), to detect images generated by VLM-driven generators. As far as we are concerned, this is the first work focusing on the forensics of text-to-image generation.
- We model the detection of VLM-generated images as an anomaly detection task and adopt one-class classification to capture features of VLM-generated images. In addition, we propose a new contrastive loss in our framework, which extends the boundary of learned features by adopting challenging VLM-edited images.
- We conducted comprehensive experiments to test the performance of our proposed methods. The experiment results suggest the effectiveness and generalization ability of our method.

2   Related Works

2.1   Vision-Language Model and Its Detection

Among all VLMs that bridge text and images, CLIP [29] is adopted by most text-to-image generators. CLIP learns visual concepts from natural language supervision with contrastive learning. Based on the large-scale dataset of paired text and image containing the same conceptions, CLIP pre-trains an image encoder and a text encoder to predict which images were paired with which texts in the dataset. With the supervision of contrastive loss, paired texts and images are optimized to have close representations in the CLIP latent space. Pre-trained CLIP works as a zero-shot classifier and enables the generation of images with text prompts, where CLIP is adopted to connect the two modalities. VAGAN-CLIP [8] used CLIP [37] to guide VQGAN [12] to produce higher visual quality outputs on a variety of tasks. Seeing the excellent generation capacity of DMs, DiffusionCLIP [21] and DALL-E [30] introduced CLIP to DMs for text-driven image editing and synthesizing, respectively. Many subsequent works including Stable diffusion [34], Imagen [37], GLIDE [27], Imagic [18], Latent Diffusion Model [33] and DreamBooth [36]. These text-to-image generators rely on CLIP to encode the text prompt to a series of tokens with the text encoder. The tokens are then used as conditions for the diffusion process of DMs so that DMs can generate images with aligned conceptions as the text tokens. The critical role of CLIP inevitably leaves some traces of generation regardless of the diverse diffusion process, which makes it possible for detectors to capture VLM traces and recognize VLM-generated images. Regardless of the promising detection accuracy reported by some detectors of AI-generated image [44,51,51], they mainly focus on artifacts within images or fingerprints left by various DM variants, which shows limited generalization ability to unseen datasets or generators.

2.2 Anomaly Detection

Anomaly detection is a crucial task in both industrial fields and academic research, aimed at identifying abnormal patterns or instances that deviate significantly from the norm within a dataset. One-class classification, often preferred for anomaly detection, operates under the assumption that anomalies are rare and usually comprise a minority of the data. Unlike traditional binary classifiers that require labelled data from both classes, one-class classification models are trained solely on examples of the normal class. During training, the detector (e.g. one-class SVM) only learns the distribution of normal data and determines a hyperplane that best separates normal data from the origin in the feature space. The trained model can then be applied to test samples by calculating the distance of a data point from the separating hyperplane. Points farther away from the hyperplane (on the side of the origin) are more likely to be anomalies. Some existing works are modelling the detection of AI-generated images as OOC: OC-FakeDect [20] adopted OOC to model the detection of fake human faces. In detail, OC-FakeDect is a VAE-based reconstruction framework, and the anomaly score can finally be calculated according to the reconstruction difference. FITYMI [25] mainly focused on the near-distribution settings, where the differences between normal and abnormal samples are subtle. FITYMI first generated near-distribution samples and then used a score-based strategy to detect unseen samples and improve the generalization ability. To learn discriminative and robust features during training, the contrastive loss has always been adopted to ensure the consistent representation of the same sample in different views (e.g. the original images, reconstructed images and perturbed images) [5,19,38]. In the era of large language models (LLM), it is more common to use pretrained models as encoders for anomaly detection. PANDA [31] improved conventional OOC and introduced feature adaption to use pre-trained features for anomaly detection better. [32] further improved contrastive loss by shifting the center of clustering by the mean of pretrained features, which significantly improved the compactness and uniformity of adapted representation in latent space.

3 OaReal: Outliers are Real

The proposed one-class classifier, **O**utliers are **Real** (OaReal), can be seen in Fig. 2. The whole framework of OaReal is trained as a one-class classifier, which regards VLM-generated images as normal samples and learns their data distribution to determine a hyperplane describing the feature space. To capture accurate and discriminative features, we first adopt VLM as the encoder, which reconstructs the representation of images in its feature space. Then, these features

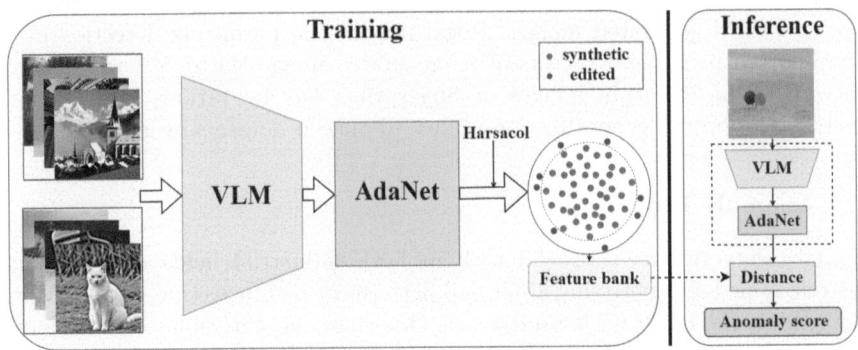

Fig. 2. The framework of our OaReal. During training, both synthetic (red frame) and edited (blue frame) images are passed to the VLM to encode images in the VLM latent space. The subsequent AdaNet re-weights the VLM features and adapts them to a more discriminative space. With the supervision of the proposed hard sample aware contrastive loss (Harsacol), the learned hyperplane is inclusive of both synthetic and edited samples. During the inference phase, the distance between the test sample and the learned feature space will be calculated as an anomaly score. Samples within the feature space will be classified as VLM-generated, while outliers are classified as real images. (Color figure online)

are re-weighted by the adaption network (AdaNet) to capture a more compact representation of the training datasets. OaReal is trained under the supervision of the proposed hard sample aware contrastive loss (Harsacol), which reinforces its capability to VLM-edited images by improving its inclusiveness.

3.1 VLM-Based Encoding

VLMs such as CLIP are proposed to align the semantic representations of the same concept from different modalities. Pre-trained with large-scale datasets of text-image pairs, VLMs are able to accurately capture the semantic representations of given images. Therefore, we adopt the powerful pre-trained VLM to encode a given input in the latent space.

Formally, given the dataset $\mathcal{D}$ consisting of n images, $\mathcal{D} = \{x_1, x_2, \ldots, x_n\}$. For a single image $x_i \in \mathcal{D}$, we aim to obtain its embeddings using a pre-trained CLIP model. The image encoder of CLIP maps an input image x_i to a feature vector.

$$\mathbf{Emb}(i) = \mathbf{CLIP^{img}}(x_i), \tag{1}$$

where the $\mathbf{CLIP^{img}}$ is the function of the CLIP image encoder, $\mathbf{Emb}(i) \in \mathbb{R}^d$ is the embedding of x_i in the CLIP latent space, d is the dimensionality of the CLIP latent space.

This function takes an image x_i as input and returns its embedding. Notable, the CLIP image encoder is frozen, and its parameters will not be updated during training, which enables us to more accurately capture the concepts of images (Fig. 3).

3.2   Feature Adaption

While CLIP-encoded embeddings represent the images well in the CLIP latent space, these embeddings are better at matching the conception with textual representations. In our task, the representations of images should have a more compact distribution so that the explored space is more discriminative for VLM-generated and real images. A more compact distribution of image representations also facilitates the generalization ability of explored space because higher compactness indicates that the representations are more VLM-specific and are more shareable among different VLM-driven generators. Therefore, we propose the Adaption Network (AdaNet) to adapt the CLIP-encoder images to fit our task better. A natural idea for the backbone of AdaNet is the attention mechanism, which assigns varying levels of importance to elements of a representation to improve the characterization ability of embeddings. An important group of attention mechanisms is self-attention. Self-attention allows the model to identify and re-weigh the importance of different parts of the input sequence by attending to itself, therefore capturing dependencies and relationships within elements of embeddings. Therefore, the adapted features are:

$$\mathbf{Emb^{ada}}(i) = \mathcal{AN}(\mathbf{Emb}(i)), \tag{2}$$

where $\mathcal{AN}$ is the function of AdaNet and $\mathbf{Emb}^{ada)}(i)$ is the embedding of image $x(i)$ re-weighted by the AdaNet.

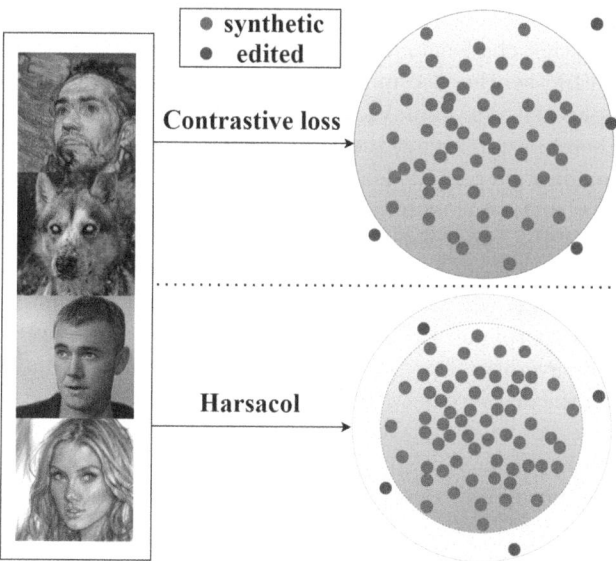

Fig. 3. Comparison of Harsacol with the conventional contrastive loss. Our Harsacol enables a fairer distribution of synthetic and edited samples according to their fully-fake and partially-fake natures.

3.3   Harsacol: Hard Sample Aware Contrastive Loss

While the pre-trained VLM is frozen in OaReal, we need to properly design the loss function to optimize AdaNet parameters. We consider three attributes while designing the contrastive loss: (1) **Uniformity** measures whether features uniformly distribute around the origin-centered sphere. (2) **Compactness**, which measures how compact the learned features are. (3) **Inclusiveness**, which measures whether the learned feature sphere includes various normal samples. Based on the traditional contrastive loss for one class classification that optimizes multi-view representations of the same sample close in the latent space, we also minimize the average cosine similarity between paired representations of the original images and the augmented images so that features can be better distributed uniformly on the unit sphere. Focusing on anomaly detection with pretrained features, [32] proposed the mean-shifted contrastive (MSC) loss, which improves uniformity and compactness by shifting the fine-tuned features by the mean of pre-trained features. While the MSC loss has successfully improved contrastive loss to fit the detection with the pre-trained model, it failed to take the inclusiveness into consideration. However, inclusiveness is an important metric for loss functions in our task because apart from VLM-synthesized images, VLM-edited images are also regarded as fake but are usually with some elements as real (i.e. pixels, textures, identities and styles). While an intuitive idea is simply to add edited samples into the training dataset and train the model with conventional contrastive loss, the real elements within the edited samples may bring confusion to the model and mislead the training process.

To overcome this challenge, we proposed to adopt VLM-edited images as hard samples (VLM-synthesized images as easy samples), pushing the boundary of normal samples (fake) so that VLM-edited images can also be learned during training and therefore improving the inclusiveness of the detector. As the hard samples are usually with partial elements edited and other elements remain real, it is intuitive that an OOC loss function should train edited samples to get closer to the boundary of the hyperplane than synthetic samples. Therefore, boundary-pushing can also be achieved by setting a shrunk hyperplane for easy samples, and the hard samples will lie between the original hyperplane (unit sphere) and the shrunk hyperplane. Both hyperplanes share the same center, which is calculated as the average of pre-trained features. We introduce a shrink factor while measuring the distance between the adapted features and the center so that VLM-synthesize and VLM-edited images are optimized to shrunk hyperplane and the original hyperplane, respectively.

Formally, the normalize representation of an image x_i is:

$$\mathbf{Emb}^{\mathbf{ada,n}}(i) = \frac{\mathbf{Emb}^{\mathbf{ada}}(i)}{\|\mathbf{Emb}^{\mathbf{ada}}(i)\|}. \tag{3}$$

The center of all samples is the average is set by $\mathbf{c}$:

$$\mathbf{c} = \mathbb{E}_{x_i \in \mathcal{D}}[\mathbf{CLIP}^{\mathbf{img}}(x_i)]. \tag{4}$$

Algorithm 1. Hard sample aware contrastive loss

Require: VLM-synthesized images $\{x_i^{synt}\}$, VLM-edited images $\{x_i^{edit}\}$, temperature τ

Ensure: Loss value L

1: Compute the center of VLM-generated data: Eq. (4).
2: Initialize total loss: $\mathcal{L} \leftarrow 0$
3: **for Batch B in** $\mathcal{D}$ **do**
4:     **for each** x_i **in B do**
5:         if $x_i \in \mathcal{D}^{edit}$ then $\alpha = 1$
6:         **Image perturbation:** $x_i^{ptb} = perturb(x_i)$
7:         **Encoding and adaption:** Eqs. (1) and (2).
8:         **Compute distance to center:** Eq. (6).
9:         Compute similarity: numerator $= \exp(sim(dist_i, dist_i^{ptb})/\tau)$
10:         Compute the Batch similarity: denominator $= \sum_{k=1}^{N} \mathbb{I}_{[k \neq i]} \exp(sim(x_i, x_k)/\tau)$
11:         **Compute loss:** $\mathcal{L}_i = numerator/denominator$, Eq. (7)
12:         Add to loss: $\mathcal{L} \leftarrow \mathcal{L} + \mathcal{L}_i$
13:     **end for**
14: **end for**
15: **return** L

It should be noticed that by introducing the VLM-edited images to the dataset, the dataset $\mathcal{D} = \mathcal{D}^{synt} \cup \mathcal{D}^{edit}$ contains both VLM-synthesized subset $\mathcal{D}^{synt}$ and the VLM-edited subset $\mathcal{D}^{edit}$.

$$\mathcal{D}^{synt} = \{x_1^{synt}, x_2^{synt}, \ldots, x_n^{synt}\}, \mathcal{D}^{edit} = \{x_1^{edit}, x_2^{edit}, \ldots, x_n^{edit}\} \quad (5)$$

The distance between x_i and the center in the latent space is:

$$\mathbf{c} = \alpha(\mathbf{Emb^{ada,n}}(i) - \mathbf{c}), \quad (6)$$

where α is the introduced shrink factor. It is a pre-set value no greater than 1. If the input during training is a VLM-edited image, α will be set to 1, meaning no shrink and the detector will optimize its representation within a unit sphere. If the input is a VLM-synthesized image, the pre-set α will lead the detector to optimize its representation within a shrunk sphere. The hard sample aware contrastive loss is:

$$\mathcal{L}(x_i, x_i^{ptb}) =$$

$$- \log \frac{\exp(sim(dist_i, dist_i^{ptb})/\tau)}{\sum_{k=1}^{2B} \mathbb{I}_{[k \neq i]} \exp(sim(dist_i, dist_k)/\tau)}, \quad (7)$$

where the superscript indicates the variable of the perturbed version of the input sample. This process is summerized in Algorithm 1.

Inference: During the inference phase, we firs extract the representation ϕ of the input sample with CLIP and AdaNet. Then, we calculate the similarity between the representation and features in the feature bank FB, which consists of features learned during the training phase (Fig. 4).

$$S_{core} = \Sigma_{x \in FB}(1 - sim(x, \phi)) \quad (8)$$

Fig. 4. Pipeline of VLM-edited image generation in our experiments. The original images are first randomly masked. Then ControlNet is adopted for inpainting.

4   Experiments

4.1   Settings

Datasets: During training, we adopt generated images either synthesized or edited by VLM-driven models. The adopted datasets containing synthesized images are:

- **DiffusionDB (DiffDB)** [45]: A large-scale text-to-image dataset generated by Stable Diffusion. DiffDB contains 14 million images generated with 1.8 million unique text prompts.
- **DiffusionForensics (DFRS)** [44]: A DM-generated image dataset with both real and generated subsets. Based on the training dataset, DFRS contains three subsets: LSUN-bedroom, ImageNet and CelebA-HQ. For each subset, we only select the images declared as generated by text prompts.
- **Synthbuster (SynBus)** [4]: A DM-generated image dataset with both real and generated subsets. DFRS consists of images generated by Stable diffusion2, IF, DALLE-2 and midjourney.
- **AI recognition Dataset (AIRD)** [22]: A dataset comprising of two parts, images generated by text-driven models such as DALL-E and midjourney, and real images known to be made by humans.
- **Deepfake StableDiffusion (DFSD)** [28]: A DM-generate human face image dataset consisting of faces generated by Stable Diffusion v1.5.

As for the edited images, we adopted text-driven image inpainting models to generate them. Because inpainted images better preserve the original elements and can effectively push the feature boundary as hard samples. In detail, we adopt real images of landscape in [2]. We masked a random region for all images and used ControlNet [48] for inpainting. We use versatile Stable Diffusion WebUI [3] for batch processing of the generation. The text prompt *"an image of beautiful nature landscape"* is applied to all images. In addition, the real images of the datasets mentioned above are integrated during the testing phase to verify the detection performance.

Baselines: In our experiments, we compare our OaReal with binary and one-class classifiers. For the binary classifier baseline, we compared our OaReal with:

- **Xception** [35]: An Deepfake detector adopting Xception as the backbone.

- **DetCNN** [42]: A Deepfake detector that improves generalization ability by various image augmentations.
- **F3Net** [7]: A Deepfake detector capturing artifacts in the frequency domain.
- **DIRE** [44]: A reconstruction-based detector for images generated by diffusion models.

For one-class classifier baselines, we compare our OaReal with:

- **OCDFD** [20]: A Deepfake detector that recognizes Deepfake images by reconstructing images with a VAE.
- **CompEmb:** [26]: A detector of GAN-Generated profile photos that use principal components analysis (PCA) to obtain compact embeddings of images. For a fair comparison, we simply replace the threshold-based classification step with a distance-measuring module of OOC.

Implementation Details: We adopt the pre-trained CLIP as the feature encoder. The CLIP parameters are frozen and will not be updated during training. Images are preprocessed using the preprocessing pipeline provided by the CLIP project. The dimension of CLIP features is 512. We adopt a multi-head self-attention mechanism as the backbone of AdaNet; the number of heads is 2. The dimension of adapted features is also 512. The adapted features are fed to the proposed loss function to explore the distribution of VLM-generated samples. The adopted perturbations include horizontal flipping and Gaussian blur. The network is optimized by Adam optimizer ($lr = 0.01$) for 50 epochs. We adopt the Area Under ROC curve (AUC) as the metric for testing the performance of the trained model.

Table 2. Comparison with state-of-the-art methods in terms of AUC. We compare our method with both binary classifiers (bi-cls) and one-class classifiers (OOC). Numbers in bold indicate the best performance.

Task	Methods	In-set	Cross-set			
		DiffDB	DFRS	SynBus	DFSD	AIRD
Bi-cls	Xception	100	52.85	53.30	49.64	54.76
	DetCNN	**100**	60.03	58.89	60.97	59.81
	F3Net	**100**	64.90	59.90	68.11	66.09
	DIRE	**100**	66.09	65.35	67.35	67.69
OOC	OCDFD	96.71	70.93	71.03	73.45	70.60
	CompEmb	86.13	65.53	62.47	60.91	58.98
	OaReal	99.97	**91.36**	**90.66**	**91.12**	**88.96**

4.2   Comparison with State-of-the-Art Methods

We compare our method with both binary classifiers and one-class classifiers. All methods are trained with the DiffDB dataset; the test results on four unseen datasets are shown in Table 2.

It can be seen that while binary classifiers achieve perfect in-set classification AUC, our OaReal also achieves close in-set performance (99.97%) but much better generalization than all baselines. Compared with our OOC framework, binary classifiers need to learn discriminative features from diverse real and fake samples, which easily leads to overfitting training samples if the scale of the dataset is not big enough. For instance, Xception simply adopts stacked CNNs to learn features for classification, and the trained model is severely overfitted to the training dataset (100% in the set) and shows poor generalization to unseen data (no more than 55%). DetCNN and F3Net introduced augmentations and frequency analysis to learn more robust features. The generalization ability is improved but no more than 70% (68.11% and 67.35% on DFSD, respectively). DIRE and OCDFD are two reconstruction-based approaches that adopt the diffusion model and VAE as reconstruction networks, respectively. Therefore, DIRE are more focused on traces left by the DM pipeline, and the classification performance drops significantly on unseen DM generators (65.35% Our OaReal achieves significantly better in-set and cross-set performance than two OOC baselines. This is owing to the accurate semantics captured by CLIP and the more discriminative adaption with AdaNet.

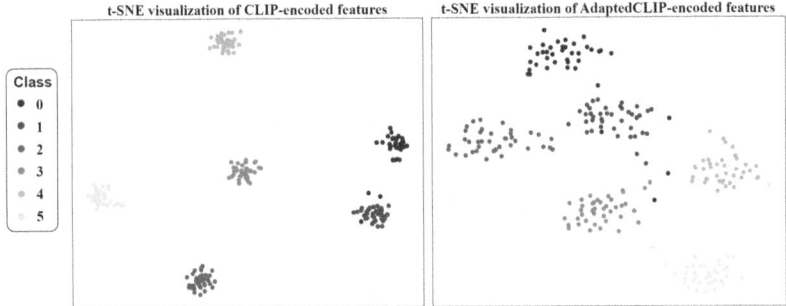

Fig. 5. t-SNE visualization of the CLIP-embedded (left) and the adapted CLIP features (right).

4.3   Ablation Study

Effectiveness of the AdaNet. To verify the effectiveness of AdaNet, we extract features from test samples in the settings with and without AdaNet adapting the CLIP features. To better show the impact of the AdaNet on optimizing the cross-class and intra-class distance, we manually select samples from the DiffDB dataset and split them into six groups (human faces, cats, dogs, fish, food and landscapes) according to the objects within the samples. The classes are given labels from 0 to 5.

The visualized results can be seen in Fig. 5. It can be seen that without feature adaption, each class is closely centred, and each class is also far from other classes. This indicates that CLIP is able to capture visual conceptions within images very well so that samples of different objects can be well discriminated. However, such characteristics are not expected in our task. Our detector aims to capture shared features among all VLM-driven models. It can be seen that after feature adaption, the cross-class distance is significantly reduced, and representations of the same class samples are significantly loosened, indicating that AdaNet is able to re-weight the features and make them more generalizable.

Effectiveness of the Loss Function. The proposed loss function Harsacol takes both synthetic and edited samples into consideration and improves the inclusiveness of our OaReal framework. To verify the effectiveness, we collected the features of synthetic and edited samples, and we visualized the distribution of their features. In detail, we train our model with vanilla contrastive loss (CL), mean-shifted contrastive loss (MSC) and our Harsacol. Then, we test the models with both synthetic and edited samples. The visualized comparison of contrastive loss and Harsacol can be seen in Fig. 6. It can be seen that our Harsacol outperforms conventional contrastive loss. This is because Harsacol uses fixed center calculated by pretrained models, improving samples' compactness and facilitating better clustering. Our loss function shows better detection accuracy and clustering and further improves the compactness of synthetic samples. This is because we regard edited as hard samples and push the feature boundary outward. Quantitative results can be seen in Table 3. While all three contrastive losses achieve close performance on the detection of synthetic datasets, our loss achieves better detection performance on edited images due to the extended boundary and improved inclusiveness.

Table 3. Effectiveness of various loss.

Encoder	DiffDB	DFRS	SynBus	DFSD	Edited
CL	99.93	90.04	90.13	90.90	87.91
MSC	99.98	91.01	90.70	91.14	89.91
Harsacol	99.97	91.36	90.66	91.12	90.70

Effectiveness of Encoder. To verify the effectiveness of CLIP as the encoder of our framework, we replace the CLIP with various commonly-used encoders,

Table 4. Effectiveness of various encoders.

Encoder	DiffDB	DFRS	SynBus	DFSD	AIRD
VGG	99.87	69.91	70.02	68.55	67.98
ResNet	99.85	68.49	69.18	71.04	70.79
Xception	99.90	70.15	71.91	72.46	71.87
ViT	**99.97**	82.45	81.49	80.31	81.80
CLIP	**99.97**	**91.36**	**90.66**	**91.12**	**88.96**

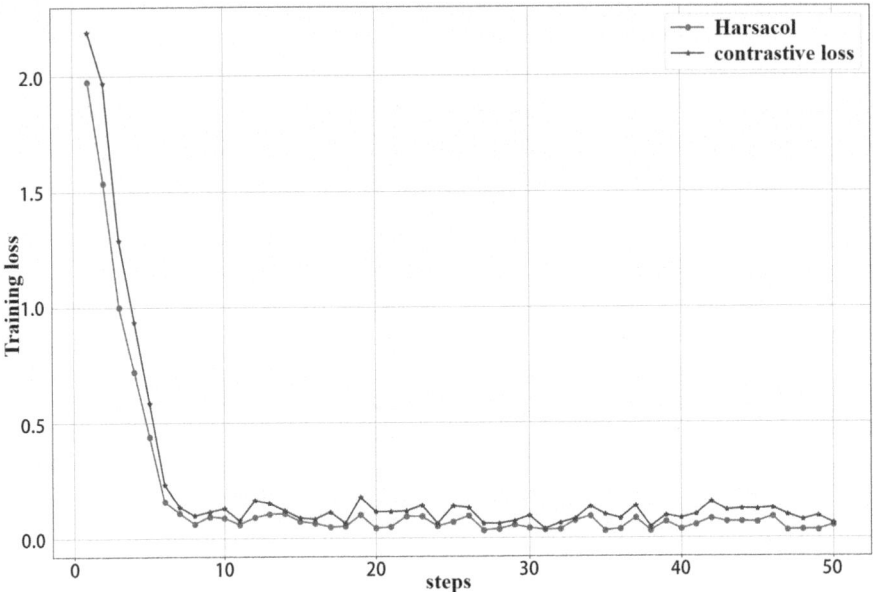

Fig. 6. Comparison of Harsacol with conventional contrastive loss (CL). Our Harsacol achieves faster and better convergence.

including VGG-19 (VGG) [39], ResNet-101 (ResNet) [14], Xception [6] and ViT-L/16 (ViT) [10].

The results can be seen in Table 4. We notice that all methods achieve over 99% in-set classification accuracy, but VGG, ResNet and Xception struggle to generalize to unseen datasets. For instance, VGG only achieves 67.98% AUC on the AIRD dataset. This is because large-scale training datasets contain very complex data distributions and can be difficult for encoders to learn. However, these encoders do not have prior knowledge of generation pipelines and can hardly obtain generalizable features from complex data distributions. This also explains why ViT and CLIP achieve better in-set and cross-set performance. CLIP outperforms ViT in cross-set settings because adopting CLIP as an encoder is like reconstructing the images in CLIP latent space. Therefore, the captured semantic representations are more accurate due to the prior knowledge of generators.

5 Conclusions and Future Work

This paper focuses on the detection of VLM-generated images. Our proposed framework, OaReal, encodes samples in latent space with pre-trained CLIP and then adapts the feature to a more discriminative space. The proposed loss function, Harsacol, further improved the inclusiveness of our framework to edited samples. The experiment results indicate our framework's discriminative capacity to VLM-synthetic and VLM-edited images. However, there are also some limitations of our work: Firstly, the proposed method is only applicable to VLM-generated samples and may struggle with samples generated by some newly-emerging generators. Secondly, due to the lack of VLMs, only CLIP is adopted as the encoder in our experiments. We also encourage further efforts to extend our work.

References

1. Afchar, D., Nozick, V., Yamagishi, J., Echizen, I.: Mesonet: a compact facial video forgery detection network. In: 2018 IEEE International Workshop on Information Forensics and Security (WIFS), pp. 1–7. IEEE (2018)
2. Arnaud58: Landscape pictures (2019). https://www.kaggle.com/datasets/arnaud58/landscape-pictures. Accessed 23 May 2024
3. AUTOMATIC1111: Stable diffusion webui (2022). https://github.com/AUTOMATIC1111/stable-diffusion-webui
4. Bammey, Q.: Synthbuster: towards detection of diffusion model generated images. IEEE Open J. Signal Process. (2023)
5. Cho, H., Seol, J., Lee, S.G.: Masked contrastive learning for anomaly detection. arXiv preprint arXiv:2105.08793 (2021)
6. Chollet, F.: Xception: deep learning with depthwise separable convolutions. In: Proceedings of the IEEE Conference on Computer Vision and Pattern Recognition (CVPR), pp. 1251–1258 (2017)
7. Corvi, R., Cozzolino, D., Zingarini, G., Poggi, G., Nagano, K., Verdoliva, L.: On the detection of synthetic images generated by diffusion models. In: ICASSP 2023-2023 IEEE International Conference on Acoustics, Speech and Signal Processing (ICASSP), pp. 1–5. IEEE (2023)
8. Crowson, K., et al.: Vqgan-clip: open domain image generation and editing with natural language guidance. In: European Conference on Computer Vision, pp. 88–105. Springer (2022)
9. Dong, X., et al.: Protecting celebrities from deepfake with identity consistency transformer. In: Proceedings of the IEEE/CVF Conference on Computer Vision and Pattern Recognition, pp. 9468–9478 (2022)

10. Dosovitskiy, A., et al.: An image is worth 16x16 words: transformers for image recognition at scale. In: International Conference on Learning Representations (2021)
11. Durall, R., Keuper, M., Keuper, J.: Watch your up-convolution: CNN based generative deep neural networks are failing to reproduce spectral distributions. In: Proceedings of the IEEE/CVF Conference on Computer Vision and Pattern Recognition, pp. 7890–7899 (2020)
12. Esser, P., Rombach, R., Ommer, B.: Taming transformers for high-resolution image synthesis. In: Proceedings of the IEEE/CVF Conference on Computer Vision and Pattern Recognition, pp. 12873–12883 (2021)
13. Goodfellow, I., et al.: Generative adversarial nets. In: Advances in Neural Information Processing Systems, vol. 27 (2014)
14. He, K., Zhang, X., Ren, S., Sun, J.: Deep residual learning for image recognition. In: Proceedings of the IEEE Conference on Computer Vision and Pattern Recognition (CVPR), pp. 770–778 (2016)
15. Ho, J., Jain, A., Abbeel, P.: Denoising diffusion probabilistic models. Adv. Neural. Inf. Process. Syst. **33**, 6840–6851 (2020)
16. Jeong, Y., Kim, D., Min, S., Joe, S., Gwon, Y., Choi, J.: Bihpf: bilateral high-pass filters for robust deepfake detection. In: Proceedings of the IEEE/CVF Winter Conference on Applications of Computer Vision, pp. 48–57 (2022)
17. Jeong, Y., Kim, D., Ro, Y., Choi, J.: Frepgan: robust deepfake detection using frequency-level perturbations. In: Proceedings of the AAAI Conference on Artificial Intelligence, vol. 36, pp. 1060–1068 (2022)
18. Kawar, B., et al.: Imagic: text-based real image editing with diffusion models. In: Proceedings of the IEEE/CVF Conference on Computer Vision and Pattern Recognition, pp. 6007–6017 (2023)
19. Keshtmand, N., Santos-Rodriguez, R., Lawry, J.: Typicality-based point OOD detection with contrastive learning. In: Northern Lights Deep Learning Conference, pp. 120–129. PMLR (2024)
20. Khalid, H., Woo, S.S.: Oc-fakedect: classifying deepfakes using one-class variational autoencoder. In: Proceedings of the IEEE/CVF Conference on Computer Vision and Pattern Recognition Workshops, pp. 656–657 (2020)
21. Kim, G., Kwon, T., Ye, J.C.: Diffusionclip: text-guided diffusion models for robust image manipulation. In: Proceedings of the IEEE/CVF Conference on Computer Vision and Pattern Recognition, pp. 2426–2435 (2022)
22. Koliha, N.: AI recognition dataset (2024). https://doi.org/10.34740/KAGGLE/DSV/7501337
23. Liu, B., Liu, B., Ding, M., Zhu, T., Yu, X.: Ti2net: temporal identity inconsistency network for deepfake detection. In: Proceedings of the IEEE/CVF Winter Conference on Applications of Computer Vision, pp. 4691–4700 (2023)
24. Liu, H., Tan, Z., Tan, C., Wei, Y., Wang, J., Zhao, Y.: Forgery-aware adaptive transformer for generalizable synthetic image detection. In: Proceedings of the IEEE International Conference on Computer Vision and Pattern Recognition (CVPR) (2024)
25. Mirzaei, H., et al.: Fake it till you make it: towards accurate near-distribution novelty detection. arXiv preprint arXiv:2205.14297 (2022)
26. Mundra, S., Porcile, G.J.A., Marvaniya, S., Verbus, J.R., Farid, H.: Exposing GAN-generated profile photos from compact embeddings. In: Proceedings of the IEEE/CVF Conference on Computer Vision and Pattern Recognition, pp. 884–892 (2023)

27. Nichol, A.Q., et al.: Glide: towards photorealistic image generation and editing with text-guided diffusion models. In: International Conference on Machine Learning, pp. 16784–16804. PMLR (2022)

28. Papa, L., Faiella, L., Corvitto, L., Maiano, L., Amerini, I.: On the use of stable diffusion for creating realistic faces: from generation to detection. In: 2023 11th International Workshop on Biometrics and Forensics (IWBF), pp. 1–6. IEEE (2023)

29. Radford, A., et al.: Learning transferable visual models from natural language supervision. In: International Conference on Machine Learning, pp. 8748–8763. PMLR (2021)

30. Ramesh, A., Dhariwal, P., Nichol, A., Chu, C., Chen, M.: Hierarchical text-conditional image generation with clip latents. arXiv preprint arXiv:2204.06125 **1**(2), 3 (2022)

31. Reiss, T., Cohen, N., Bergman, L., Hoshen, Y.: Panda: adapting pretrained features for anomaly detection and segmentation. In: Proceedings of the IEEE/CVF Conference on Computer Vision and Pattern Recognition, pp. 2806–2814 (2021)

32. Reiss, T., Hoshen, Y.: Mean-shifted contrastive loss for anomaly detection. In: Proceedings of the AAAI Conference on Artificial Intelligence, vol. 37, pp. 2155–2162 (2023)

33. Rombach, R., Blattmann, A., Lorenz, D., Esser, P., Ommer, B.: High-resolution image synthesis with latent diffusion models. In: Proceedings of the IEEE/CVF Conference on Computer Vision and Pattern Recognition, pp. 10684–10695 (2022)

34. Rombach, R., Blattmann, A., Lorenz, D., Esser, P., Ommer, B.: High-resolution image synthesis with latent diffusion models (2021)

35. Rossler, A., Cozzolino, D., Verdoliva, L., Riess, C., Thies, J., Nießner, M.: Face-forensics++: learning to detect manipulated facial images. In: Proceedings of the IEEE/CVF International Conference on Computer Vision, pp. 1–11 (2019)

36. Ruiz, N., Li, Y., Jampani, V., Pritch, Y., Rubinstein, M., Aberman, K.: Dreambooth: fine tuning text-to-image diffusion models for subject-driven generation. In: Proceedings of the IEEE/CVF Conference on Computer Vision and Pattern Recognition, pp. 22500–22510 (2023)

37. Saharia, C., et al.: Photorealistic text-to-image diffusion models with deep language understanding. Adv. Neural. Inf. Process. Syst. **35**, 36479–36494 (2022)

38. Seifi, S., Reino, D.O., Chumerin, N., Aljundi, R.: OOD aware supervised contrastive learning. In: Proceedings of the IEEE/CVF Winter Conference on Applications of Computer Vision, pp. 1956–1966 (2024)

39. Simonyan, K., Zisserman, A.: Very deep convolutional networks for large-scale image recognition. arXiv preprint arXiv:1409.1556 (2014)

40. Tan, C., et al.: Rethinking the up-sampling operations in CNN-based generative network for generalizable deepfake detection (2023)

41. Wang, J., et al.: M2TR: multi-modal multi-scale transformers for deepfake detection. In: Proceedings of the 2022 International Conference on Multimedia Retrieval, pp. 615–623 (2022)

42. Wang, S.Y., Wang, O., Zhang, R., Owens, A., Efros, A.A.: CNN-generated images are surprisingly easy to spot... for now. In: Proceedings of the IEEE/CVF Conference on Computer Vision and Pattern Recognition, pp. 8695–8704 (2020)

43. Wang, Y., Yu, K., Chen, C., Hu, X., Peng, S.: Dynamic graph learning with content-guided spatial-frequency relation reasoning for deepfake detection. In: Proceedings of the IEEE/CVF Conference on Computer Vision and Pattern Recognition, pp. 7278–7287 (2023)

44. Wang, Z., et al.: Dire for diffusion-generated image detection. arXiv preprint arXiv:2303.09295 (2023)
45. Wang, Z.J., Montoya, E., Munechika, D., Yang, H., Hoover, B., Chau, D.H.: Diffusiondb: a large-scale prompt gallery dataset for text-to-image generative models. arXiv preprint arXiv:2210.14896 (2022)
46. Wu, C., Yin, S., Qi, W., Wang, X., Tang, Z., Duan, N.: Visual chatgpt: talking, drawing and editing with visual foundation models. arXiv preprint arXiv:2303.04671 (2023)
47. Yang, T., Huang, Z., Cao, J., Li, L., Li, X.: Deepfake network architecture attribution. In: Proceedings of the AAAI Conference on Artificial Intelligence, vol. 36, pp. 4662–4670 (2022)
48. Zhang, L., Rao, A., Agrawala, M.: Adding conditional control to text-to-image diffusion models (2023)
49. Zhao, C., Wang, C., Hu, G., Chen, H., Liu, C., Tang, J.: ISTVT: interpretable spatial-temporal video transformer for deepfake detection. IEEE Trans. Inf. Forensics Secur. **18**, 1335–1348 (2023)
50. Zhao, H., Zhou, W., Chen, D., Wei, T., Zhang, W., Yu, N.: Multi-attentional deepfake detection. In: Proceedings of the IEEE/CVF Conference on Computer Vision and Pattern Recognition, pp. 2185–2194 (2021)
51. Zhu, M., et al.: Genimage: a million-scale benchmark for detecting AI-generated image. In: Advances in Neural Information Processing Systems, vol. 36 (2024)

TSformer: A Transformer-Based Model Focusing Specifically on the Fusion of Temporal-Spatial Features for Traffic Forecasting

Yuquan Chu[1], Peng Liu[2(✉)], Jin Fan[1], Haocheng Ye[1], and Tianfan Jiang[1]

[1] HDU-ITMO Joint Insistute, Hangzhou Dianzi University, Hangzhou 310018, China
{232320008,fanjin,yehaocheng,232320005}@hdu.edu.cn
[2] School of Computer Science and Technology, Hangzhou Dianzi University, Hangzhou 310018, China
perryliu@hdu.edu.cn

Abstract. With the rapid development of Intelligent Transportation Systems (ITS), how to accurately obtain traffic data predictions has become a key challenge. In recent years, many neural networks with complex structures have been designed to meet this challenge; however, these models often use the method of extracting temporal and spatial features independently and then fusing them, which ignores the intrinsic interconnectivity of spatio-temporal features. To address this problem, we propose TSformer, a Transformer-based model specifically designed to extract important features in both time and space. Our innovation lies in integrating Transformer's cross-attention mechanism to pre-learn the model in the temporal dimension in phase before the spatial features are extracted, seamlessly fusing temporal and spatial dimensional features. We refer to this novel approach as Temporal-Spatial Cross Attention Fusion (TSCAF), which improves the model's ability to capture the intrinsic connection between temporal and spatial features. Our experimental results on six traffic prediction datasets show the state-of-the-art performance of the model.

Keywords: Traffic forecasting · Attention Mechanisms · Intelligent Transportation Systems · Convolutional Neural Networks

1 Introdutcion

The goal of Traffic forecasting [1] is to predict future traffic data information for a road network based on historical observations. In recent years, deep learning-based Traffic forecasting models have achieved remarkable success. Among them, spatio-temporal graphical neural networks (STGNNs) [2,3] and Transformer-based models [4,5] have become very popular due to their outstanding performance. In addition to these, researchers have developed many sophisticated and

© The Author(s), under exclusive license to Springer Nature Singapore Pte Ltd. 2025
T. Zhu et al. (Eds.): ICA3PP 2024, LNCS 15251, pp. 309–318, 2025.
https://doi.org/10.1007/978-981-96-1525-4_17

complex models for traffic flow prediction, including graph convolution models [6,7], learning graph structure models [8–12], efficient attention mechanism models [13–17] and other models [18–20].

Feature extraction is a crucial step in Traffic forecasting. STGNNs mainly use feature embedding E_f. With E_f, the original input can be projected to the high-latitude hidden space to capture spatio-temporal features. Since the attention mechanism cannot preserve the location features of the time series, models based on the attention mechanism require additional embedding structures such as temporal location encoding E_{tpe} and periodic (daily, weekly, monthly) embedding E_p. Models in recent years, including PDFormer [4], GMAN [5], STID [21] and STAE-former [22], all use spatial embedding E_S to better capture spatial features. Among these, STID [21] has a very innovative embedding model, which employs both spatial embedding and temporal periodic embedding and achieves significant performance with a simple Multi-layer Perceptron (MLP). STAEformer [22] pioneered the application of a spatio-temporal adaptive embedding layer, which utilizes adaptive embedding E_a to capture features in both the temporal and spatial dimensions.

However, all these methods have the obvious shortcoming which is extracting temporal and spatial features independently and ignoring the intrinsic correlation between temporal and spatial information. We are inspired by the Transformer [25] architecture, especially the way its encoder and decoder are connected to each other, and use the cross-attention mechanism to realize the fusion of temporal and spatial features. We fuse complex spatio-temporal features more effectively by allowing the spatial tensor to pre-capture features on the temporal tensor. The contributions of this paper are mainly as follows:

- We propose a TSformer based on Transformer, which utilizes the cross-attention mechanism to allow the model to capture temporal features before learning spatial features, thus realizing the fusion of temporal and spatial features.
- For the purpose of improving the efficiency of pre-learning, we propose a Feature Pre-extraction Block (FPB), which applies convolutional layers in the spatio-temporal cross-attention layer for the pre-capture of temporal features.
- Experiments and analyses on six real-world traffic datasets prove that our model can achieve state-of-the-art performance (SOTA) for traffic forecasting using the transformer.

2   Related Work

2.1   Attention Mechanism

Attention Mechanism [25] has been widely used in various domains because of its efficiency and flexibility in modeling dependencies. Its core idea is to adaptively focus on the most relevant features based on the input data. In our method, we incorporate the cross-attention mechanism [25] into modeling the intrinsic temporal-spatial dependencies, thus better integrating temporal and spatial features. Furthermore, inspired by COTattention [23], we apply convolutional layers

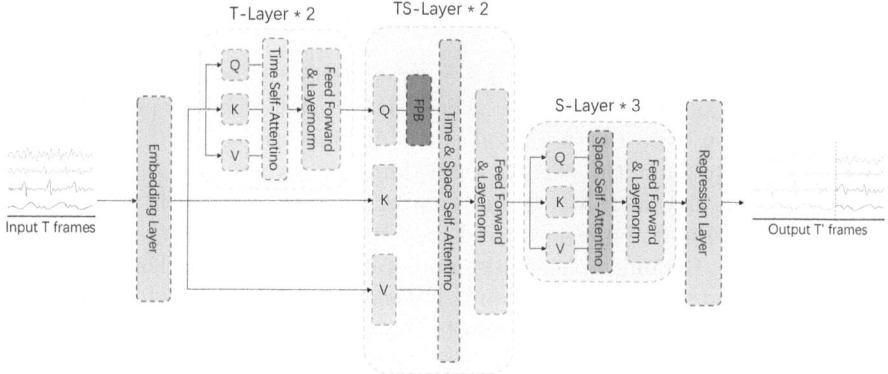

Fig. 1. The Architecture of Temporal-Spatial former (**TSformer**).

in the temporal-spatial cross-attention layer to pre-capture temporal features, aiming to enhance the fusion of temporal and spatial features.

2.2   Traffic Forecasting

Traffic forecasting has been studied extensively over the past few decades. For example, DCRNN [2] models the dynamics of traffic flow as a diffusion process and proposes a diffusion convolution operation to capture spatial dependencies. AGCRN [24] considering the specificity of traffic patterns at different nodes, separate weights and biases are used for each node to learn the traffic patterns specific to each node in order to capture the characteristics of different observed nodes. STGCN [6] uses a purely convolutional structure to build the model, using fewer parameters in exchange for faster training. There are also models that do not require the aid of graph structures, such as STID [21] and STAEformer [22], the former applies spatial embedding and temporal periodic embedding, and the latter applies an adaptive spatio-temporal embedding layer on this basis, both achieve excellent performance through a simple network structure.

The purpose of Traffic forecasting is to predict the future traffic time series in road networks based on historical observations. Specifically, the historical observations $X_{t-T+1:t}$ includes the traffic data of the previous T time nodes. Traffic prediction aims to infer the traffic data of the next T' time points by training a model $F(\cdot)$ with parameters Θ, which can be expressed as:

$$[X_{t-T+1:}, ..., X_t] \xrightarrow{F(\Theta)} [X_{t+1}, ..., X_{(t+T')}] \tag{1}$$

where the information for each time node $X_i \in \mathbb{R}^{N \times d}$, N is the number of spatial nodes, and d is the dimension of the input feature, which equals 1 in our case, representing traffic flow.

3   Methodology

As shown in Fig. 1, our model consists of three parts, the embedding layer, the attention layer and the regression layer. First, considering the periodicity of the real world, we use two temporal embedding matrices: $T_D \in \mathbb{R}^{N_D \times d}$ and $T_W \in \mathbb{R}^{N_W \times d}$ to form the periodic embedding E_p, where N_D is the number of samples in a day and N_W is the number of days in a week. Second, we project along the spatial dimension and the temporal dimension: $f_s \in \mathbb{R}^{N \times d}$ and $f_t \in \mathbb{R}^{N \times d}$, thus obtaining the spatio-temporal embedding E_{ts}. For each node, the temporal and spatial information is learned adaptively through a randomly initialized matrix $E_a \in \mathbb{R}^{N \times d_a}$, aiming to capture the unique traffic patterns of the nodes. Finally, by concatenating, we obtain the output of the Embedding Layers $Z \in \mathbb{R}^{T \times N \times d_h}$:

$$Z = E_p || E_a || E_{ts} \tag{2}$$

The most complex part of them is the attention layer. The attention layer consists of three sub-layers: the first is a temporal self-attention layer, the second is a temporal-spatio cross-attention layer (which incorporates Feature Pre-extraction Block(FPB)), and the third is a spatial self-attention layer. The input data is mapped to a high dimensional hidden space through the embedding layer, after which it is successively passed through the temporal self-attention layer, the spatio-temporal cross-attention layer and the spatial self-attention layer for feature extraction, and the obtained data is processed through the regression layer to obtain the final prediction results.

3.1   Temporal-Spatial Attention and Regression Layer

We applied the basic Transformer model along the temporal and spatial axes to capture complex traffic relationships. The input to the spatial self-attention layer is the hidden vector $Z \in \mathbb{R}^{T \times N \times d_h}$, which contains T temporal nodes and N spatial nodes. First, we obtain the query matrix Q^{te}, key matrix K^{te}, and the value matrix Vte through the self-temporal self-attention layer:

$$Q^{te} = ZW_Q^{te}, \quad K^{te} = ZW_K^{te}, \quad V^{te} = ZW_V^{te} \tag{3}$$

Here $W_Q^{te}, W_K^{te}, W_V^{te} \in \mathbb{R}^{d_h \times d_h}$ are the learnable parameters. We calculate the self-attention weights as:

$$A^{te} = \mathrm{Softmax}\left(\frac{Q^{te}K^{te^T}}{\sqrt{d_h}}\right) \tag{4}$$

where $A^{te} \in \mathbb{R}^{N \times T \times T}$ captures the temporal relationships in different nodes. Finally, we obtain the output of temporal self-attention:

$$Z^{te} = A^{te}V^{te} \tag{5}$$

Next, we input the Z^{te} into the Temporal Cross Attention layer. Here, we draw on the cross-attention mechanism from the Transformer [25], linking the output

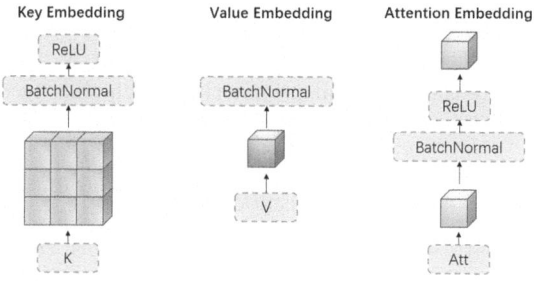

Fig. 2. The components of Feature Pre-Extraction Block (FPB).

of the temporal self-attention layer Z^{te} with the initial input Z, and performing cross-attention calculations along the spatial axis. The purpose is to pre-learn temporal features before extracting spatial features, with the learning process as follows:

$$Z_n^{ts} = \text{CrossAttention}(Z^{te}, Z_{n-1}^{ts}, Z_{n-1}^{ts}) \qquad (6)$$

where n represents the index of the cross attention sublayer. When $n = 1$, Z_{n-1}^{ts} is the initial input Z. After that, we input the Z_n^{ts} into the spatial self-attention layer and continue the self-attention computation along the spatial dimension to obtain the result that incorporates spatio-temporal features Z', the process is shown below:

$$Z' = \text{SelfAttention}(Z_n^{ts}) \qquad (7)$$

Finally, we take the output of the spatial self-attention layer $Z' \in \mathbb{R}^{(T \times N \times d_h)}$ through the regression layer to generate the predicted values. The detailed process can be represented as follows:

$$\hat{Y} = FC(Z') \qquad (8)$$

where $\hat{Y} \in \mathbb{R}^{T' \times N \times d_h}$ is the predicted value, and T' is the time span of the prediction, and d_h is the dimension of the output feature, which in our case is equal to 1. Thus, the regression layer takes the tensor dimension from $T \times N$ in the Z' to the $T' \times (d = 1)$ in the $\hat{Y}$.

3.2   Feature Pre-extraction Block (FPB)

To better enable the model to pre-learn temporal features before capturing spatial features, we apply a Feature Pre-Extraction Block to enhance the efficiency of pre-learning. The structure of our Feature Pre-Extraction Block is shown in Fig. 2. It consists of three key components: Key Embedding, Value Embedding, and Attention Embedding. The Key Embedding primarily comprises a convolution kernel $Cov_3 \in \mathbb{R}^{3 \times 3}$, a Batch Normalization layer, and a ReLU activation layer. The Value Embedding consists of a convolution kernel $Cov_1 \in \mathbb{R}^{1 \times 1}$ and a Batch Normalization layer. The Attention Embedding is made up of two convolution kernels $Cov_1 \in \mathbb{R}^{1 \times 1}$, a Batch Normalization layer, and a ReLU activation layer. We assign the original data X as input to K, Q, and V. We use Key Embedding with K to generate the key feature matrix K_1, use Value Embedding and V to generate the value feature matrix V', which is used for subsequent attention weighting.

$$K_1 = \text{KeyEmbedding}(K) \tag{9}$$

$$V' = \text{ValueEmbedding}(V) \tag{10}$$

After that, the key feature matrix K_1 is concatenated with the original input X along the feature dimension to obtain Y'. Y' is then processed through Attention Embedding to obtain the attention distribution Att. Next, the computed Attention Distribution Att and Value Feature Matrix V' undergo dot product processing to obtain the weighted feature matrix K_2.

$$Att = \text{Concat}(K_1, Q) \tag{11}$$

$$K_2 = \text{DotProduct}(Att, V') \tag{12}$$

Finally, the values at corresponding positions of the key feature matrix K_1 and the weighted feature matrix X' are added together to obtain the final output feature map X':

$$X' = \text{Add}(K_1, K_2) \tag{13}$$

Table 1. Performance on METR-LA and PEMS-BAY.

Datasets		Metric	HP[2021]	GWNet[2019]	DCRNN[2018]	AGCRN[2020]	STGCN[2018]	GTS[2021]	MTGNN[2020]	STNorm[2021]	GMAN[2020]	PDFormer[2023]	STID[2022]	STAEformer[2023]	TSformer
METR-LA	Horizon 3 (15 min)	MAE	6.80	2.69	2.67	2.85	2.75	2.75	2.69	2.81	2.80	2.83	2.82	2.65	2.62
		RMSE	14.21	5.15	5.16	5.53	5.29	5.27	5.16	5.57	5.55	5.45	5.53	5.11	5.04
		MAPE	16.72%	6.99%	6.86%	7.63%	7.10%	7.12%	6.89%	7.40%	7.41%	7.77%	7.75%	6.85%	6.71%
	Horizon 6 (30 min)	MAE	6.80	3.08	3.12	3.20	3.15	3.14	3.05	3.18	3.12	3.20	3.19	2.97	2.95
		RMSE	14.21	6.20	6.27	6.52	6.35	6.33	6.13	6.59	6.49	6.46	6.57	6.00	5.98
		MAPE	16.72%	8.47%	8.42%	9.00%	8.62%	8.62%	8.16%	8.47%	8.73%	9.19%	9.39%	8.13%	8.09%
	Horizon 12 (60 min)	MAE	6.80	3.51	3.54	3.59	3.60	3.59	3.47	3.57	· 3.44	3.62	3.55	3.34	3.33
		RMSE	14.20	7.28	7.47	7.45	7.43	7.44	7.21	7.51	7.35	7.47	7.55	7.02	6.99
		MAPE	10.15%	9.96%	10.32%	10.47%	10.35%	10.25%	9.70%	10.24%	10.07%	10.91%	10.95%	9.70%	9.68%%
PEMS-BAY	Horizon 3 (15 min)	MAE	3.06	1.30	1.31	1.35	1.36	1.37	1.33	1.33	1.35	1.32	1.31	1.31	1.30
		RMSE	7.05	2.73	2.76	2.88	2.88	2.92	2.80	2.82	2.90	2.83	2.79	2.78	2.77
		MAPE	6.85%	2.71%	2.73%	2.91%	2.86%	2.85%	2.81%	2.76%	2.87%	2.78%	2.78%	2.76%	2.73%
	Horizon 6 (30 min)	MAE	3.06	1.63	1.65	1.67	1.70	1.72	1.66	1.65	1.65	1.64	1.64	1.62	1.60
		RMSE	7.04	3.73	3.75	3.82	3.84	3.86	3.77	3.77	3.82	3.79	3.73	3.68	3.65
		MAPE	6.84%	3.73%	3.71%	3.81%	3.79%	3.88%	3.75%	3.66%	3.74%	3.71%	3.73%	3.62%	3.61%
	Horizon 12 (60 min)	MAE	3.05	1.99	1.97	1.94	2.02	2.06	1.95	1.92	1.92	1.91	1.91	1.88	1.87
		RMSE	7.03	4.60	4.60	4.50	4.63	4.50	4.50	4.45	4.49	4.43	4.42	4.34	4.30
		MAPE	6.83%	4.71%	4.68%	4.55%	4.72%	4.88%	4.62%	4.46%	4.52%	4.51%	4.55%	4.41%	4.42

4   Experiments

4.1   Experimental Setup

Datasets, Metrics. We conducted experiments on six traffic prediction benchmark datasets, namely METR-LA, PEMS-BAY, PEMS03, PEMS04, PEMS07, and PEMS08. The first two datasets were introduced by DCRNN [2], and the latter four by STSGCN [16] proposed. The time sampling interval of these six datasets is 5 min, thus there are 12 time points per hour. We used three widely adopted metrics in the traffic prediction task, namely MAE, RMSE, and MAPE. Following prior work, we chose to evaluate the average performance over the 12 predicted time steps for the PEMS03, PEMS04, PEMS07, and PEMS08 datasets. To assess performance on the METR-LA and PEMS-BAY datasets, we compared the performance over time horizons of 3, 6, and 12 (15, 30, and 60 min).

Implementation. We implement the model using the PyTorch toolkit on a Windows server equipped with a GeForce RTX 4070 Ti GPU. For METR-LA and PEMS-BAY, we divided them into training, validation, and testing sets with a ratio of 7:1:2, while PEMS03, PEMS04, PEMS07, and PEMS08 were divided according to a 6:2:2 ratio. More specifically, the embedding dimensions d_a is 84, and d is 24. The number of layers of temporal self-attention layer and temporal-spatio cross-attention layer is 2, the spatial self-attention layer is 3, and the number of heads is 4. We set the input and prediction lengths to 1 h, i.e. $T = T' = 12$. We choose Adam as the optimizer with a learning rate decaying from 0.001.

4.2   Performance Evaluation

As shown in Table 1 and Table 2, our method outperforms on most metrics across the six datasets, where TSformer significantly surpasses various STGNNs models without any graph modeling. STNorm and STID also perform well, and other Transformer-based models demonstrate their superiority in capturing complex spatio-temporal relationships.

Table 2. Performance on PEMS03, 04, 07, and 08.

Dataset	PEMS03			PEMS04			PEMS07			PEMS08		
Metric	MAE	RMSE	MAPE	MAE	RMSE	MAPE	MAE	RMSE	MAPE	MAE	RMSE	MAPE
HI[2021]	32.62	49.89	30.60%	42.35	61.66	29.92%	49.03	71.18	22.75%	36.66	50.45	21.63%
GWNet[2019]	14.59	25.24	15.52%	18.53	9.92	12.89%	20.47	33.47	8.61%	14.40	23.39	9.21%
DCRNN[2018]	15.54	27.18	15.62%	19.63	31.26	13.59%	21.16	34.14	9.02%	15.22	24.17	10.21%
AGCRN[2020]	15.24	26.65	15.89%	19.38	31.25	13.40%	20.57	34.40	8.74%	15.32	24.41	10.03%
STGCN[2018]	15.83	27.51	16.13%	19.57	31.38	13.44%	21.74	35.27	9.24%	16.08	25.39	10.60%
GTS[2021]	15.41	26.15	15.39%	20.96	32.95	14.66%	22.15	35.10	9.38%	16.49	26.08	10.54%
MTGNN[2020]	14.85	25.23	14.55%	19.17	31.70	13.37%	20.89	34.06	9.00%	15.18	24.24	10.20%
STNorm[2021]	15.32	25.93	14.37%	18.96	30.98	12.69%	20.50	34.66	8.75%	15.41	24.77	9.76%
GMAN[2020]	16.87	27.92	18.23%	19.14	31.60	13.19%	20.97	34.10	9.05%	15.31	24.92	10.13%
PDFormer[2023]	14.94	25.39	15.82%	18.36	30.03	12.00%	19.97	32.95	8.55%	13.58	23.41	9.05%
STID[2022]	15.33	27.40	16.40%	18.38	29.95	12.04%	19.61	32.79	8.30%	14.21	23.28	9.27%
STAEformer[2023]	15.35	27.55	15.18%	18.22	30.18	11.98%	19.14	32.60	8.01%	13.46	23.25	8.88%
TSformer	14.75	25.75	15.02%	18.05	29.90	12.36%	19.11	32.36	7.96%	13.34	23.11	8.75%

4.3   Ablation Study

To evaluate the effectiveness of each component in TSformer, we conducted ablation studies, including two variants of our model:

- **w/o FPB:** Remove the FPB block.
- **w/o TS:** Remove the temporal-spatio cross-attention layer.

Table 3 shows the impact of various modules on the performance of our model.

Table 3. Ablation Study on PEMS03, PEMS04 and PEMS08.

Dataset	PEMS03			PEMS04			PEMS08		
Metric	MAE	RMSE	MAPE	MAE	RMSE	MAPE	MAE	RMSE	MAPE
w/o FPB	14.89	25.85	15.47%	18.26	30.15	12.50%	13.45	23.25	8.88%
w/o TS	15.11	26.10	15.63%	18.39	30.21	12.61%	13.66	23.13	8.87%
TSformer	14.72	25.75	15.02%	18.05	29.90	12.36%	13.34	23.11	8.75%

When analyzing model performance comparisons, We examined the differences between the models with only the temporal self-attention layer, only the spatial self-attention layer, and the original model on PEMS03, PEMS04, and PEMS08 datasets (see Fig. 3). The analysis results indicate that the model performance significantly deteriorates when only the temporal attention layer is retained. Based on this observation, we choose to adopt the strategy of prioritizing the extraction of temporal features, which helps to improve the training effectiveness of the model.

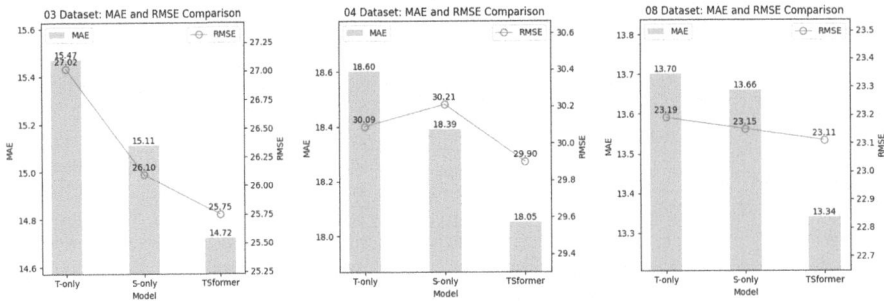

Fig. 3. Comparison of TSformer variants: (1) TSformer with only temporal attention layer (T-only), (2) TSformer with only spatial attention layer (S-only), and (3) Original TSformer, evaluated on PEMS03, PEMS04, and PEMS08 datasets.

5   Conclusion

By applying a temporal-spatial cross-attention fusion (TSCAF) mechanism, we have achieved remarkable success in traffic prediction. Our research demonstrates significant advantages in handling complex spatio-temporal patterns and overcoming the performance bottlenecks of traditional neural networks. Experimental results indicate that our model achieves state-of-the-art performance across six traffic prediction datasets, particularly excelling in simulating intricate spatio-temporal relationships. This novel approach offers an effective solution to traffic forecasting challenges, achieving satisfactory results in experiments.

References

1. Jiang, R., et al.: DL-Traff: survey and benchmark of deep learning models for urban traffic prediction. In: Proceedings of the 30th ACM International Conference on Information & Knowledge Management, pp. 4515–4525 (2021)
2. Li, Y., Yu, R., Shahabi, C., Liu, Y.: Diffusion convolutional recurrent neural network: data-driven traffic forecasting. In: International Conference on Learning Representations (2018)
3. Yu, B., Yin, H., Zhu, Z.: Spatio-temporal graph convolutional networks: a deep learning framework for traffic forecasting. In: Proceedings of the 27th International Joint Conference on Artificial Intelligence, pp. 3634–3640 (2018)
4. Jiang, J., Han, C., Zhao, W.X., Wang, J.: PDFormer: propagation delay-aware dynamic long-range transformer for traffic flow prediction. In: AAAI. AAAI Press (2023)
5. Zheng, C., Fan, X., Wang, C., Qi, J.: GMAN: a graph multi-attention network for traffic prediction. In: Proceedings of the AAAI Conference on Artificial Intelligence, vol. 34, pp. 1234–1241 (2020)
6. Song, C., Lin, Y., Guo, S., Wan, H.: Spatial-temporal synchronous graph convolutional networks: a new framework for spatial-temporal network data forecasting. In: Proceedings of the AAAI Conference on Artificial Intelligence, vol. 34, pp. 914–921 (2020)
7. Wang, X., et al.: Traffic flow prediction via spatial temporal graph neural network. In: Proceedings of the Web Conference 2020, pp. 1082–1092 (2020)
8. Jiang, R., et al.: Spatio-temporal meta-graph learning for traffic forecasting. In: Proceedings of the AAAI Conference on Artificial Intelligence, vol. 37, pp. 8078–8086 (2023)
9. Shang, C., Chen, J., Bi, J.: Discrete graph structure learning for forecasting multiple time series. In: International Conference on Learning Representations (2021)
10. Wu, Z., Pan, S., Long, G., Jiang, J., Chang, X., Zhang, C.: Connecting the dots: multivariate time series forecasting with graph neural networks. In: Proceedings of the 26th ACM SIGKDD International Conference on Knowledge Discovery & Data Mining, pp. 753–763 (2020)
11. Wu, Z., Pan, S., Long, G., Jiang, J., Zhang, C.: Graph wavenet for deep spatial-temporal graph modeling. In: Proceedings of the 28th International Joint Conference on Artificial Intelligence, pp. 1907–1913 (2019)
12. Zhang, Q., Chang, J., Meng, G., Xiang, S., Pan, C.: Spatio-temporal graph structure learning for traffic forecasting. In: Proceedings of the AAAI Conference on Artificial Intelligence, vol. 34, pp. 1177–1185 (2020)

13. Cirstea, R.-G., Yang, B., Guo, C., Kieu, T., Pan, S.: Towards spatio-temporal aware traffic time series forecasting–full version. arXiv preprint arXiv:2203.15737 (2022)
14. Liu, S., et al.: Pyraformer: low-complexity pyramidal attention for long-range time series modeling and forecasting. In: International Conference on Learning Representations (2021)
15. Wu, H., Xu, J., Wang, J., Long, M.: Autoformer: decomposition transformers with auto-correlation for long-term series forecasting. In: Advances in Neural Information Processing Systems 34, pp. 22419–22430 (2021)
16. Zhou, H., et al.: Informer: beyond efficient transformer for long sequence time-series forecasting. In: Proceedings of the AAAI Conference on Artificial Intelligence, vol. 35, pp. 11106–11115 (2021)
17. Zhou, T., Ma, Z., Wen, Q., Wang, X., Sun, L., Jin, R.: FEDformer: frequency enhanced decomposed transformer for long-term series forecasting. In: International Conference on Machine Learning. PMLR, pp. 27268–27286 (2022)
18. Cirstea, R.-G., Kieu, T., Guo, C., Yang, B., Pan, S.J.: EnhanceNet: plugin neural networks for enhancing correlated time series forecasting. In: 2021 IEEE 37th International Conference on Data Engineering (ICDE), pp. 1739–1750. IEEE (2021)
19. Lee, H., Jin, S., Chu, H., Lim, H., Ko, S.: Learning to remember patterns: pattern matching memory networks for traffic forecasting. In: International Conference on Learning Representations (2022)
20. Pan, Z., et al.: AutoSTG: neural architecture search for predictions of spatio-temporal graph. In: Proceedings of the Web Conference 2021, pp. 1846–1855 (2021)
21. Shao, Z., Zhang, Z., Wang, F., Wei, W., Xu, Y.: Spatial-temporal identity: a simple yet effective baseline for multivariate time series forecasting. In: Proceedings of the 31st ACM International Conference on Information & Knowledge Management, pp. 4454–4458 (2022)
22. Liu, H., Dong, Z., Jiang, R., Deng, J., Chen, Q., Song, X.: Spatio-temporal adaptive embedding makes vanilla transformer SOTA for traffic forecasting. In: Proceedings of the 32nd ACM International Conference on Information and Knowledge Management (CIKM) (2023)
23. Li, Y., Yao, T., Pan, Y., Mei, T.: Contextual transformer networks for visual recognition. IEEE Trans. Pattern Anal. Mach. Intell. **45**(2), 1489–1500 (2023)
24. Bai, L., Yao, L., Li, C., Wang, X., Wang, C.: Adaptive graph convolutional recurrent network for traffic forecasting. In: Advances in Neural Information Processing Systems 33, pp. 17804–17815 (2020)
25. Vaswani, A., Shazeer, N., Parmar, N., et al.: Attention is all you need. In: Advances in Neural Information Processing Systems 30 (2017)

Temperature-Based Watermarking and Detection for Large Language Models

Weitong Chen[1], Zhenxin Zhang[2], Huali Ren[3], Pei-Gen Ye[4(✉)],
Zhengdao Li[1], and Shanshan Huang[1]

[1] Institute of Artificial Intelligence, Guangzhou University, Guangzhou, China
[2] School of Cyber Engineering, Xidian University, Xi'an, China
[3] School of Cyberspace Security, Guangzhou University, Guangzhou, China
[4] School of Cyberspace Science and Technology, Beijing Institute of Technology,
Beijing, China
pgmhxy@gmail.com

Abstract. With the wide application of Large Language Models
(LLMs), protecting the copyright of generated content and preventing
its misuse becomes important. This paper proposes a temperature-based
watermark embedding algorithm that embeds watermarks in text using
the Softmax function and polynomial sampling techniques. Meanwhile,
this paper also discusses a watermark detection technique based on statistical testing, which can effectively identify and verify watermarks embedded in text. By applying these techniques to different LLMs and computing environments, including OPT series, Llama series, BLOOM series
and GPT-2, this paper analyses the scenarios, evaluates the key parameters in the algorithms and proposes solutions to ensure the integration
of watermarks without compromising on the performance of the model
or the naturalness of the generated text.

Keywords: Large Language Models · Text Watermarking · Machine
Learning Security · Copyright Protection

1 Introduction

Recently, Large Language Models (LLMs) have made a profound impact on
the field of AI [1,2]. Particularly in the field of natural language processing
(NLP) [3–5], as the number of parameters increases, LLMs have shown significant improvement in capabilities in fields such as information retrieval (IR)
[6,7]. However, using LLMs also presents some challenges [8,9], such as their
ability to quickly generate high-quality text, which may accelerate the spread of
misinformation [10]. Additionally, LLMs involve important intellectual property
issues, including copyright issues related to training datasets [11] and the right
to extract knowledge from LLMs [12]. To address these challenges, watermarking
has emerged as a vital technique. It embeds hidden marks in text generated by
LLMs for content tracking and source attribution, aiding in copyright protection,
document authentication, and preventing plagiarism.

© The Author(s), under exclusive license to Springer Nature Singapore Pte Ltd. 2025
T. Zhu et al. (Eds.): ICA3PP 2024, LNCS 15251, pp. 319–329, 2025.
https://doi.org/10.1007/978-981-96-1525-4_18

In this paper, we propose a temperature-based watermark embedding algorithm to enhance copyright protection and content originality by adjusting token probability vectors. We utilize Softmax and multinomial sampling for efficient watermark training and a statistical test for detection effectiveness. The algorithm is applied to 11 LLMs (OPT series, Llama series, BLOOM series, and GPT-2) across different types and parameter scales, analyzing watermark effectiveness across various parameter variations. Our main contributions are:

- We propose a rational temperature-integrated watermark embedding method that effectively embeds watermarks in text, ensuring content originality and recognizability.
- To effectively detect the embedded watermark, we present a watermark detection technique based on statistical tests.
- We validated the proposed watermarking technique's applicability and effectiveness across various LLMs through extensive experiments, including 11 different LLMs, and analyzed the impact of different key parameter settings on watermark effectiveness and text quality.

2   Related Work

In terms of adding watermarks to text, Brassil et al. [13,14], Por et al. [15], Rizzo et al. [16], and Sato et al. [17] focus on altering the text's appearance. Munyer and Zhong [18], WordNet [19], Topkara et al. [20], Yang et al. [21], and Yoo et al. [22] use semantic modifications, while Atallah et al. [3] and Topkara et al. [23] enhance robustness by changing grammatical structure. Abdelnabi and Fritz [24], Zhang et al. [25,26] embed watermarks during text generation.

With advancements in LLMs, researchers are developing methods to watermark LLM-generated text. Current techniques include watermarking during training, logits generation, and token sampling. Watermarking during training integrates information directly into the model, allowing specific watermarked text generation. Liu et al. [27] use modified inputs and labels for triggers, and Sun et al. [28] employ adversarial learning, though retraining is required. Watermarking during logits generation embeds watermarks in outputs before the final layer, as shown by Kirchenbauer et al. [29]. Watermarking during token sampling uses pseudorandom numbers for tamper-proofing, with Kuditipudi et al. and Hou et al. [30] improving this by randomizing sequences and embedding watermarks at the sentence level.

3   Watermark Embedding and Detection

3.1   Temperature Watermarking

This paper discusses watermarking algorithms for large language models, which embed watermarks while generating log vectors. Based on the KGW watermarking technique proposed by Kirchenbauer et al. [29], the vocabulary is divided into two lists: a green list and a red list, with words randomly assigned to each. We introduce a temperature parameter to modify the model's log vectors, further affecting the sharpness of the final probability distribution.

Given a vocabulary V ($V \geq 50,000$) and a prompt consisting of token sequence $s^{(-\tilde{N})}, \cdots, s^{(-1)}$, the model generates a sequence $s^{(0)}, \cdots, s^{(T)}$. For each predicted token $s^{(t)}$, the model outputs a log vector $l^{(t)}$, which is converted into a probability distribution $p_k^{(t)}$. Then, a sampling algorithm is used to obtain the token with the highest probability at position t from this distribution. To ensure consistency and security, the hash value of the previous token sequence is used as a seed for a random number generator. This maintains consistency in the generated sequence under the same conditions while ensuring randomness. Next, use this random number generator to generate a sequence of random numbers and randomly divide the vocabulary into a green list (G) and a red list (R). If the size of the green list $|S|_G$ is $\mu|V|$, then the size of the red list $|S|_R$ is $(1-\mu)|V|$, where $\mu \in (0,1)$.

Adjust the log vector of each token output by the model using the temperature parameter φ to obtain

$$\tilde{l}_k^{(t)} = \frac{l_k^{(t)}}{\varphi} \tag{1}$$

When $\varphi > 1$, high temperature makes the probability distribution more uniform, increasing the diversity and randomness of the text. When $\varphi < 1$, low temperature sharpens the probability distribution, making the generated text more deterministic but potentially decreasing diversity and innovation. When $\varphi = 1$, the probability distribution remains unchanged, reflecting the model's original predictions and preserving its inherent characteristics.

The log vectors of tokens $\tilde{l}_k^{(t)}$ are converted to probability distributions $p_k^{(t)}$ on the word list using the classification function Softmax,

$$p_k^{(t)} = \frac{e^{l_k^{(t)}}}{\sum_{i=1}^{|V|} e^{l_k^{(t)}}} \tag{2}$$

To increase the probability that the generated tokens are included in the green list, constant θ ($\theta > 0$) is introduced that is added to the log vectors of the words $\tilde{l}_k^{(t)}$ ($k \in G$) in the green list, from which the probability distributions of the tokens $\tilde{p}_k^{(t)}$ in the different lists can be obtained.

$$\tilde{p}_k^{(t)} = \begin{cases} \dfrac{e^{\left(\tilde{l}_k^{(t)}+\theta\right)}}{\sum_{i \in G} e^{\left(\tilde{l}_i^{(t)}+\theta\right)}+\sum_{i \in R} e^{\tilde{l}_i^{(t)}}}, & k \in G \\[4mm] \dfrac{e^{\tilde{l}_k^{(t)}}}{\sum_{i \in G} e^{\left(\tilde{l}_i^{(t)}+\theta\right)}+\sum_{i \in R} e^{\tilde{l}_i^{(t)}}}, & k \in R \end{cases} \tag{3}$$

Based on the probability distribution of the tokens $\tilde{p}_k^{(t)}$ in the word list, the next token $s^{(t)}$ is sampled using polynomial sampling to estimate the statistical properties of the predictive distribution, and the predicted sequence of tokens is generated step-by-step.

3.2 Watermark Detection

When detecting watermarks as opposed to generating watermark text, one only needs to understand the third-party hash function and random number generator to regenerate a red list for each token and calculate the frequency of rule violations without accessing the language model.

For this watermarking algorithm, the null hypothesis H_0 is set as follows: the token sequence $\{s^{(t)}\}(t = 0, 1, \cdots, T)$ is generated without knowledge of the red and green list rules.

Given the total number of tokens T to be generated, the proportion of green-list words appearing in the generated token sequence $p = |s|_G/T$, and the pre-determined proportion of green-list words $p_0 = \mu$. The standard error (SE) and the z-statistic value are calculated as follows:

$$SE = \sqrt{\frac{\mu(1 - \mu)}{T}} \tag{4}$$

$$z = \frac{|s|_G - \mu T}{SE \cdot T} \tag{5}$$

Assume that the threshold for the z value is z_0, which determines the significance level of the detection process. The effectiveness of the watermark can be assessed by comparing the z-value with the threshold z_0. Suppose the z-value produced by the watermarking algorithm exceeds the assumed threshold. In that case, it can be considered that the watermark is significant, and its presence can be detected efficiently and accurately.

4 Experiments and Results

Applying the watermarking algorithm to different LLMs such as the OPT series (OPT-125M, OPT-1.3B, OPT-2.7B, OPT-6.7B), LLAMA series (LLAMA-7B, LLAMA2-7B, CHINESE-LLAMA2-7B), BLOOM series (BLOOM-560M, BLOOM-3B, BLOOM-7B1), and the GPT-2. By studying four key variables in the algorithm, we investigate the effect of the watermarking algorithm on different large language models. The null hypothesis is rejected when the confidence level is significant at $z = 4$, indicating that the text generated by large language models has a watermark. Therefore, the threshold for z is set to 4.

4.1 The Impact of Predicted Number of Tokens T on Watermarking

This section primarily discusses the watermarking effectiveness when using 11 different LLMs with embedded watermarking algorithms under a unified standard $\mu = 0.25, \theta = 2, \varphi = 0.7$ when generating different numbers of tokens $T = 25, 50, 75, 100, 125, 150, 175, 200$.

Table 1. The z-values for LLMs with different numbers of predicted tokens T

T	opt-125m	opt-1.3b	opt-2.7b	opt-6.7b	llama-7b	llama2-7b	chinese-llama2-7b	bloom-560m	bloom-3b	bloom-7b1	gpt2
25	4.71	3.30	3.77	2.36	1.89	2.83	2.83	2.83	3.77	2.36	5.66
50	6.33	3.55	5.53	2.56	3.22	4.00	5.20	4.87	4.87	4.28	8.17
75	8.85	6.31	6.04	3.09	5.24	5.50	6.58	6.04	6.04	4.97	8.99
100	10.10	8.88	6.56	5.86	6.32	6.56	7.25	7.49	7.02	6.56	9.81
125	11.50	9.33	7.47	7.67	7.67	7.88	7.88	7.47	7.67	7.67	9.75
150	12.30	9.41	8.28	8.47	8.66	8.84	9.75	8.10	8.84	8.84	10.90
175	12.80	9.72	8.84	8.67	9.54	9.37	10.40	8.32	9.37	9.72	12.20
200	13.50	9.37	9.21	9.04	10.00	9.54	11.30	8.72	9.21	10.40	13.00

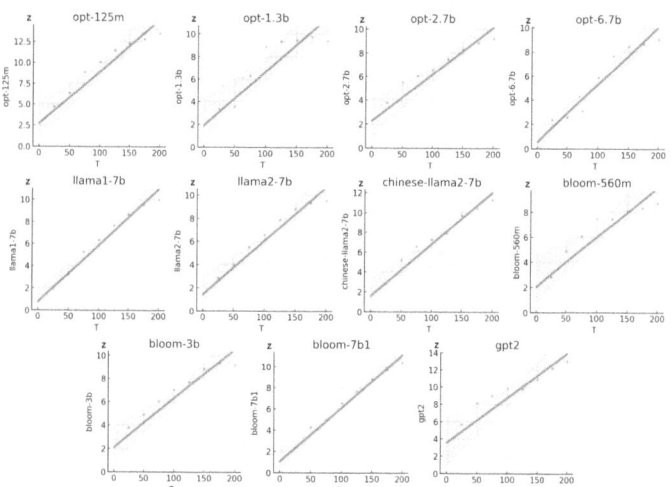

Fig. 1. Scatter plot of T and z-values

For the descriptive statistics of T and z-scores (from Table 1 and Fig. 1), it can be concluded that the z-values for all models increase as T increases. OPT-125M and GPT-2 reach thresholds earlier, showing more efficient watermark embedding, while larger models like OPT-6.7B need more tokens, indicating varied efficiency by model size. The z-values and the value of T have a strong linear relationship, with correlation coefficients between 0.90 and 0.99, suggesting a reliable prediction of z-value changes by token count.

Overall, all models show a strong positive linear relationship between the z-value and T, with T being a strong predictor of the z-value. When $T > 100$, all text sequences generated by these models can be identified as machine-generated with embedded watermarks, indicating that the watermark embedding algorithm has achieved good effectiveness.

4.2 The Impact of List Classification Parameters μ on Watermarking

This section discusses the classification scenarios with varying proportions of greenlisted items. Through experiments, data for different μ are obtained when $T = 200, \theta = 2, \varphi = 0.7$.

Observations from Table 2 and Fig. 2 show that around $\mu = 0.25$, the z-values of most models reach their peak, indicating that when approximately 25% of the words are labeled as greenlisted, the watermark embedding effect is optimal for most models. However, excessive greenlisted items are detrimental to watermarking, as the declining z-values with increasing μ values indicate. Thus, optimizing the watermarking algorithm is crucial at $\mu = 0.25$. Adjusting list classification parameters impacts model performance, with different models showing varying sensitivities. This insight facilitates optimizing watermarking strategies for improved model performance in practical applications.

Table 2. The z-values for LLMs with different list classification parameters μ

μ	opt-125m	opt-1.3b	opt-2.7b	opt-6.7b	llama1-7b	llama2-7b	chinese-llama2-7b	bloom-560m	bloom-3b	bloom-7b1	gpt2
0.10	9.76	8.29	5.46	7.82	5.46	5.22	6.64	5.22	4.99	7.11	8.06
0.15	10.40	8.37	6.38	8.96	7.18	5.39	9.56	6.18	6.58	6.98	10.40
0.20	11.40	9.78	8.19	9.25	7.66	9.78	9.43	8.01	8.54	10.70	11.90
0.25	13.50	9.37	9.21	9.04	10.00	9.54	11.30	8.72	9.21	10.4	13.00
0.30	11.60	7.63	8.25	9.17	9.02	7.94	10.6	8.09	7.47	8.40	11.20
0.35	10.10	7.04	7.19	7.78	7.93	6.29	9.71	7.48	7.19	7.48	9.71
0.40	9.11	6.57	6.57	6.71	6.71	5.12	8.60	6.71	7.44	6.71	8.45
0.45	8.84	6.62	5.76	6.19	5.76	3.91	7.47	7.33	6.48	5.76	8.04
0.50	7.68	5.74	5.17	5.46	5.32	4.18	6.73	6.03	6.17	5.74	7.73
0.55	7.16	4.64	3.78	4.78	4.50	6.21	6.06	5.64	5.21	5.49	7.06
0.60	6.41	4.72	4.14	5.01	3.85	6.45	6.45	5.3	4.57	5.15	6.02
0.65	6.15	3.81	3.81	5.00	3.37	5.45	5.89	4.85	4.11	5.89	5.30
0.70	5.02	4.75	4.28	5.37	2.89	4.59	4.75	4.44	3.2	5.21	4.28
0.75	4.51	4.54	3.56	4.87	1.60	3.4	4.05	3.07	2.41	4.54	4.22
0.80	3.48	4.57	2.98	4.22	0.85	2.98	3.15	1.56	2.27	4.04	2.98
0.85	3.32	3.74	2.75	3.54	1.96	3.15	2.35	1.96	1.76	4.34	2.15
0.90	2.56	2.58	1.16	2.58	1.39	2.58	1.16	1.16	2.10	3.52	1.63

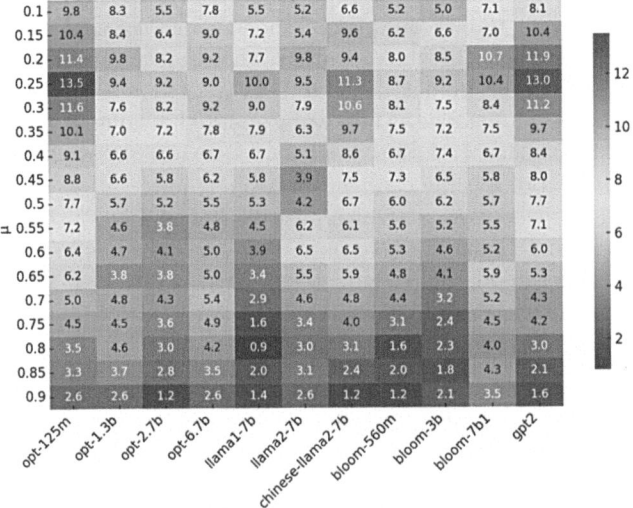

Fig. 2. Heatmap of z-values for each model under different μ

4.3   The Impact of Positive Deviation Amount θ on Watermarking

This section mainly discusses the watermarking effects of LLMs when adding different positive bias amounts θ to increase the probability of generating tokens as greenlisted items. Through experiments, data for different θ values ($\theta = 0, 0.1, 0.5, 1, 2, 5, 10$) are obtained when $T = 20, \mu = 0.25, \varphi = 0.7$.

Table 3. The z-values for LLMs with different positive deviation amount θ

θ	opt-125m	opt-1.3b	opt-2.7b	opt-6.7b	llama1-7b	llama2-7b	chinese-llama2-7b	bloom-560m	bloom-3b	bloom-7b1	gpt2
0	2.01	0.37	−0.12	−1.60	0.04	2.01	2.55	0.61	1.19	0.86	2.82
0.1	0.86	0.53	0.21	2.33	0.7	0.69	5.44	−0.78	2.01	1.84	1.84
0.5	5.77	0.04	1.84	1.51	4.3	2.66	4.00	2.82	4.62	2.82	1.35
1	7.74	5.44	5.28	4.13	5.12	2.99	10.40	8.55	5.61	6.26	5.28
2	13.50	9.37	9.21	9.04	10.00	9.54	11.30	8.72	9.21	10.40	13.00
5	23.10	15.60	18.00	21.00	20.70	6.14	20.60	16.30	23.50	16.90	22.80
10	24.40	24.30	23.70	23.80	24.00	24.40	20.80	16.32	24.30	24.30	24.40

With the increase in positive bias amount (see Table 3), the z-values of most models show growth, especially in certain models (such as CHINESE-LLAMA2-7B and GPT-2), where the z-values are higher, indicating sensitivity to watermarking. However, interpreting text sequences generated by large language models reveals a significant decrease in text quality and readability when $\theta > 5$. For example, considering OPT-1.3B, the perplexity of the generated text is calculated when $\theta = 0, 1, 2, 5, 10$ (Fig. 3).

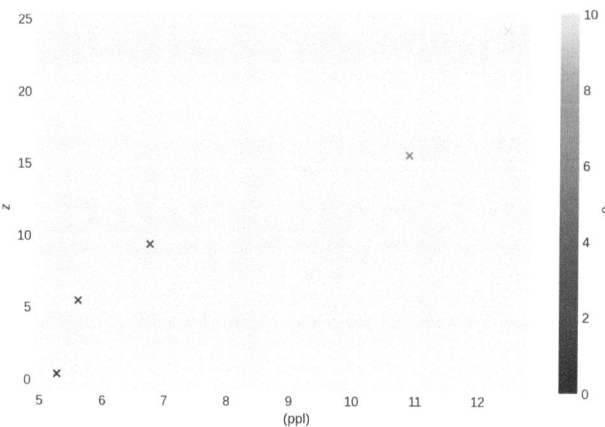

Fig. 3. Scatter plot of θ, z-values, and perplexity (ppl)

As θ increases, the z-values also increase, while perplexity (ppl) exhibits a certain degree of upward trend. This suggests that in this dataset, improving watermarking effectiveness comes at the expense of model-generated text quality. Therefore, selecting appropriate parameters to balance watermarking effectiveness and text quality is crucial.

4.4   The Impact of Temperature Parameters φ on Watermarking

This section mainly discusses the impact of watermarking effects on LLMs when using polynomial sampling with different temperature parameters φ for token sampling. Data for different temperature parameter values ($\varphi = 0.1, 0.2, 0.3, 0.4, 0.5, 0.6, 0.7$) are obtained through experiments.

Table 4. The z-values for LLMs with different temperature parameters φ

φ	opt-125m	opt-1.3b	opt-2.7b	opt-6.7b	llama1-7b	llama2-7b	chinese-llama2-7b	bloom-560m	bloom-3b	bloom-7b1	gpt2
0.1	7.41	7.41	6.92	10.4	9.86	9.7	11.7	10.7	13.8	14.4	11.5
0.2	14.8	9.6	7.41	11.2	11.8	8.55	9.86	11.3	9.21	8.72	15.3
0.3	14.1	9.7	5.77	12.8	5.28	8.72	10.5	13	11.5	8.06	15.3
0.4	14.3	11.8	10.4	6.75	7.24	8.72	11.2	11	12.5	14.8	17.9
0.5	14.3	9.86	9.7	10.8	13	9.21	8.88	12.5	12.3	11.2	14.9
0.6	11.5	10.8	8.55	7.24	9.7	10.8	8.95	13.1	12.3	13	12
0.7	13.5	9.37	9.21	9.04	10	9.54	11.3	8.72	9.21	10.4	13
0.8	13.3	10.5	8.72	8.72	9.54	10.5	6.59	10.5	9.21	17.6	10.2
0.9	13	11.3	9.21	8.94	8.72	5.93	6.43	5.11	8.23	12.8	11.3
1	12.2	9.54	7.24	7.24	8.06	8.72	12	10	11.2	9.05	10

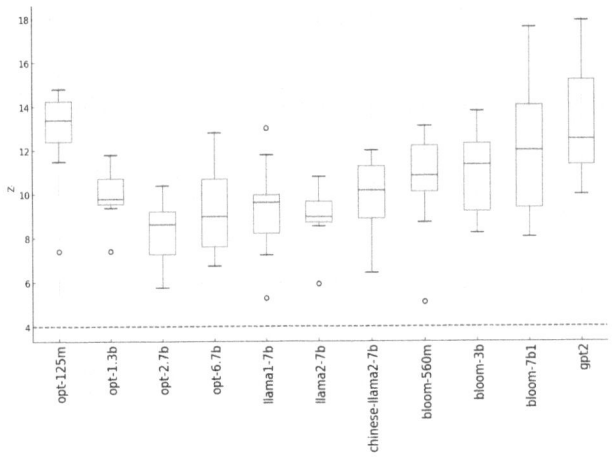

Fig. 4. Boxplot of z-values for each model under different φ

The z-values of all models under all coefficients exceed the set threshold of 4, further validating the effectiveness and consistency of these models in watermark embedding. Additionally, different models exhibit varying sensitivities of z-values to temperature coefficients φ, indicating significant fluctuations in z-values under different coefficients (from Table 4 and Fig. 4). Therefore, the coefficient of variation (CV) is introduced to understand the relative fluctuations of z-values under different temperature coefficients. Therefore, the temperature coefficient impacts watermark embedding in LLMs, making its optimization crucial for improving watermarking effectiveness and stability.

5   Conclusions

This paper comprehensively studies the development, implementation, and effectiveness of watermark embedding and detection techniques in LLMs. In particular, this study proposes a temperature-based embedding algorithm using Softmax and polynomial sampling, which not only enhances copyright protection but also ensures the originality and recognizability of the content. Additionally, the paper introduces a statistical test-based watermark detection technique that can effectively detect the presence and integrity of watermarks without directly accessing the language model. We apply the watermarking algorithm to 11 different LLMs (OPT series, Llama series, BLOOM series, and GPT-2), including various types of LLMs and models of the same type with different parameter scales, and analyzes the effectiveness of the watermark added to all models under four key parameter (predicted number of tokens T, list classification parameters μ, positive deviation amount θ and temperature parameters φ). Through many experiments, it can be concluded that the optimal parameters setting suggestions of the watermarking algorithm are as $T = 200$, $\mu = 0.25$, $\theta = 2$, $\varphi = 0.7$. However, further research is still needed regarding text generation accuracy and readability. These insights and data support further research and optimization of watermarking techniques, with significant implications for model optimization, advancing NLP technologies, and protecting digital content copyrights.

Acknowledgments. This work was supported by the National Natural Science Foundation of China (No. U23A20307, No. 62272118). We extend our gratitude to the Foundation for their financial support, our collaborators and team members for their contributions, and the editors and reviewers for their valuable feedback.

References

1. Bubeck, S., et al.: Sparks of artificial general intelligence: early experiments with GPT-4. arXiv preprint arXiv:2303.12712 (2023)
2. Turing, A.M.: Computing Machinery and Intelligence. Springer (2009)
3. Thede, S.M., Harper, M.: A second-order hidden Markov model for part-of-speech tagging. In: Proceedings of the 37th Annual Meeting of the Association for Computational Linguistics, pp. 175–182 (1999)
4. Bahl, L.R., Brown, P.F., De Souza, P.V., Mercer, R.L.: A tree-based statistical language model for natural language speech recognition. IEEE Trans. Acoust. Speech Signal Process. **37**(7), 1001–1008 (1989)
5. Brants, T., Popat, A., Xu, P., Och, F.J., Dean, J.: Large language models in machine translation. In: Proceedings of the 2007 Joint Conference on Empirical Methods in Natural Language Processing and Computational Natural Language Learning (EMNLP-CoNLL), pp. 858–867 (2007)
6. Liu, X., Croft, W.B.: Statistical language modeling for information retrieval. Annu. Rev. Inf. Sci. Technol. **39**(1), 1–31 (2005)
7. Mikolov, T., et al.: Statistical language models based on neural networks. Presentation at Google, Mountain View, 2 April 2012, vol. 80, no. 26 (2012)
8. Mirsky, Y., et al.: The threat of offensive AI to organizations. Comput. Secur. **124**, 103006 (2023)

9. Bergman, A.S., et al.: Guiding the release of safer e2e conversational AI through value sensitive design. In: Proceedings of the 23rd Annual Meeting of the Special Interest Group on Discourse and Dialogue. Association for Computational Linguistics (2022)
10. Megías, D., Kuribayashi, M., Rosales, A., Mazurczyk, W.: Dissimilar: towards fake news detection using information hiding, signal processing and machine learning. In: Proceedings of the 16th International Conference on Availability, Reliability and Security, pp. 1–9 (2021)
11. Tang, R., Feng, Q., Liu, N., Yang, F., Hu, X.: Did you train on my dataset? Towards public dataset protection with cleanlabel backdoor watermarking. ACM SIGKDD Explor. Newsl. **25**(1), 43–53 (2023)
12. Zhao, X., Wang, Y.-X., Li, L.: Protecting language generation models via invisible watermarking. In: International Conference on Machine Learning. PMLR, pp. 42187–42199 (2023)
13. Brassil, J.T., Low, S., Maxemchuk, N.F., O'Gorman, L.: Electronic marking and identification techniques to discourage document copying. IEEE J. Sel. Areas Commun. **13**(8), 1495–1504 (1995)
14. Begum, M., Uddin, M.S.: Digital image watermarking techniques: a review. Information **11**(2), 110 (2020)
15. Por, L.Y., Wong, K., Chee, K.O.: UniSpaCh: a text-based data hiding method using Unicode space characters. J. Syst. Softw. **85**(5), 1075–1082 (2012)
16. Rizzo, S.G., Bertini, F., Montesi, D.: Content-preserving text watermarking through Unicode homoglyph substitution. In: Proceedings of the 20th International Database Engineering & Applications Symposium, pp. 97–104 (2016)
17. Sato, R., Takezawa, Y., Bao, H., Niwa, K., Yamada, M.: Embarrassingly simple text watermarks. arXiv preprint arXiv:2310.08920 (2023)
18. Munyer, T., Zhong, X.: DeepTextMark: deep learning based text watermarking for detection of large language model generated text. arXiv preprint arXiv:2305.05773 (2023)
19. Topkara, U., Topkara, M., Atallah, M.J.: The hiding virtues of ambiguity: quantifiably resilient watermarking of natural language text through synonym substitutions. In: Proceedings of the 8th Workshop on Multimedia and Security, pp. 164–174 (2006)
20. Fellbaum, C.: WordNet: An Electronic Lexical Database. MIT Press, Cambridge (1998)
21. Yang, X., et al.: Tracing text provenance via context-aware lexical substitution. In: Proceedings of the AAAI Conference on Artificial Intelligence, vol. 36, no. 10, pp. 11613–11621 (2022)
22. Yoo, K., Ahn, W., Jang, J., Kwak, N.: Robust multi-bit natural language watermarking through invariant features. In: Proceedings of the 61st Annual Meeting of the Association for Computational Linguistics (Volume 1: Long Papers), pp. 2092–2115 (2023)
23. Topkara, M., Topkara, U., Atallah, M.J.: Words are not enough: sentence level natural language watermarking. In: Proceedings of the 4th ACM International Workshop on Contents Protection and Security, pp. 37–46 (2006)
24. Abdelnabi, S., Fritz, M.: Adversarial watermarking transformer: towards tracing text provenance with data hiding. In: 2021 IEEE Symposium on Security and Privacy (SP), pp. 121–140. IEEE (2021)
25. Zhang, R., Hussain, S.S., Neekhara, P., Koushanfar, F.: Remark-LLM: a robust and efficient watermarking framework for generative large language models. arXiv preprint arXiv:2310.12362 (2023)

26. Jang, E., Gu, S., Poole, B.: Categorical reparameterization with gumbel-softmax. arXiv preprint arXiv:1611.01144 (2016)
27. Liu, Y., Hu, H., Zhang, X., Sun, L.: Watermarking text data on large language models for dataset copyright protection. arXiv preprint arXiv:2305.13257 (2023)
28. Sun, Z., Du, X., Song, F., Ni, M., Li, L.: CoProtector: protect open-source code against unauthorized training usage with data poisoning. In: Proceedings of the ACM Web Conference 2022, pp. 652–660 (2022)
29. Kirchenbauer, J., Geiping, J., Wen, Y., Katz, J., Miers, I., Goldstein, T.: A watermark for large language models. In: International Conference on Machine Learning. PMLR, pp. 17061–17084 (2023)
30. Iqbal, M.M., Khadam, U., Han, K.J., Han, J., Jabbar, S.: A robust digital watermarking algorithm for text document copyright protection based on feature coding. In: 2019 15th International Wireless Communications & Mobile Computing Conference (IWCMC), pp. 1940–1945. IEEE (2019)

CPAKE: Dynamic Batch Authenticated Key Exchange with Conditional Privacy

Axin Xiang[1,2] , Youliang Tian[1,2(✉)] , Jinbo Xiong[3(✉)] , Zuobin Ying[4] , and Changgen Peng[1,2]

[1] State Key Laboratory of Public Big Data, College of Computer Science and Technology, Guizhou University, Guiyang 550025, China
youliangtian@163.com
[2] Guizhou Provincial Key Laboratory of Cryptography and Blockchain Technology, Guiyang 550025, China
[3] Industrial School of Joint Innovation, Quanzhou Vocational and Technical University, Quanzhou 362268, China
jbxiong@qvtu.edu.cn
[4] Faculty of Data Science, City University of Macau, Macau 999078, China

Abstract. Authenticated key exchange (AKE) needs to be designed for realizing point-to-multipoint secure communications in blockchain networks (BNet). However, since BNet is open, untrusted and decentralized, traditional certificateless AKE schemes are difficult to overcome two technical bottlenecks, i.e., private distribution of partial keys and accurate identification of anonymous traitors. This paper proposes a conditionally private and dynamic batch AKE scheme (CPAKE) in BNet. We first develop a privacy-preserving certificateless key generation algorithm to prevent the disclosure of nodes' privacy during partial key distribution. Then, we design a simple anonymous traitor tracing mechanism against Byzantine adversaries to accurately identify and trace the traitor's identity. Finally, we build a dynamic batch authentication protocol with decentralized node management to ensure the final validity of honest nodes. Moreover, performance analysis indicates that our scheme reduces the signing and batch-verifying time by approximately 193.10 ms and 10.93 ms respectively over state-of-the-art.

Keywords: Blockchain · Authenticated Key Exchange · Conditional Privacy

1 Introduction

Blockchain is a novel distributed paradigm and can bootstrap decentralized secure communications without trusted center [7]. Authenticated key exchange (AKE) [1,8] allows n users to agree on a shared session key, ensuring security of point-to-multipoint (P2MP) communications in blockchain networks (BNet). However, AKE schemes in BNet face two main challenges: the one is to authenticate a message transmitted in open BNet comes from a legitimate anonymous

© The Author(s), under exclusive license to Springer Nature Singapore Pte Ltd. 2025
T. Zhu et al. (Eds.): ICA3PP 2024, LNCS 15251, pp. 330–340, 2025.
https://doi.org/10.1007/978-981-96-1525-4_19

node [12]. The other implies that the trusted authority (TA) in decentralized BNet is multi-centered or cannot corrupt the overall AKE process [14].

Unlike hash-based message authentication [12] and factor-based identity authentication [16], digital signature can reliably link a message to its originating node, ensuring authenticity and integrity. However, since there is no a single TA in BNet to provide the registration and authentication services for public keys, adversaries can replace or forge keys and corrupt all signed AKE messages, threatening P2MP security and node privacy [1,7]. Naturally, certificateless aggregate signature provide security and efficiency guarantee in decentralized AKE by introducing batch verification mechanism and certificateless cryptography, avoiding the issues of key escrow [13], single point of failure [18], and certificate management [8]. Ulteriorly, a well-designed certificateless AKE scheme [3] can effectively prevent the illegal leakage or tampering of AKE materials [14].

Currently, some certificateless AKE schemes [9,17] use private channels to secure partial private key distribution, but their high design costs make them unsuitable for resource-limited scenarios like Internet of Vehicles (IoV). In the open wireless BNet, nodes communicate directly without access control [10], posing the risk of privacy disclosure when requesting partial private keys from key generation center (KGC). One reason is that if the keys are transmitted in plaintext, any adversary can directly exploit the eavesdropping attacks to compromise data privacy. Another reason is that KGC is typically required to authenticate node's identity, so the privacy protection of identity, i.e., anonymity mechanism [3], must be considered in such AKE schemes to resist the privacy theft by adversaries. Therefore, it is crucial to design a privacy-preserving and certificateless AKE scheme that ensures the dual privacy of data and identity.

While BNet often claims to be anonymous, this feature can be a double-edged sword in IoV [8]. The anonymity mechanism can protect nodes' identity and avoid frequent address changes [4], but it has spawned numerous uncontrollable fund-raising and fraud activities, resulting in chaos in blockchain market and highlighting the need for government oversight [2]. In a distributed AKE scenario, nodes authenticate via blockchain address to avoid exposing real identities in unknown environments [12]. However, if a malicious node (i.e., traitor [14]) disrupts the AKE process, service provider may need to trace and identify the real culprit, especially during the event, thus balancing privacy and traceability in BNet remains an open question [11]. Moreover, BNet's dynamic nature and potential rogue-key attacks [6] in batch verification mechanism complicate node and public key management in certificateless AKE schemes [5].

Briefly, we designs a conditionally private AKE scheme (CPAKE) with dynamic node management. The main contributions are summarized as follows.

- We propose a privacy-preserving certificateless key generation algorithm to achieve partial private key distribution without disclosing on-chain identity.
- We design a simple anonymous traitor tracing mechanism against Byzantine adversaries to achieve the accurate detection of malicious behavior.
- We build a decentralized and dynamic batch authentication protocol to realize fast on-chain registration and revocation of nodes and their public keys.

2   CPAKE: Dynamic Batch Key Exchange with Conditional Privacy

As shown in Fig. 1, our CPAKE scheme involves five entities: Key Generation Center (KGC), Identity Manager (IM), Blockchain (B), Smart Contract (SC), Blockchain Node (BN). It is divided into four phases with a total of 10 steps.

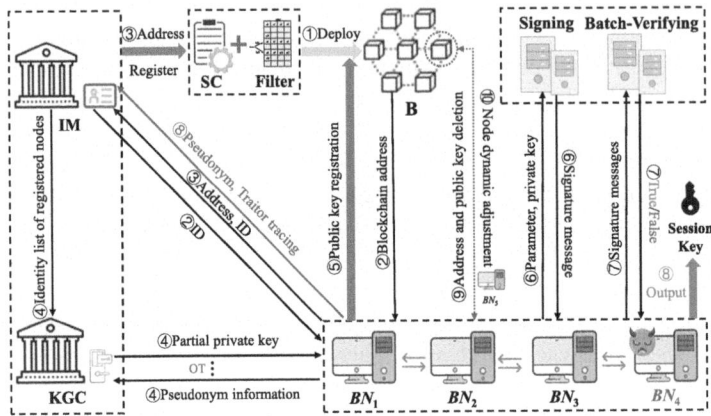

Fig. 1. Overview of the CPAKE scheme with 4 BNs where BN_4 is malicious.

2.1   System Initialization

Given a security parameter λ, TA, i.e., KGC and IM, selects an additive cyclic group $\mathbb{G}$ of prime order q over an elliptic curve E, a generator P of $\mathbb{G}$, a minimum number n of nodes participating in a single AKE and four hash functions $H_1 : \mathbb{G} \times \{0,1\}^* \rightarrow \mathbb{Z}_q^*$, $H_2 : \mathbb{G} \times \mathbb{G} \times \{0,1\}^* \times \{0,1\}^* \rightarrow \mathbb{Z}_q^*$, $H_3 : \mathbb{G} \times \mathbb{G} \times \{0,1\}^* \times \{0,1\}^* \times \mathbb{G} \times \{0,1\}^* \rightarrow \mathbb{Z}_q^*$, $H_4 : \{\mathbb{G}, \times^n\} \times \{\mathbb{G}, \times^n\} \times \{\{0,1\}^*, \times^n\} \times \mathbb{G} \rightarrow \mathbb{Z}_q^*$, where $\{\circ, \times^n\}$ denotes that there are n elements $\circ$ existing side by side in the form of symbol "$\times$". Then, KGC randomly selects $\alpha \in \mathbb{Z}_q^*$ as the system private key sk_{KGC} and computes the system public key $pk_{KGC} = sk_{KGC} \cdot P$. Next, IM randomly selects $\beta \in \mathbb{Z}_q^*$ as its private key and computes its public key $pk_{IM} = \beta \cdot P$. Finally, TA publishes the system public parameters $Params = \{E, \mathbb{G}, q, P, pk_{KGC}, pk_{IM}, n, H_1, H_2, H_3, H_4\}$. Furthermore, TA pre-deploys SC with functions of *NodeRegister*, *NodePkRegister*, *NodePkValidate*, *TraitorTracingValidate*, and *Noderevoke*. Given on-chain private key $ask_{IM} \in \mathbb{Z}_q^*$, IM obtains blockchain address $addr_{IM}$ and then is pre-written immutably into SC with pk_{IM}, where $addr_{IM}$ is the last 20 bytes of Keccak-256 hash [10] of the ask_{IM}-related on-chain public key, and SC can enhance anti-center dependence and on-chain information reliability [5].

2.2   Node and Public Key Registration

Node Registration: after obtaining the off-chain identity BID_i from IM, each BN_i with the on-chain private key $ask_i \in \mathbb{Z}_q^*$ and blockchain address $addr_i$ sends

---

Algorithm 1: *NodeRegister* Function

Input: $addr_i$
Output: $true/false$

1 ras is an address array for IM.
2 **if** $msg.sender$ *not in* ras **then**
3 | **return** $false$
4 **end**
5 $users$ is a mapping type.
6 **if** $users[msg.sender][addr_i] == true$ **then**
7 | **return** $false$
8 **end**
9 **else**
10 | $users[msg.sender][addr_i] \leftarrow true$
11 | **return** $true$
12 **end**

---

the message $\langle addr_i, BID_i \rangle$ to IM via private channel, completing the binding registration of BID_i to $addr_i$, and importantly, it provides the trusted address support for dynamic public key management. Specifically, if BID_i is legitimate, with the given transaction $TX^1_{IM} = sign_{ask_{IM}}(addr_i)$ using ECDSA, IM calls Algorithm 1 to determine and send TX^1_{IM} to BNet. Finally, IM locally stores a registered node identity list $list_{ID} = \{BID_1, BID_2, ..., BID_m\}$.

Public Key Registration: after registering the BN_i's address, other nodes $\{BN_j\}_{j \neq i}$ also need to confirm the validity of BN_i's public keys. Here, the registration details of BN_i's public keys are described as follows.

(1) Identity anonymization. Each node that wants to participate in an AKE randomly selects $r_i \in \mathbb{Z}_q^*$ and computes $pse_{i,1} = r_i \cdot P$, $pse_{i,2} = BID_i \oplus H_1((r_i \cdot pk_{IM})\|VT_i)$. Finally, the BN_i's pseudonym information $pse_i = \{pse_{i,1}\|pse_{i,2}\|VT_i\}$, where VT_i is the timestamp.
(2) Partial private key generation with privacy. IM first shares $list_{ID}$ with KGC. Notably, $list_{ID}$ covers all registered nodes' identity, but it cannot accurately determine which identity corresponds to which address, thus effectively ensuring the on-chain privacy of a certain BN except IM. Then, the detailed process of KGC generating the BN_i's partial private key is as follows.

- After receiving pse_i, KGC first randomly selects $s_{i,j} \in \mathbb{Z}_q^*$ for each identity BID_j in $list_{ID}$ and computes $S_{i,j} = s_{i,j} \cdot P$, $Q_{i,j} = H_2(S_{i,j}\|pse_i)$, $psk_{i,j} = (s_{i,j} + sk_{KGC} \cdot Q_{i,j}) \bmod q$. Then, by using an encoding function $f(x)$, KGC maps $S_{i,j}$ and $psk_{i,j}$ to elliptic curve points $W_{i,j}$ and $W'_{i,j}$ respectively. Next, KGC randomly selects $t_i, c_{i,j} \in \mathbb{Z}_q^*$ and computes $x_{i,j,1} = W_{i,j} + c_{i,j} \cdot t_i \cdot pse_{i,1}$, $x_{i,j,2} = W'_{i,j} + c_{i,j} \cdot t_i \cdot pse_{i,1}$, $x_{i,j,3} = c_{i,j} \cdot P$. Finally, KGC sends $X = \{x_{i,j,1}, x_{i,j,2}, x_{i,j,3}\}_{1 < j < m}$ to BN_i.

Algorithm 2: *NodePkRegister* Function

Input: $addr_{IM}$, $pklist_i$
Output: $true/false$

1 $K \leftarrow 2$ is the number of bucket arrays.
2 **if** $cuckoofilter == NULL$ **then**
3       $cuckoofilter.buckets \leftarrow new()[K]$ // Two filters mark two items.
4       $cuckoofilter.counter \leftarrow 0$
5 **end**
6 **if** $users[addr_{IM}][msg.sender] == true$ **then**
7   **for** pse_i and each npk_i^x in $pklist_i$ **do**
8   **if** $cuckoofilter.counter > MAX$ **then**
9               $cuckoofilter.empty()$ // Set two filters to initial state.
10   **end**
11          $cuckoofilter.buckets[0].add(pse_i)$ // pse_i is only inserted once.
12          $cuckoofilter.buckets[1].add(npk_i^x)$
13          $cuckoofilter.counter \leftarrow cuckoofilter.counter + 1$
14   **end**
15   **return** $true$
16 **end**
17 **else**
18   **return** $false$
19 **end**

- After receiving X, BN_i randomly select $\lambda_i \in \mathbb{Z}_q^*$ and computes $R_i = x_{l_i,3} + \lambda_i \cdot P$, where $x_{l_i,3}$ denotes the third element in X corresponding to the position l of BID_i in $list_{ID}$. Finally, BN_i sends R_i to KGC.
- After receiving R_i, KGC computes $R_i' = t_i \cdot R_i$ and $Z_i = t_i \cdot pse_{i,1}$. Finally, KGC returns $\{pse_i, Z_i, R_i'\}$ to BN_i.
- Based on $\{pse_i, Z_i, R_i'\}$ and $X_{i,j}$, BN_i computes $W_{l_i} = x_{l_i,1} - r_i \cdot R_i' + \lambda_i \cdot Z_i$, $W_{l_i}' = x_{l_i,2} - r_i \cdot R_i' + \lambda_i \cdot Z_i$. Then, based on an inverse encoding function $f^{-1}(x)$, BN_i maps W_{l_i} and W_{l_i}' to S_{l_i} and psk_{l_i} respectively, and obtains $S_i = S_{l_i}$, $psk_i = psk_{l_i}$. Finally, if the equation $psk_i \cdot P \overset{?}{=} S_i + H_2(S_i \| pse_i) \cdot pk_{KGC}$ is true, BN_i locally obtains partial private key $\{pse_i, S_i, psk_i\}$. Otherwise, the process terminates.

(3) Public key registration. BN_i randomly selects $\gamma_i \in \mathbb{Z}_q^*$ as his/her private key $nsk_i = \gamma_i$ and computes public key $npk_i = \gamma_i \cdot P$. To register multiple public keys, each BN_i prepares a list $pklist_i = \{pse_i, \{npk_i^x\}_{1 \leq x \leq M}\}$ and completes the registration operation in one step, where M is maximum number of public keys. Due to the anonymity of pse_i in BNet, the *NodePkRegister* function based on the cuckoo filter (CF) can only determine that npk_i is valid for $addr_i$, but cannot determine which identity the npk_i belongs to. Here, with the given transaction $TX_i^1 = sign_{ask_i}(addr_{IM}, pklist_i)$, BN_i calls Algorithm 2 to determine and send TX_i^1 to BNet, where only verified BNs can register their public keys. Two CFs are used to efficiently store

---

Algorithm 3: *NodePkValidate* Function

Input: $addr_{IM}$, $addrlist_i$, $npklist_i$
Output: $true/false$

1 **for** *each $addr_j \in addrlist_i$ and $npk_j \in npklist_i$* **do**
2   **if** $users[addr_{IM}][addr_j] == false$ **then**
3   **return** $false$
4   **end**
5   **if** $cuckoofilter.buckets[1].check(npk_j) == true$ **then**
6   **return** $true$
7   **end**
8 **end**
9 **return** $false$

---

elements in *pklisti*. If the number of items in a CF exceeds the threshold MAX, no new elements can be added, thus ensuring space efficiency. Based on the above design, BN_i can accurately check whether a certain item exists, facilitating the publication of the public key list.

2.3 Message Authentication and Key Exchange

In an authenticated BNet, we consider a complex but more appropriate P2MP application, in which a set $\{BN_1, BN_2, ..., BN_n\}$ containing $n \ll m$ BNs wants to agree a shared session key SK.

Message Signing: based on pseudonym information $pse_i = \{pse_{i,1} \| pse_{i,2} \| VT_i \}$, partial private key $\{pse_i, S_i, psk_i\}$, timestamp NT_i and public-private key pair $\{npk_i, nsk_i\}$, each BN_i randomly selects $f_i \in \mathbb{Z}_q^*$ and computes $v_i = f_i \cdot P$, $h_i = H_3(v_i \| pse_i \| npk_i \| NT_i)$, $\sigma_i = (psk_i + h_i \cdot nsk_i) \bmod q$. Finally, BN_i broadcasts the signed AKE message $msg_i = \{npk_i, pse_i, S_i, v_i, \sigma_i, NT_i\}$ to other nodes BN_j, where $1 < j < n$ and $j \neq i$.

P2P Key Exchange: upon receiving msg_j from BN_j with $addr_j$, BN_i verifies the freshness by checking $NT_{curr} - NT_j \leq \Delta T$, where NT_{curr} is the current timestamp. If this fails, BN_i terminates the process. Otherwise, BN_i computes $Q_j = H_2 (S_j \| pse_j)$ and $h_j = H_3 (v_j \| pse_j \| npk_j \| NT_j)$, and checks if $\sigma_j \cdot P \stackrel{?}{=} S_j + Q_j \cdot pk_{KGC} + h_j \cdot npk_j$. If not, BN_i rejects msg_j. Otherwise, BN_i computes two-party session key $SK_i^j = H_4((pse_i \| pse_j) \| (npk_i \| npk_j) \| (NT_i \| NT_j) \| f_i \cdot v_j)$.

P2MP Key Exchange: after receiving and verifying $n-1$ messages msg_j from BN_j with $addr_j$, given the IM's address $addr_{IM}$, node address list $addrlist_i = \{addr_j\}_{1<j<n, j \neq i}$ and the public key list $npklist_i = \{npk_j\}_{1<j<n, j \neq i}$, each BN_i builds a transaction $TX_i^2 = sign_{ask_i}(addr_{IM}, addrlist_i, npklist_i)$ and calls Algorithm 3 to determine and send TX_i^2 to BNet to check whether each BN_j's public key exists, mitigating rogue-key attacks [6]. If the output is $false$, BN_i terminates the process. Otherwise, for each valid msg_j, BN_i computes $Q_j =$

Algorithm 4: *TraitorTracingValidate* Function

Input: $addrlist_{IM}$, $pselist_{IM}$
Output: $true/false$

1 **if** $msg.sender\ not\ in\ ras$ **then**
2 | **return** $false$
3 **end**
4 **for** $each\ addr_z \in addrlist_{IM}\ and\ pse_z \in pselist_{IM}$ **do**
5 | **if** $users[msg.sender][addr_z] == false$ **then**
6 | | **return** $false$
7 | **end**
8 | **if** $cuckoofilter.buckets[0].check(pse_z) == true$ **then**
9 | | **return** $true$
10 | **end**
11 **end**
12 **return** $false$

$H_2\left(S_j \| pse_j\right)$ and $h_j = H_3\left(v_j \| pse_j \| npk_j \| NT_j\right)$, and then checks if the batch verification equation $\left(\sum_{j=1,j\neq i}^{n} \sigma_j\right) \cdot P \overset{?}{=} \sum_{j=1,j\neq i}^{n}\left(S_j + Q_j \cdot pk_{KGC} + h_j \cdot npk_j\right)$ holds. If not, BN_i implements traitor tracing (see Sect. 2.4 for details). Otherwise, BN_i accepts the $n-1$ messages and computes multi-party session key $SK = H_4((pse_1\| \cdots \| pse_n)\|(npk_1\| \cdots \| npk_n)\|(NT_1\| \cdots \| NT_n)\|\theta)$, where $\theta = f_i \cdot v_1 \cdot v_2 \cdot \ldots \cdot v_{i-1} \cdot v_{i+1} \cdot \ldots \cdot v_n$.

2.4   Traitor Tracing and Revocation

If the batch verification fails, traitors are identified by detecting invalid signatures regardless of network failures and tracing their identities through a pseudonym mechanism, thus updating the set of honest nodes participating in an AKE. The specific operations are as follows.

Invalid Signature Identifying: based on the fast binary invalid signature identification technology [15] and n signed messages $\{msg_1, msg_2, ..., msg_n\}$, each BN_i builds a binary tree of fast identification with a height of $\lfloor log_2 n \rfloor + 1$. The tree is built by splitting in half and batch-verifying the message set layer by layer until the leaf nodes only contain invalid signature messages. The reason why the number of invalid signatures $z \leq n/3$ is to ensure the reliability and validity of the AKE messages in BNet, i.e., Byzantine fault tolerance.

Traitor Tracing: in the Byzantine BNet, since BNs locally batch-verify all signed messages, there may be a case where some BNs maliciously initiate the traitor tracing requests. To address this, a fault-tolerant determination mechanism against $n/3$ Byzantine adversaries is designed to ensure consistent outputs across all honest nodes. Here, the tracing details are described as follows.

Algorithm 5: *Noderevoke* Function

Input: $addr_z$, $Trlist_i$

Output: $true/false$

1 **if** $msg.sender$ not in ras **then**
2 | **return** $false$
3 **end**
4 **if** $users[msg.sender][addr_z] == true$ **then**
5 | $users[msg.sender][addr_z] \leftarrow false$ // Invalidate node address. **for** pse_z and each npk_z^x in $Trlist_i$ **do**
6 | | $cuckoofilter.buckets[0].delete(pse_z)$ // pse_z is only revoked once.
7 | | $cuckoofilter.buckets[1].delete(npk_z^x)$
8 | **end**
9 | **return** $true$
10 **end**
11 **else**
12 | **return** $false$
13 **end**

- After obtaining $\{msg_z\}_{1 \leq z \leq n/3}$, each BN_i constructs and securely sends a traitor tracing request packet $ttrpacket_i = \{\{pse_z\}_{1 \leq z \leq n/3}, \{npk_z\}_{1 \leq z \leq n/3}, T_i^{IM}\}$ to IM, where T_i^{IM} is the timestamp of sending the packet to IM.

- After receiving and verifying $2n/3$ packets $ttrpacket_i$, IM re-counts the number $count$ for the entire set $\{pse_z\}_{1 \leq z \leq n/3}$. If $count \geq n/3$, given the node address list $addrlist_{IM} = \{addr_z\}_{1 \leq z \leq n/3}$ and pseudonym list $pselist_{IM} = \{pse_z\}_{1 \leq z \leq n/3}$, IM builds a transaction $TX_{IM}^2 = sign_{ask_{IM}}(addrlist_{IM}, pselist_{IM})$ and calls Algorithm 4 to determine and send TX_{IM}^2 to BNet to check the validity of each traitor's pseudonym information. If the output is $false$, IM terminates the process. Otherwise, for each valid pse_z, IM computes $BID_z = pse_{z,2} \oplus H_1((pse_{z,1} \cdot sk_{IM}) \| VT_z)$. Finally, IM obtains all traitors' identities $\{BID_z\}_{1 \leq z \leq n/3}$.

Traitor Revoking: after receiving $\{BID_z\}_{1 \leq z \leq n/3}$, given the traitor revocation list $Trlist_i = \{pse_z, \{npk_z^x\}_{1 \leq x \leq M}\}$, IM builds a transaction $TX_{IM}^3 = sign_{ask_{IM}}(addr_z, Trlist_i)$ and calls Algorithm 5 to determine and send TX_{IM}^3 to BNet to revoke $addr_z$, pse_z, and all npk_z^x of BN_z, in which the public information of BN_z will be removed from the two CF bucket arrays, and the state of the addresses will be set to invalid.

Dynamic Update: after revoking the traitors $\{BID_z\}_{1 \leq z \leq n/3}$, the remaining honest nodes $\{BID_1, BID_2, ..., BID_n\} \backslash \{BID_z\}$ exchange AKE materials with other registered nodes $\{BID_1, BID_2, ..., BID_m\} \backslash \{BID_n\}$ to replace the removed nodes, and then repeat the previous operations cyclically until the batch verification is passed, thereby ensuring that all final nodes are trusted in the current round of P2MP secure communication.

3 Evaluation

This section uses the JPBC library [7] to evaluate the computational cost of our CPAKE scheme and the existing certificateless AKE schemes [3,9,18] on a Windows 11 system with Intel(R) Xeon(R) W-1390 2.80 GHz processor and 64.0 GB RAM, in which we use Type A bilinear pairings with field width $q = 512$ and curve order $r = 160$ to construct the elliptic curve $y^2 = x^3 + x$.

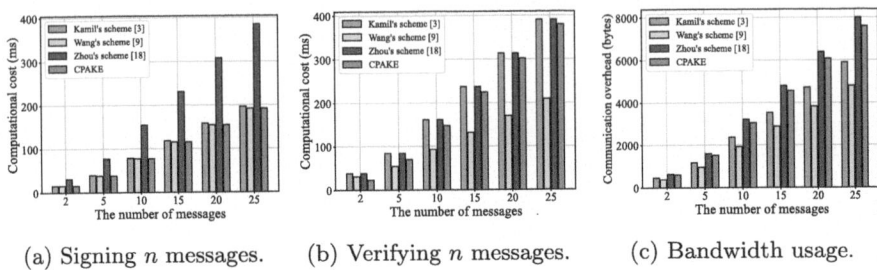

(a) Signing n messages.   (b) Verifying n messages.   (c) Bandwidth usage.

Fig. 2. Execution overhead of existing schemes

Figure 2 shows computation/communication cost of all AKE schemes when the number of messages $n \in [2, 25]$. As depicted in Fig. 2a, when signing n messages, the running time of our scheme always overlaps with that of the scheme [9], reducing by up to approximately 193.10 ms at $n = 25$. Figure 2b shows that our scheme reduces batch-verifying time by 2.8% to 39.75% compared to [3,18], with a 10.93 ms reduction at $n = 25$. Figure 2c depict that our scheme has lower communication overhead than the scheme [18] and the maximum difference is 400 bytes at $n = 25$, but has higher communication than [3,9]. Altogether, the execution overhead of our TP-AKA scheme is overall acceptable.

4 Conclusion

In this paper, we proposed a novel conditionally private and batch AKE scheme in BNet. The proposed CPAKE scheme first uses an oblivious transfer protocol to effectively address the privacy disclosure problem of partial private key. Then, based on a binary tree of fast verification, CF and SC, CPAKE scheme effectively achieves accurate identification, anonymous tracing, and public revocation of all traitors, thereby dynamically adjusting the set of honest nodes. Finally, performance analyses indicate that our scheme has relatively acceptable efficiency advantages over the existing certificateless AKE schemes.

Acknowledgements. This work was supported in part by National Key Research and Development Program of China under Grant 2021YFB3101100; National Natural Science Foundation of China under Grant 62272123, 62272102, 62462012; Science and

Technology Program of Guiyang under Grant [2022]2-4; Science and Technology Program of Guizhou Province under Grant [2022]065; Natural Science Foundation of Fujian Province under Grant 2023J02014; Open Research Fund of State Key Laboratory of Integrated Services Networks under Grant ISN24-16.

References

1. Badshah, A., Waqas, M., Muhammad, F., et al.: AAKE-BIVT: anonymous authenticated key exchange scheme for blockchain-enabled internet of vehicles in smart transportation. IEEE Trans. Intell. Transp. Syst. **24**(2), 1739–1755 (2022)
2. Houy, S., Schmid, P., Bartel, A.: Security aspects of cryptocurrency wallets-a systematic literature review. ACM Comput. Surv. **56**(1), 1–31 (2023)
3. Kamil, I.A., Ogundoyin, S.O.: A lightweight certificateless authentication scheme and group key agreement with dynamic updating mechanism for LTE-V-based internet of vehicles in smart cities. J. Inf. Secur. Appl. **63**, 102994 (2021). https://doi.org/10.1016/j.jisa.2021.102994
4. Li, L., Liu, J., Chang, X., et al.: Toward conditionally anonymous bitcoin transactions: a lightweight-script approach. Inf. Sci. **509**, 290–303 (2020)
5. Liu, S., Chen, L., Chen, L., et al.: Integrated and accountable data sharing for smart grids with fog and dual-blockchain assistance. IEEE Trans. Ind. Inf. **20**(3), 4940–4952 (2024)
6. Shim, K.A.: On the security of a certificateless aggregate signature scheme. IEEE Commun. Lett. **15**(10), 1136–1138 (2011)
7. Singh, A.K., Kumar, S.: An efficient and secure CLAKA protocol for blockchain-aided wireless body area networks. Expert Syst. Appl. **242**, 122740 (2024). https://doi.org/10.1016/j.eswa.2023.122740
8. Vallent, T.F., Hanyurwimfura, D., Kim, H., et al.: Certificate-less authenticated key agreement scheme with anonymity for smart grid communications. J. Intell. Fuzzy Syst. **43**(2), 1859–1869 (2022)
9. Wang, Z., Zhou, Y., Qiao, Z., et al.: An anonymous and revocable authentication protocol for vehicle-to-vehicle communications. IEEE Internet Things J. **10**(6), 5114–5127 (2023)
10. Wei, L., Cui, J., Zhong, H., et al.: A decentralized authenticated key agreement scheme based on smart contract for securing vehicular ad-hoc networks. IEEE Trans. Mob. Comput. **23**(5), 4318–4333 (2024)
11. Wen, B., Wang, Y., Ding, Y., et al.: Security and privacy protection technologies in securing blockchain applications. Inf. Sci. **645**, 119322 (2023)
12. Xiang, A., Gao, H., Tian, Y., et al.: DBKEM-AACS: a distributed key escrow model in blockchain with anonymous authentication and committee selection. Sci. China Inf. Sci. **66**(3), 139102 (2023)
13. Xiang, A., Gao, H., Tian, Y., et al.: H2CT: asynchronous distributed key generation with high computational efficiency and threshold security in blockchain network. IEEE Internet Things J. 1 (2024)
14. Xie, X., Wu, B., Hou, B.: BEPHAP: a blockchain-based efficient privacy-preserving handover authentication protocol with key agreement for internet of vehicles. J. Syst. Architect. **138**, 102869 (2023)
15. Xiong, H., Jin, C., Alazab, M., et al.: On the design of blockchain-based ECDSA with fault-tolerant batch verification protocol for blockchain-enabled IoMT. IEEE J. Biomed. Health Inform. **26**(5), 1977–1986 (2022)

16. Yang, K., Zhang, Z., Youliang, T., et al.: A secure authentication framework to guarantee the traceability of avatars in metaverse. IEEE Trans. Inf. Forensics Secur. **18**, 3817–3832 (2023)
17. Wei, J., Zhang, J., Zhang, J., et al.: CKAA: certificateless key-agreement authentication scheme in digital twin telemedicine environment. Trans. Emerg. Telecommun. Technol. **35**(1), e4922 (2024)
18. Zhou, J., Luo, M., Song, L., et al.: A dynamic group key agreement scheme for UAV networks based on blockchain. Perv. Mob. Comput. **95**, 101844 (2023). https://doi.org/10.1016/j.pmcj.2023.101844

Certificate-Based Transport Layer Security Encrypted Malicious Traffic Detection in Real-Time Network Environments

Yiran Suo[1], Jingfeng Xue[1($\boxtimes$)], Wenjie Guo[1], Wenbiao Du[1], Weijie Han[2($\boxtimes$)], and Chang Xu[1]

[1] Beijing Institution of Technology, Beijing, China
{3120230921,jfxue,3120215536,3220231346,3220231277}@bit.edu.cn
[2] Space Engineering University, Beijing, China
bit_hwj2016@126.com

Abstract. Encryption technology has become ubiquitous in network communication and encrypted malicious traffic detection becomes an important part of malware detection and cyber attack detection. Existing machine learning models and deep learning models are mainly trained based on packet length sequence information and time series information. Recent studies have shown that these models perform poorly in real network environments. In response to this challenge, this paper proposes a novel malicious traffic detection method based on certificate information extracted during the TLS (Transport Layer Security) encrypted handshake protocol. Our approach demonstrates that certificate information exhibits a strong correlation with the maliciousness of traffic, while remaining unaffected by the complexities of the real network environment. The experimental results illustrate that our method has high accuracy and low time overheading.

Keywords: Encrypted Malicious Traffic Detection · TLS · Certificate

1 Introduction

With the continuous advancement of network devices, monitoring traffic has become increasingly accessible to third parties. In addition, to safeguard the security and privacy of network transmissions, encryption technologies such as SSL/TLS are widely adopted. According to the Google Transparency Report, approximately 96% of the services provided by Google are encrypted [10]. The Transport Layer Security (TLS) protocol, as a crucial component of modern cybersecurity, facilitates secure communication over the internet. It has evolved from its predecessor, Secure Sockets Layer (SSL), and is instrumental in encrypting data exchanged between clients and servers, thereby ensuring the confidentiality and integrity of transmitted information.

© The Author(s), under exclusive license to Springer Nature Singapore Pte Ltd. 2025
T. Zhu et al. (Eds.): ICA3PP 2024, LNCS 15251, pp. 341–350, 2025.
https://doi.org/10.1007/978-981-96-1525-4_20

Nevertheless, malicious users are leveraging encryption technologies to shield their nefarious activities. The 2022 Encrypted Attacks Report [20] reveals that over 85% of attacks were encrypted in 2022, with the total attack volume exceeding that of 2021 by 20%. Some attackers employ Denial of Service (DoS) attacks to control botnets, aiming to circumvent DNS request detection. Additionally, hackers utilize the TLS protocol for communication post successful malware deployment. Traditional traffic packet detection methods like Deep Packet Inspection (DPI) are rendered inadequate for encrypted traffic inspection due to traffic encryption. Consequently, classifying encrypted traffic and promptly detecting abnormal information pose significant challenges that require resolution.

Recent research has introduced various methods for encrypted traffic analysis. Shen et al. [17] classify encrypted decentralized application traffic using graph neural networks. Fu et al. [8] detect malicious traffic and mitigate false positives through point cloud analysis. However, Xie and Cao et al. [19] observed that DL and ML models trained on features influenced by the network environment exhibit poor performance in real-world scenarios. Torroledo et al. [18] discovered that certificates harbor valuable information conducive to encrypted traffic detection.

Motivated by the findings of Xie et al. and the work of Torroledo et al., this work propose a certificate-based method for malicious traffic detection. Unlike methods reliant on packet length and timing features, our approach operates independently of the network environment. Furthermore, our paper curate public datasets on malicious traffic detection and amalgamate them into a comprehensive dataset encompassing multiple scenarios. Notably, our method exhibits high accuracy on this dataset and is suitable for real-time network deployment since it solely relies on certificate information, which can be obtained before TLS connection establishment.

The contributions are as follows:

1. This paper consolidate mainstream malicious traffic datasets to create a comprehensive dataset characterized by broad coverage, substantial data volume, and heightened complexity.
2. This paper introduce a novel approach for detecting encrypted malicious traffic, leveraging certificate information exclusively during TLS encrypted transmissions.
3. This paper conduct extensive validation experiments to substantiate the effectiveness, real-time performance, and low time complexity of our proposed method.

The remainder of the paper is structured as follows: Sect. 2 illustrates the motivation behind our work and related research. Section 3 describes dataset, our method and model. In Sect. 4, extensive experiments validate the efficacy of our method in malicious traffic detection. Finally, Sect. 5 concludes this paper.

2   Related Work

2.1   Encrypted Traffic Detection Based on ML and DL

As traditional DPI methods become inadequate for detecting encrypted traffic, researchers are increasingly turning to ML and DL techniques in this domain. The ML and DL-based approach typically includes three main stages: Data Collection, Feature Engineering, and Model Selection. With the continuous innovation in the field of deep learning, more and more studies are using deep learning rather than machine learning for detection. At the same time, many studies in recent years have shown that graph-based deep learning models can understand the relationships between packages and thus produce better training results.

In the Data Collection stage, researchers select datasets based on their research domain. Shen et al. [16] have categorized the research domain in encrypted traffic analysis, yet in the realm of malicious encrypted traffic analysis, two main domains emerge: malware detection and network anomaly detection. In our study, this paper gather four public datasets pertinent to malicious encrypted traffic analysis, as detailed in Sect. 3.A.

Regarding Feature Engineering, Shen et al. [15] have compiled a list of features in encrypted traffic, including packet ordering, packet length, and packet timing. However, our research observed many studies utilizing packet header contents as features. Consequently, our reasearch incorporate packet content features alongside those summarized by Shen et al.

In Model Selection, DL methods are increasingly favored due to their capability to perform feature selection without relying heavily on expert knowledge. Nonetheless, it's crucial to note that ML methods offer better interpretability and significantly reduce training time. Researchers must weigh these factors carefully when selecting an appropriate model for their specific application.

2.2   Transport Layer Security

Transport Layer Security (TLS) is a cryptographic protocol designed to ensure secure communication over computer networks. Serving as the successor to Secure Sockets Layer (SSL), TLS is extensively employed to safeguard web browsing and email communications by encrypting data exchanged between clients and servers. By upholding privacy, integrity, and authenticity of data, TLS effectively mitigates risks associated with eavesdropping, tampering, and forgery.

Behavioral information characterizing TLS traffic typically includes three primary aspects: Information contained in the Client Hello package, Information in the Server Hello package, and Information in the Certificate package. Notably, John Althouse et al. have introduced ja3 and ja4 as network fingerprinting methods for representing TLS traffic [4,11]. These techniques hold promise in detecting malicious traffic; however, they are susceptible to evasion by attackers who can manipulate traffic information.

To counteract the limitations posed by potentially manipulable traffic information, our research focuses on leveraging certificate information. Unlike traffic header data and time information, certificate information remains immutable, making it a reliable source for detecting malicious activities.

3   Methodology

This chapter focuses on our approach based on certificate information. This chapter will be divided into four parts: Framework, Data Collection, Feature Engineering and Model Selection.

3.1   Framework

As shown in Fig. 1, our work first collected Pcap files of all the datasets and we extracted the TLS encrypted traffic in these datasets using Wireshark. After that our work extracted the certificate information during each session using Brim. Finally our reasearch integrated all the extracted certificate data into one dataset and trained machine learning models to help us detect it.

In the feature extraction part, we just choose the certificate information as the feature for model training. The reason for this is that most of the models used in past studies are based on package ordering, package length, package timing. On the one hand, this information is easily influenced by the network environment. In the context of a high-performance network, these features may have a high degree of consistency with the data in the dataset. However, in a network environment with network congestion and performance limitations, packet loss, retransmission, and many other phenomena may occur. On the other hand, this information can also be modified by an attacker to evade model detection.

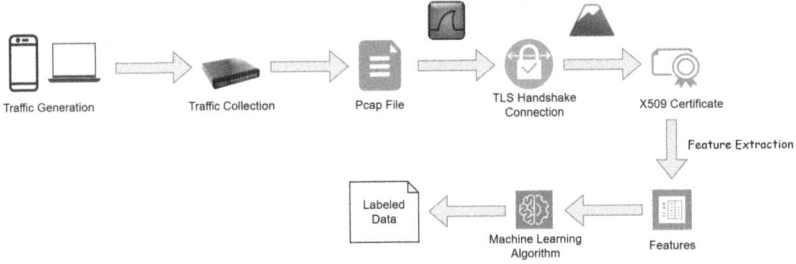

Traffic Generation   Traffic Collection   Pcap File   TLS Handshake   X509 Certificate
                                                                          Connection

Feature Extraction

Labeled Data   Machine Learning   Features
                    Algorithm

Fig. 1. Framework of certificate-based encrypted traffic detection

3.2   Data Collection

The study utilized four open-source datasets, each representing distinct malicious traffic scenarios:

1. CICIDS-2017 Dataset [14]: This dataset comprises traffic samples extracted from cyber attacks. The CIC-IDS-2017 dataset comprises both benign traffic and samples of the latest common attacks, mimicking real-world network traffic scenarios captured in Packet Capture (PCAP) files.
2. MTA Malware Dataset [3]: The MTA Malware Dataset contains traffic generated by numerous contemporary malware specimens. By including traffic data from recent malware, this dataset offers insights into evolving malware behaviors and tactics.
3. DataCon2020-Encrypted Malicious Traffic Dataset [1]: This dataset focuses on encrypted malicious traffic and traffic samples originating from sophisticated cyber threats. Analyzing this dataset provides valuable insights into encrypted traffic patterns associated with malicious activities.
4. CTU13 Dataset [9]: The CTU13 Dataset captures data emanating from botnet communications. By studying botnet traffic patterns, researchers can gain a deeper understanding of botnet behaviors and their implications for network security.

Leveraging these datasets allows for comprehensive analysis and evaluation of detection methods across diverse malicious traffic scenarios.

3.3   Feature Engineering

This paper extract the certificate information from the dataset in Sect. 3.2. Our work chose features that were valuable for the study. The features are shown in Table 1. The certificate subject and certificate issuer are the core fields of this certificate. In the traditional TLS connection establishment process, the mainstream operating system uses a whitelisting mechanism to determine whether the certificate issuer is trustworthy or not. The versions of the TLS protocol include TLS v1.0, TLS v1.1, TLS v1.2, and TLS v1.3. The TLS protocol has been upgraded to become more and more secure. The encryption method, signature method and key length of the certificate also indicate the security of this certificate in the issuance process. Alternative name of the subject of the certificate also indicate the domains supported by this certificate. Correlation studies have shown a correlation between this field and the maliciousness of encrypted traffic.

Based on the features shown in Table 1, we extracted more detailed features. The first feature our work selected was the certificate level. Certificates for SSL/TLS are categorized into three levels of verification: Domain Validation (DV), Organization Validation (OV), and Extended Validation (EV). Domain Validation Certificate are designed to verify the ownership of a domain. Organization Validation Certificate offer a higher level of verification, including the

Table 1. Certificate Features extracted from the traffic features

No.	Feature	Description
1	subject	Certificate subject information
2	issuer	Certificate issuer information
3	certificate.version	X509 certificate version
4	certificate.key_alg	Encryption algorithm used by the certificate
5	certificate.sig_alg	Sign algorithm used by the certificate
6	certificate.key_length	The length of the key used by the certificate
7	san_ip_len	The number of IPs included in the Extension Alternative Name of the certificate
8	san_dns_len	The number of DNS contained in the Extended Alternative Name of the certificate

verification that the website owner is a legitimate organization. Extended Validation Certificate provide the most stringent verification process. In addition to verifying domain and organizational details, it also verifies the legal status, operational location, and identity of the organization. It is worth noting that in the actual process of applying for certificates, EV certificates are much more expensive than OV and DV certificates, so it is reasonable to believe that these three levels of certificates reflect different levels of malice.

The second feature selected for our work focuses on the granularity of the subject and issuer details present in certificates. Our work observed that both the subject and issuer fields contain numerous attributes. Initially, this work quantified the level of detail by summarizing the number of fields within each.

Moreover, to assess certificate security, our research utilized the root certificates recognized by the Windows operating system. Our research determined whether the root certificate authorities associated with these certificates were included in the Windows trusted certificates list. This criterion served as an indicator of certificate legitimacy and trustworthiness.

3.4   Model Selection

This research chose traditional machine learning models for our tests. These include KNN [7], Decision Tree [13], Random Forest [5], lightgbm [12] and xgboost [6]. Each of these models has its unique strengths and is widely used in various applications. In this work, we are more concerned with the validity of the features this work have selected.

For our tests, this research aim to evaluate the performance of these models in detecting malicious encrypted traffic.

4   Experiments

In this section, this research evaluate our framework using the encrypted traffic from real malware. Firstly, this research describe our dataset. Secondly, this research analyze the effectiveness of each classifier and the effectiveness after ensemble.

4.1   Dataset

This paper have collected 4 open source datasets: CICIDS-2017 Dataset [14], MTA Malware Dataset [3], DataCon2020-Encrypted Malicious Traffic Dataset [1] and CTU13 Dataset [9]. This work extracted the TLS traffic in these four datasets as sessions by Brim [2].

Table 2. Distribution of Malicious Traffic Between Datasets

Ref	Dataset	Benign	Malicious
[14]	CICIDS-2017	92973	901
[3]	MTA	0	82540
[9]	DataCon-2020	27825	90585
[2]	CTU13	74644	3634

The total number of traffic extracted from the above four datasets using the TLS encryption protocol is 373,102, of which the data distribution of each dataset is shown in Table 2.

It can be seen that in these four datasets, the data distribution is unbalanced. But this work combined all of the above data and created a dataset with a rich set of scenarios.

Table 3. Accurancy and F1 score based on ML models

Model	Accuracy	F1 Score
KNN	93.2%	0.91
XGBoost	94.3%	0.92
Decision Tree	94.6%	0.92
LightGBM	93.3%	0.91
Random Forest	94.6%	0.93
Average	94.2%	0.918

4.2   Detection Results

Our work start by dividing all Pcap packet traffic into sessions by using the quin-tuple (source IP address, source port, destination IP address, destination port, and transport layer protocol) and extracting TLS encrypted sessions from them. Our work extractes time-related information, packet sequence-related informa-tion, and certificate information from each TLS encrypted session, and summa-rized this information to form our original dataset by means of table connection.

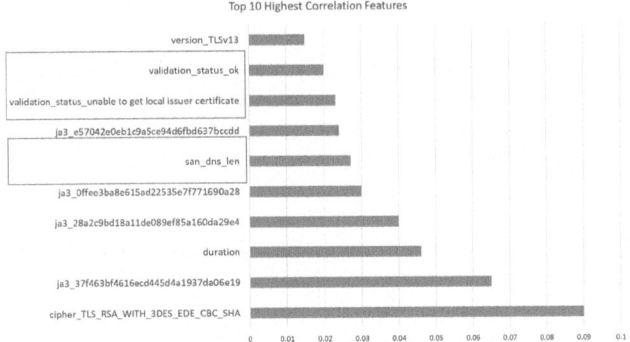

Fig. 2. Correalation

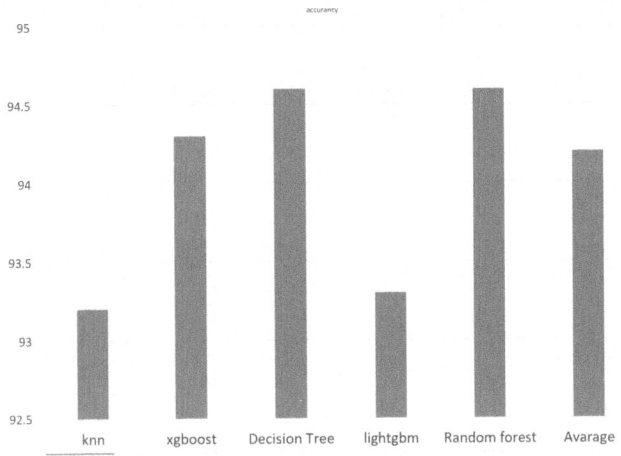

Fig. 3. Accurancy based on ML models

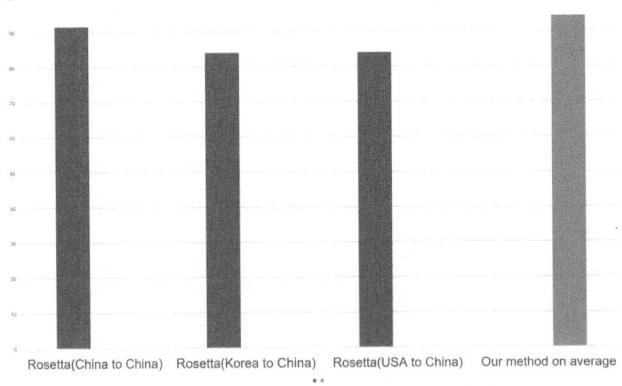

Fig. 4. Accurancy of our method compared to Rosetta

By One-Hot coding the text fields of this dataset, our research obtain a dataset with 230-dimensional features.

The first thing our work do is test whether the information associated with the certificate is relevant to malignity. Our work calculate the correlation between each feature and traffic malignancy based on a decision tree modeling. After that our model selects the 10 feature fields with the highest relevance. As Fig. 2 shows, The validity of TLS certificates and the Subject Alternative name in the TLS extension field have correlation coefficients of 2.7 and 2.3, respectively, with the maliciousness of traffic (Fig. 3).

This result shows that traffic certificate related information can be used for traffic maliciousness detection. Based on the above experimental results, we chose the certificate information and did the feature extraction on the certificate according to the method in Sect. 3.3.

After stating the value of the certificate information, this research compare our method with the with state-of-the-art method Rosetta [19]. This research test dataset using the model KNN, Decision Tree, Random Forest, lightgbm and xgboost. As Table 3 shows, our method has 94.2% accurancy on average.

As Fig. 4 shows, our method have a improvement on accurancy compared with Rosetta. Considering that our feature selection process selects only 19 features, our method is more efficient. Because the features this research chose is unmodifiable, so our method can be used in real network. Besides, our method has a low time complexity and our method can detect the maliciousness during the TLS handshake protocol. So our method has real-time and real-world deployment value.

4.3   Discussion

The experimental results illustrate that our method has high accuracy and practicality. Despite the challenges of dataset imbalance, the method shows robust generalization across various models, indicating its potential for real-world application. With a focus on certificate information, the approach outperforms existing methods like Rosetta in terms of accuracy and training time, highlighting its advantages for immediate and practical deployment in network security environments. Future work could further explore strategies to address data imbalance and optimize feature selection to enhance model performance even further.

5   Conclusion

This paper introduces an innovative method for detecting malicious traffic through the analysis of certificate information during TLS handshakes. Our approach offers high accuracy and efficiency, overcoming the challenges faced by traditional models in real network conditions. The method's effectiveness is supported by extensive experiments, showcasing its potential for immediate application in real-time cybersecurity. Future work could explore the integration of this method with existing network security infrastructures and further enhance its scalability and adaptability to evolving encryption techniques and cyber threats.

Acknowledgement. This work was supported by Major Scientific and Technological Innovation Projects of shandong Province (2020CXGC010116) and the National Natural Science Foundation of China (No. 62172042).

References

1. DataCon 2020 dataset (2020). https://datacon.qianxin.com/opendata
2. Brim software (2022). https://github.com/brimdata/brimcap
3. MTA dataset (2023). https://malware-traffic-analysis.net
4. Althouse, J.: ja4 (2024). https://github.com/FoxIO-LLC/ja4
5. Breiman, L.: Random forests. Mach. Learn. **45**, 5–32 (2001)
6. Chen, T., Guestrin, C.: XGBoost: a scalable tree boosting system. In: Proceedings of the 22nd ACM SIGKDD International Conference on Knowledge Discovery and Data Mining, pp. 785–794 (2016)
7. Cover, T., Hart, P.: Nearest neighbor pattern classification. IEEE Trans. Inf. Theory **13**(1), 21–27 (1967)
8. Fu, C., Li, Q., Xu, K., Wu, J.: Point cloud analysis for ML-based malicious traffic detection: reducing majorities of false positive alarms. In: Proceedings of the 2023 ACM SIGSAC Conference on Computer and Communications Security, pp. 1005–1019 (2023)
9. Garcia, S., Grill, M., Stiborek, J., Zunino, A.: An empirical comparison of botnet detection methods. Comput. Secur. **45**, 100–123 (2014)
10. Google: Google transparency report (2024). https://transparencyreport.google.com/
11. Althouse, J., Atkinson, J., Atkins, J.: ja3 (2017). https://github.com/salesforce/ja3
12. Ke, G., et al.: LightGBM: a highly efficient gradient boosting decision tree. In: Advances in Neural Information Processing Systems 30 (2017)
13. Quinlan, J.R.: C4. 5: Programs for Machine Learning. Elsevier (2014)
14. Sharafaldin, I., Lashkari, A.H., Ghorbani, A.A., et al.: Toward generating a new intrusion detection dataset and intrusion traffic characterization. In: ICISSp, vol. 1, pp. 108–116 (2018)
15. Shen, M., Liu, Y., Zhu, L., Xu, K., Du, X., Guizani, N.: Optimizing feature selection for efficient encrypted traffic classification: a systematic approach. IEEE Netw. **34**(4), 20–27 (2020)
16. Shen, M., et al.: Machine learning-powered encrypted network traffic analysis: a comprehensive survey. IEEE Commun. Surv. Tutor. **25**(1), 791–824 (2022)
17. Shen, M., Zhang, J., Zhu, L., Xu, K., Du, X.: Accurate decentralized application identification via encrypted traffic analysis using graph neural networks. IEEE Trans. Inf. Forensics Secur. **16**, 2367–2380 (2021)
18. Torroledo, I., Camacho, L.D., Bahnsen, A.C.: Hunting malicious TLS certificates with deep neural networks. In: Proceedings of the 11th ACM Workshop on Artificial Intelligence and Security, pp. 64–73 (2018)
19. Xie, R., et al.: Rosetta: enabling robust TLS encrypted traffic classification in diverse network environments with TCP-aware traffic augmentation. In: Proceedings of the ACM Turing Award Celebration Conference-China 2023, pp. 131–132 (2023)
20. Zscaler: 2022 encrypted attacks report (2024). https://www.zscaler.com/blogs/security-research/2022-encrypted-attacks-report/

MAP-SIM: A Performance Model for Shared-Memory Heterogeneous Systems with Mapping Awareness

Yuhang Li, Mei Wen$^{(\boxtimes)}$, Junzhong Shen, Zhaoyun Chen, and Yang Shi

Key Laboratory of Advanced Microprocessor Chips and Systems, National University of Defense Technology, Changsha, China
{liyuhang,meiwen}@nudt.edu.cn

Abstract. As performance demands continue to rise, shared-memory heterogeneous systems (SMHSs) have been widely adopted for their ability to enable efficient communication and data sharing between different heterogeneous cores. However, existing SMHS face challenges in uneven workload distribution among heterogeneous cores and suboptimal mapping schemes, preventing them from fully leveraging their architectural advantages. To address these issues, this paper introduces a performance model for SMHS named MAP-SIM. By performing performance modeling for CPUs and systolic array structures, and considering rational schemes for the partition and mapping of computational tasks, MAP-SIM aims to evaluate and optimize the computational performance of heterogeneous multicore architectures. The experimental results show that compared to previous schemes, MAP-SIM can lead to a 1.5 to 3.5 times increase in computational performance for SMHS.

Keywords: Performance Model · Systolic Array · Shared Memory Heterogeneous System · Mapping Scheme

1 Introduction

The increasing demand for improved performance has driven the development of shared-memory heterogeneous systems, making them widely used in on-chip systems. The state-of-the-art SMHS connects CPU cores and accelerator cores through shared memory on the same chip to facilitate efficient communication and data sharing among heterogeneous cores. However, mainstream SMHS still face the following two key issues when in use, making it difficult to fully leverage their inherent architectural advantages. **The first problem is that many systems oriented towards accelerators offload most of their computations to the accelerator cores, resulting in idle CPU cores while the accelerator cores are computing.** Researchers opt for this strategy primarily because accelerator cores exhibit significant performance advantages over CPU cores in highly parallel computing tasks. However, as instruction sets expand, exemplified

© The Author(s), under exclusive license to Springer Nature Singapore Pte Ltd. 2025
T. Zhu et al. (Eds.): ICA3PP 2024, LNCS 15251, pp. 351–361, 2025.
https://doi.org/10.1007/978-981-96-1525-4_21

by ARM's NEON instruction set [2], CPU cores have demonstrated enhanced performance in vector computations. Therefore, in heterogeneous architectures, the computing power of CPU cores should not be underestimated. **The second problem is that some frameworks have implemented task schedulers that distribute tasks across CPU cores and accelerator cores [3], but such schemes map the same tasks to different processing units.** Due to the varying computational throughput of different heterogeneous cores, this allocation of workload can still lead to inefficient utilization of available resources.

To overcome the aforementioned limitations, this work introduces a performance model named MAP-SIM, designed to discern and assess various partitioning and mapping schemes effectively. By conducting performance modeling for both CPUs and the classic parallel computing architecture, the systolic array, MAP-SIM enables precise evaluations of computational performance for SMHS. It employs a partitioning strategy for workload distribution, maximizing the use of CPU cores and accelerator cores within the system. This approach dynamically adjusts the partition ratio, considering the performance variances across heterogeneous cores and workload shape characteristics, to achieve an optimal workload distribution. Furthermore, MAP-SIM rigorously examines the impact that distinct mapping schemes have on overall system performance, employing a heuristic method to explore potential mappings. This culminates in the recommendation of a mapping strategy adept at minimizing computational latencies. In summary, this work makes the following contributions:

(1) We propose a novel performance model, MAP-SIM, which simulates the computational process of workloads being partitioned and mapped to different heterogeneous cores on SMHS and gives the corresponding computational time overhead.

(2) We have used methods for dynamically adjusting the workload partition ratio and heuristically exploring the mapping space to evaluate the impact of various partitioning and mapping schemes on the performance of SMHS, providing an implementation approach that achieves the shortest computation time.

(3) Experimental results show that MAP-SIM can offer efficient partitioning and mapping schemes for various types of workloads. Compared to previous schemes, the use of MAP-SIM can result in a 1.5 to 3.5 times performance improvement.

2   Background and Motivation

SMHS can integrate various types of processing units and enhance hardware resource utilization by appropriately allocating workloads, offering significant advantages in handling complex computational tasks. However, several common heterogeneous computing paradigms have certain shortcomings, and Fig. 1 shows these examples in detail. In Fig. 1, the blue sections represent the time overhead for off-chip and on-chip data communication, while the grey sections indicate the time overhead for data processing on the corresponding computing cores.

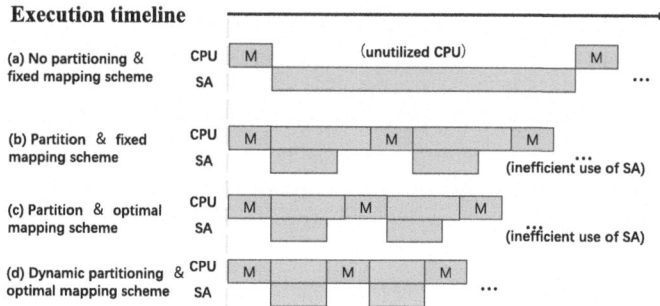

Fig. 1. Different partitioning and mapping schemes. (Color figure online)

In heterogeneous architectures, a common scheme involves offloading most computations to accelerators, as shown in Fig. 1(a). This approach makes full use of the accelerator's powerful computing capabilities but can also lead to underutilization of the CPU. To address this issue, a common approach is to partition the workload and deploy it across different cores. Figure 1(b) illustrates a strategy that involves partitioning the workload and distributing it across multiple cores. This method can prevent the issue of underutilization in individual cores but does not take into account the impact of different mappings on computing performance. Figure 1(c) depicts the strategy of partitioning the workload while considering optimal mapping. This method further enhances the overall performance of the heterogeneous system but still has shortcomings. Current task partitioning schemes divide computational tasks based on the peak performance of heterogeneous cores. However, in practice, the heterogeneous cores may be limited by factors such as bandwidth and buffer size, making it difficult to consistently maintain peak performance. Therefore, there is still room for improvement in this approach. Figure 1(d) showcases a strategy for dynamically adjusting the workload partitioning ratio while taking optimal mapping into account. This approach balances the computational time overhead of different heterogeneous cores by dynamically adjusting the workload distribution ratio, ultimately aiming to further enhance the overall efficiency of the system.

3   Related Work

Heterogeneous architectures are extensively utilized in a variety of compute-intensive and data-intensive scenarios, primarily because they can offer higher energy efficiency and computational performance.

Previous work has been centered on the issue of partitioned collaboration on task loads across heterogeneous cores. Stealing [3] introduced a cross-core distribution scheme for heterogeneous systems, allowing different cores to be more fully utilized during computation. However, this method overlooks the performance disparities between heterogeneous cores, distributing the same computational tasks among different cores, which results in overall computing efficiency still not being optimal. STREAM [6] proposes an optimized scheduling scheme for heterogeneous multicore scenarios; however, STREAM's approach to heterogeneous multicore is limited to different types of accelerator cores and does not consider distributing the workload to the CPU. Other research employs a specialized heterogeneous decomposition, selecting different heterogeneous cores for task mapping based on computational efficiency of the tasks. This thoroughly considers the characteristics of heterogeneous cores, but a drawback is the difficulty in flexibly decomposing work so that the workload can be evenly distributed across different heterogeneous cores.

4   MAP-SIM Performance Model

4.1   Systolic Array Performance Model

MAP-SIM is based on the fact that the patterns of data movement within the systolic array during computation are largely deterministic. It mathematically models the computation process and derives the corresponding formulas.

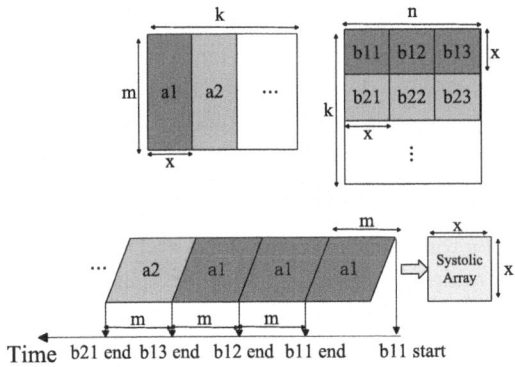

Fig. 2. Computation process of GEMM with IS dataflow.

Figure 2 illustrates the computation process of general matrix multiplication with input-stationary dataflow. During computation, a1 is computed with b11, b12, and b13 in sequence, after which a2 is computed with b21, b22, and b23 in sequence. It can be observed intuitively from the figure that when the pipeline technology is adopted, the time overhead of most matrix computations being run at full capacity on the systolic array is m.

T1 corresponds to the computational time overhead of the first block, which is determined by the size of the systolic array and the size of the matrix. In the formula, pe_latency represents the number of cycles needed for a processing element (PE) to complete a single multiply-accumulate operation. Thus, we can obtain T1 as follows:

$$T1 = m - 1 + (n - 1)\%x + (k - 1)\%x \times pe\_latency \tag{1}$$

T2 represents the time cost of the same color matrix computation in Fig. 2. Taking b21, b22, and b23 as examples, the time overhead of each matrix computation is m, while the number of matrices is determined by n and the size of the systolic array, from which we can obtain T2:

$$T2 = (n - 1)/x \times m \tag{2}$$

T3 describes the number of matrices of different colors in Fig. 2, which is in turn determined by k and the size of the systolic array. Accordingly, we can obtain T3 as follows:

$$T3 = m \times (n/x) \times (k - 1)/x \tag{3}$$

In summary, we can get the computation latency T_{SA} as follows. For WS and OS dataflows, we can also derive the corresponding computation latency in a similar way.

$$T_{SA} = T1 + T2 + T3 \tag{4}$$

4.2 CPU Performance Model

When constructing a CPU performance model, MAP-SIM takes into account multiple factors, including workload size, CPU peak performance, memory access latency, cache hit rate, etc. Taking floating-point computation as an example, assuming the total number of floating-point operations required by the workload to be computed is num_flops, and the corresponding number of memory accesses is num_access, then the overall execution time can be estimated as T_{CPU}:

$$T_{CPU} = num\_flops \times (FLOP\_latency + num\_access \times AMAT) \times \alpha \tag{5}$$

FLOP_latency represents the average execution time of floating-point operations, determined by the number of floating-point units in the CPU and its clock frequency. If the CPU's clock frequency is f and it can perform k floating-point operations per cycle, then FLOP_latency is $1/(f \times k)$. AMAT [8] represents the Average Memory Access Time, which is determined by the access delays and hit rates of each level of cache in the CPU. The access delays for the CPU's various cache levels can be acquired through specific lmbench [5] testing, while the access hit rates of the various cache levels can be obtained through appropriate measurement tools, such as AMD's uProf [1] and the perf tool [7] within the Linux system. α represents the coefficient of efficiency improvement through pipeline optimization during instruction execution by the CPU, with this parameter being determined by the actual measurement results of the hardware.

4.3   Dynamic Partitioning and Mapping Space Exploration

MAP-SIM partitions the workload and maps it to the corresponding processing cores while also invoking performance models for CPUs and systolic arrays to estimate the time overhead. Figure 3 showcases an example of the partitioning and mapping for the GEMM workload. MAP-SIM initially employs a technique proposed by Huang [4], partitioning the workload based on the number of heterogeneous cores and their maximum computational capabilities (indicated by the red line in Fig. 3). On this basis, MAP-SIM further examines the impact of different mapping schemes on data scheduling and storage, leading to a second partitioning (corresponding to the process denoted by the black line in Fig. 3).

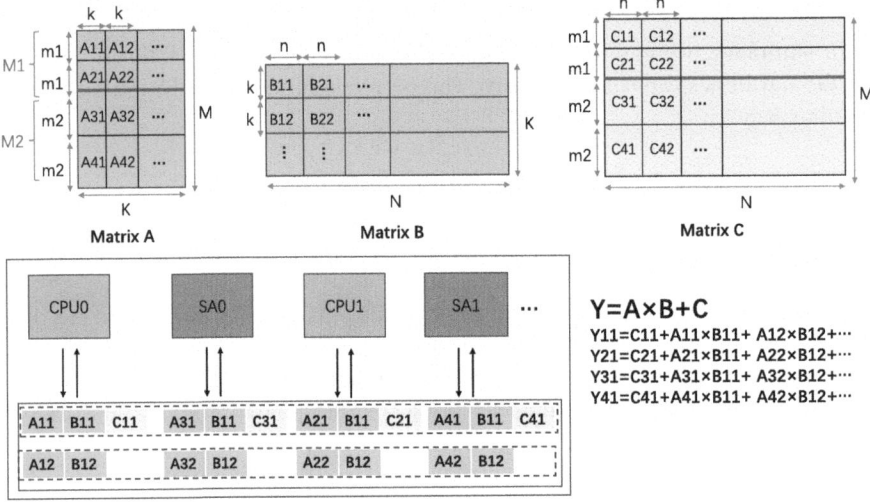

Fig. 3. Example of partitioning and mapping for GEMM workload. (Color figure online)

Taking SA0 in Fig. 3 as an example, the hardware first moves A31, B11, and C31 to the on-chip shared memory for computation. Once the data transfer is complete, SA0 starts the computation process, while the hardware continues to move A32 and B12 to the on-chip shared memory. The size of the sub-block is limited by the scale of the shared memory to ensure that the computation time of the heterogeneous cores can overlap with the data transfer time of the shared memory. Taking the range of the black dashed box of shared memory in Fig. 3 as granularity, assuming the computational time overhead for a given sub-block by the CPU and SA are t_{CPU} and t_{SA}, respectively, and the time overhead for moving data from off-chip to the shared memory within a single black box is t_{move}, then the total time overhead for this part of the sub-block should be the maximum of the three:

$$t_{cost} = \max\{t_{CPU}, t_{SA}, t_{move}\} \tag{6}$$

Finally, the number of black dashed boxes is jointly determined by the scale of the workload and the size of the sub-blocks. Therefore, the complete time evaluation metric is as follows, where x and y represent the numbers of CPUs and SAs, respectively.

$$\min\{t_{cost} \times \lceil \frac{K}{k} \rceil \times \lceil \frac{N}{n} \rceil \times \lceil \frac{M}{x \times m1 + y \times m2} \rceil\} \tag{7}$$

When MAP-SIM finds a suitable mapping scheme, it cannot yet determine whether this solution is optimal, as the initial partition based on the peak computational power of heterogeneous cores may not necessarily be the best choice. In actual hardware computation, the real computational power is often affected by various factors, leading to a difference from the peak computational power. Therefore, after obtaining a mapping scheme through a heuristic algorithm, MAP-SIM continues to invoke the performance models for CPU and SA to perform a time evaluation of the sub-block sizes provided by the mapping scheme. Based on the evaluated time ratios, it adjusts the initial partitioning ratio of the workload and then conducts a new heuristic search for mapping schemes. This process is repeated until the time overhead ratio of the sub-blocks aligns with the initial workload partitioning ratio, at which point it stops.

5   Evaluation

5.1   Experimental Settings

Current mainstream work employs a diverse array of heterogeneous cores in SMHS. To emphasize the impact of workload partitioning and mapping on the overall efficiency of SMHS, which is the focus of this paper, we use an internally developed simulator for experimentation. The simulator takes system configurations and workload information as input and outputs the corresponding execution time.

In the experiments, the L1 cache size of the CPU cores is 64 KB, and the L2 cache size is 1 MB. Each buffer in the SA cores is 32 KB, and dual-buffer technology is employed. The L3 cache size is 16 MB, which is shared by all the CPU and SA cores. Each SA core has 16 MAC units for computation, while the computing unit of a CPU core is one-fourth of that of an SA core.

The workloads selected for the experiment include the large language model Llama, the convolutional neural network AlexNet, and several representative layers from widely used contemporary natural language processing models: GNMT, DeepSpeech2 (DB), and Neural Collaborative Filtering (NCF).

5.2   Evaluations on Partitioning and Mapping Schemes

We implemented different partitioning and mapping schemes on a simulator and conducted experimental comparisons across a variety of workloads. The baseline used in the experiments assigns all workloads to the Systolic Array for computation and employs a fixed mapping scheme. This approach is referred to as SA+FM in the experiments. For the scheme that only uses SA for computation and considers the impact of mapping on performance, it is called SA+BM in the experiments. Correspondingly, for the schemes that consider workload partitioning, they are named CPU+SA+FM and CPU+SA+BM based on whether they take into account the impact of mapping on performance. Finally, this paper proposes the MAP-SIM scheme, which utilizes dynamic partitioning and optimal mapping.

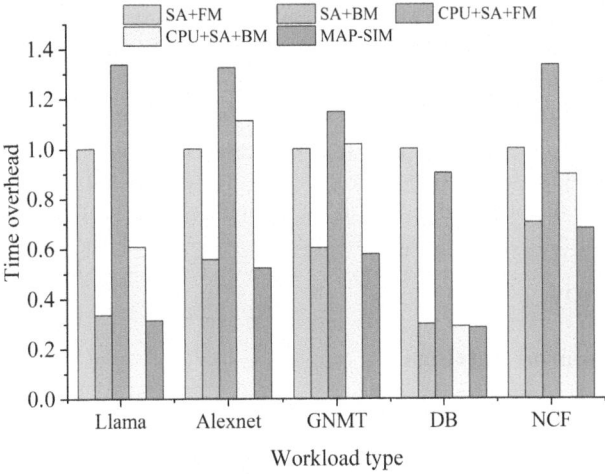

Fig. 4. Performance comparison of different partitioning and mapping schemes under 4-core CPU and 4-core systolic array environments.

Figure 4 presents the performance comparison of different partitioning and mapping schemes across various workloads. For the sake of clarity, we set the time overhead of the baseline as 1, and the time overhead of the other schemes is converted proportionally. First, it is intuitively evident from the figure that the MAP-SIM scheme offers the lowest execution time overhead across all workloads. Using the MAP-SIM scheme can bring a performance improvement of 1.5 to 3.5 times, which proves the effectiveness of the MAP-SIM model. Secondly, regardless of whether the CPU is involved in the computation, the optimal mapping schemes significantly enhance the overall performance of the SMHS compared to fixed mapping schemes, with the corresponding reduction in time overhead reaching up to 74%. Moreover, the experimental data also indicate that using peak computing power as the basis for partitioning when employing CPU cores for joint computation does not necessarily lead to performance improvement; in fact, this scheme can result in a decrease in overall system performance in the computation of workloads like Alexnet and GNMT.

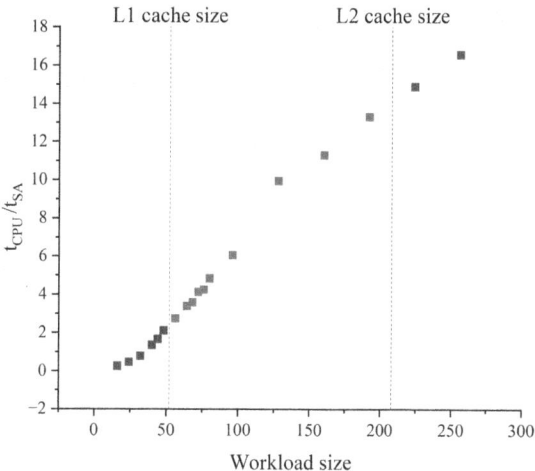

Fig. 5. Comparison of computational time overhead between CPU and SA.

The fundamental reason for this kind of situation lies in the different hardware structures and computing methods of the CPU and SA, which ultimately lead to varying computational performance of the two types of processing cores depending on the shape and size of the workloads. To more intuitively demonstrate the impact of this issue, we also conducted tests on different sizes of workloads separately on a single-core environment for both CPU and SA, and Fig. 5 presents the detailed results of these experiments.

Overall, as the size of the workload increases, the ratio of computational time overhead between CPU and SA gradually increases, and the speed of this ratio's growth also accelerates. Due to the difference in actual peak computing power, ideally, the ratio of computational time overhead between CPU and SA should be 4. However, in actual computation, this ratio fluctuates significantly, for the following reasons. Firstly, when the size of the workload is small, the L1 cache of the CPU has enough space to accommodate the data, thus the memory access overhead of the CPU is lower, leading to higher computational efficiency. In contrast, the SA experiences a reduction in average computational efficiency because the systolic array is not fully loaded (corresponding to the computation process of b11 in Fig. 2). In this case, the ratio of computational time overhead between CPU and SA can be less than 4. As the size of the workload increases, the average computational efficiency of the systolic array gradually increases, so the time overhead ratio gradually grows to around 4. When the size of the workload continues to increase until it exceeds the capacity of CPU's L1 cache, the growth rate of the time overhead ratio begins to accelerate. During this time, the CPU frequently experiences L1 cache misses during computations, necessitating data retrieval from the L2 cache. This leads to additional data access time, thereby reducing its computational efficiency. Similarly, as the workload size continues to increase beyond the L2 cache capacity, the CPU will increasingly access data from the L3 cache, resulting in a further decline in efficiency. Since data retrieval from lower-level caches incurs greater time overhead, the rate at which the time overhead ratio between the two increases also accelerates.

6   Conclusion

Overall, this paper introduces MAP-SIM, a performance model specially designed for shared-memory heterogeneous systems, focusing on improving hardware computational performance through efficient workload partitioning and mapping schemes. By addressing the issues of uneven workload distribution and suboptimal mapping in existing SMHS, MAP-SIM achieves good results in leveraging the architectural advantages of heterogeneous multicore architecture. The model conducted a comprehensive performance evaluation for heterogeneous systems composed of CPUs and Systolic Arrays, considering the impact of factors such as the computational throughput difference between heterogeneous cores and workload partitioning on system performance. Experimental results highlight MAP-SIM's ability to significantly improve computational performance, proving its effectiveness compared to previous works.

Acknowledgement. This work was supported by NUDT Foundation (No. ZK20 23-16) and National Natural ScienceFoundation of China (Grant No, U23A20301).

References

1. Advanced Micro Devices: AMD uProf User Guide (2023). https://developer.amd.com/amd-uprof/
2. ARM Holdings: ARM NEON Programming Guide. https://developer.arm.com/architectures/instruction-sets/simd-isas/neon
3. Cheng, L.: Intelligent scheduling for simultaneous CPU-GPU applications (2017). https://api.semanticscholar.org/CorpusID:50189368
4. Huang, S., et al.: Analysis and modeling of collaborative execution strategies for heterogeneous CPU-FPGA architectures. In: Proceedings of the 2019 ACM/SPEC International Conference on Performance Engineering (2019). https://api.semanticscholar.org/CorpusID:102347963
5. McVoy, L.W., Staelin, C.: lmbench: portable tools for performance analysis. In: USENIX Annual Technical Conference (1996). https://api.semanticscholar.org/CorpusID:10094752
6. Symons, A., Mei, L., Colleman, S., Houshmand, P., Karl, S., Verhelst, M.: Stream: a modeling framework for fine-grained layer fusion on multi-core DNN accelerators. In: 2023 IEEE International Symposium on Performance Analysis of Systems and Software (ISPASS), pp. 355–357 (2023). https://doi.org/10.1109/ISPASS57527.2023.00051

7. The Linux Foundation: perf: Linux profiling with performance counters. Online. https://training.linuxfoundation.org/resources/webinars/linux-performance-analysis-with-perf/

8. Wulf, W.A., McKee, S.A.: Hitting the memory wall: implications of the obvious. SIGARCH Comput. Archit. News **23**, 20–24 (1995). https://api.semanticscholar.org/CorpusID:263893286

A Vectorized Sequence-to-Graph Alignment Algorithm

Chenchen Peng[1], Shengbo Tang[1], Yifei Guo[1], Zeyu Xia[1],
Canqun Yang[1,2,3], and Yingbo Cui[1(✉)]

[1] College of Computer Science, National University of Defense Technology,
Changsha, China
yingbocui@nudt.edu.cn
[2] National Supercomputer Center in Tianjin, Tianjin, China
[3] Haihe Lab of ITAI, Tianjin, China

Abstract. A pangenome graph can represent the genomes of multiple individuals simultaneously. It provides a more comprehensive reference and overcomes the allele bias caused by linear reference genome, which is becoming a powerful new source of information in research. Sequence-to-graph alignment is the computational core of many pangenome related tasks, which aligns sequences to a graph to find the best matches. It serves as a crucial link for upstream and downstream research applications. However, existing sequence-to-graph alignment algorithms can not handle large-scale sequences efficiently. In this paper, we propose a vectorization version of sequence-to-graph alignment algorithm. We vectorize the core of Gwfa with SIMD instructions to improve the speed of per sequence alignment. Experiments on real and simulated datasets showed that, our algorithm achieved speedups of about 1.5×, while maintaining the alignment result consistent.

Keywords: Sequence-to-graph alignment · Vectorization · Bioinformatics · Read alignment

1 Introduction

Biological research often relies on inferring relationships between genomes, where the functionality of genes is closely linked to their genomic locations [11,33]. Reference genome provides a positional coordinate system for genomic analysis, which is an important resource of modern genomics [7]. The current human reference is an individual's genome (T2T-CHM13) [23] or a linear composite of genome segments from more than 20 people (GRCh38) [8,27,29,30]. These linear reference genomes can represent at most one haplotype. However, the genomes of human beings contain multiple alleles and genetic variants, which could not be fully represented by a linear reference. The sequences that do not appear by the reference genome may fail to align and will not be available for subsequent functional analysis. This phenomenon is called reference bias [34].

© The Author(s), under exclusive license to Springer Nature Singapore Pte Ltd. 2025
T. Zhu et al. (Eds.): ICA3PP 2024, LNCS 15251, pp. 362–371, 2025.
https://doi.org/10.1007/978-981-96-1525-4_22

These limitations have driven the development of reference genomes towards greater diversity and flexibility. The emergence of pangenomes has, to some extent, addressed the aforementioned issues [34]. A pangenome refers to the set of genomes from multiple individuals within a population, with most genes shared within the same population [4,6]. These shared regions are called homologous regions, while individuals within the population have personalized gene expression in other regions. Due to the typically large volume of genomic data, in practical applications, homologous regions among different individuals are often stored only once to form a single sequence (although methods involving multiple storage instances also exist [15]). Variations among individuals are stored separately from the homologous regions and are linked together, forming branches extending from the main backbone. This representation naturally forms a pangenome graph, which is emerging as a research focus in the field, regarding pangenome graphs as graph reference genomes [5,16,19].

The emergence of more comprehensive and diverse graph reference genomes has led to higher accuracy in many applications such as genotyping [3,31], functional annotation of variants [19], and long-read sequencing error correction [1]. Sequence alignment is the computational core of many pangenome related tasks. However, aligning sequences to a graph introduces greater complexity than aligning to linear sequences. Classical sequence alignment algorithms such as Smith-Waterman [32] and Needleman-Wunsch algorithms [22] have quadratic time complexity, not to mention their graph extension versions [12,21]. Existing sequence-to-graph alignment tools are primarily developed based on these algorithms [6,10,25,31].

With the advancement of sequencing technology, sequencing reads have evolved from high-throughput sequencing to long-read sequencing [35]. As the length of sequencing reads increases, the issue of the time required for sequence alignment has become a pressing problem that needs to be addressed. Over the past decade, many tools have been proposed to reduce the time for sequence alignment [9,14,28]. The main heuristic methods used include selecting characteristic substrings for localization [13] and setting thresholds to terminate the computation early when calculating the dynamic programming matrix for exact matches [39].

Myer proposed a time complexity $O(ND)$ difference algorithm [20], which efficiently searches for the longest common substring between strings by leveraging the similarity between them, where N is the length of the pattern and D is the edit distance between strings, where edit distance refers to the number of mismatched characters between two sequences. Santiago Marco-Sola et al. introduced the gap-affine penalty model into Myer's difference algorithms, and proposed a fast sequence alignment algorithm called wavefront alignment algorithm (WFA) [17]. Leveraging the similarity between sequences, WFA exhibits high performance in sequence alignment [18]. The graph extension of WFA algorithm, Gwfa, outperforms many existing sequence-to-graph alignment tools, particularly aligning to target objects with smaller edit distances [38].

In addition to algorithmic optimizations, some tools employ parallel acceleration strategies to improve efficiency. This is especially common in the era of high-throughput sequencing, where the large number of sequences makes it practical to divide tasks based on the number of sequences to achieve parallelism [37]. Other methods include partitioning the dynamic programming matrix for computation [36]. Sequence-to-graph alignment methods have also adopted these acceleration strategies. Currently, well-known alignment tools like vg [6], GraphAligner [26], and Vargas [2] mainly rely on heuristic algorithms or Single Instruction, Multiple Data (SIMD) acceleration. However, they have not yet explored the potential for application on larger-scale systems.

To further enhance the efficiency of alignment algorithms, various strategies have been explored, providing us with insights. Gwfa aims to improve overall program speed by converting some computations into bitwise operations and making significant efforts in memory management overhead. However, we observe that the program structure is entirely serial. Additionally, due to the complexity of the problem being addressed and the presence of numerous branches within the algorithm, this part offering opportunities for optimization.

In this paper, we propose an optimization version sequence-to-graph alignment algorithm. We tailor the data structure of Gwfa and used SIMD instructions to accelerate the computational cores and bottleneck branches, which are time-consuming parts.

The remainder of this paper is organized as follows. Section 2 describes the architecture and design of the algorithm. Section 3 presents the experiments of our algorithm on both simulated and real datasets. Section 4 concludes the paper and discusses the future directions of work.

2   Methods

2.1   Graph Wavefront Alignment Algorithm

Existing sequence-to-graph alignment algorithms typically inherit the ideas of sequence alignment algorithms. One of the fast sequence alignment algorithms is the WFA algorithm, which combines the dynamic programming approach of calculating the longest common substring along the matrix diagonals iteratively. This combination achieves at least a $10\times$ speedup compared to other sequence alignment methods [17,18]. Extending this idea to sequence-to-graph alignment allows for faster computation of the edit distance between sequences and graphs.

Before elaborating on the specific algorithm, it is necessary to clarify the concepts involved and formulate the problem that the algorithm aims to solve. In sequence-to-graph alignment algorithms, the query sequence $q[i...j]$ typically refers to a DNA fragment detected by sequencing technologies. Depending on the sequencing technique used, the query sequence may vary in length. A sequence can generally be understood as a string based on the alphabet $\Sigma = \{A, G, C, T\}$.

A sequence graph is a directed graph $G(V, E)$ constructed from multiple genomes, comprising nodes and edges. Each node represents a variable-length string based on the alphabet Σ, and directed edges denote their connectivity.

w_{v_i,v_j} represents a walk from the start node v_i to the end node v_j. Concatenating all the sequences represented by the nodes it passes through yields the sequence represented by this walk. Based on the concepts mentioned above, the sequence-to-graph alignment problem can be formulated as finding a walk in the graph that typically has the minimum edit distance with the query sequence.

Using dynamic programming (DP) algorithm, Gwfa compute the sequence-to-string labeled vertex sequence graph according to the following iterative rules as Eq. (1) shown. Here, $H_{i,v,j}$ represents the minimum edit distance between the prefix of the query sequence consisting of the first i characters and a path ending at the jth character of vertex v in the sequence graph. If $q[i] = v[j]$, then $\Delta_{i,v,j} = 0$; otherwise, $\Delta_{i,v,j} = 1$. When searching for the best alignment, it sets the starting node as v_s and the ending node as v_e. The initial condition is set as $H_{0,v_s,0} = 0$, and the final minimum edit distance obtained is $H_{|q|,v_e,|v_e|}$.

$$H_{i,v,j} = \min \begin{cases} H_{i-1,v,j} + 1, & i \geq 1 \\ H_{i,v,j-1} + 1, & j \geq 1 \\ H_{i-1,v,j-1} + \Delta_{i,v,j}, & i \geq 1, j \geq 1 \\ H_{i,u,|u|}, & j = 0, \forall u, (u,v) \in E \end{cases} \tag{1}$$

In practical computation, with an initial edit distance of 0, Gwfa first seeks the next potential starting position for the edit distance by computing the farthest mismatch along the current diagonal. This primarily involves two processes. The first is extending, which continuously extends the wavefront by matching characters until no further matches occur. The second is expanding, which keeps track of the current mismatch positions for use in the next extending process. However, in real-world scenarios, situations arise where there are gaps between nodes. This complicates the cases the algorithm must handle, necessitating the introduction of numerous conditional statements and making vectorization optimization significantly more challenging.

To effectively optimize vectorization, we employ program timing to pinpoint the most time-consuming section. As indicated in Table 1, The performance bottleneck of Gwfa lies within the alignment phase from sequences to graphs. Performance optimization can be further achieved through vectorization techniques for batch data computation and simplifying branching.

Table 1. Table showing the runtime duration and percentage of each stage of Gwfa.

Stages	Time(s)	Proportion
Total Time	26.11×10^3	100.00%
Calculate LCP	2.52×10^3	9.65%
Expand	18.02×10^3	69.02%
Sort	2.92×10^2	1.11%
Others	5.29×10^3	20.26%

2.2   Vectorization Algorithm of Gwfa

Vectorization is a parallel computing technique that accelerates calculations by processing multiple data elements simultaneously, leveraging SIMD instruction sets. This allows multiple operations to be executed in a single cycle, improving efficiency. Originating in the 1980s with supercomputers, vectorization has become integral to modern processors. Instruction sets such as SSE, AVX, and NEON support vectorized operations with widths like 128-bit and 256-bit, offering greater parallelism and performance. The main advantages include enhanced computational efficiency, reduced memory access overhead, and improved energy efficiency. By processing multiple elements at once, vectorization significantly speeds up execution, minimizes data transfer, and lowers overhead.

The current Gwfa algorithm is developed for computing edit distances. During the process of expanding along the diagonal, only the next mismatch position needs to be recorded. This position represents the starting element of all cells with an edit distance incremented by one. Moreover, due to the large number of nodes in the graph, it's impractical to store computation matrices for each node. Therefore, the algorithm's implementation doesn't actually compute matrices; instead, it maintains two queues. Queue A stores the positions of all cells with the current edit distance, using values from both forward and backward diagonals to locate positions within the matrix. Additionally, for each element in queue A, the algorithm calculates the Longest Common Prefix (LCP) to find the next mismatch position. When a mismatch position is found, the cell is added to queue B. This process continues until all elements in queue A have been expanded, treating queue B as the new queue of cells to be expanded, repeating the aforementioned steps. The process of calculating LCP can be understood as continuously matching and moving diagonally downwards to the right, which is referred to as extending. Finding mismatch positions and adding them to queue B is called expanding. In understanding the algorithm, we realize that the various diagonals stored in queue A can theoretically be computed independently, providing a foundation for vectorized algorithms.

Eliminating Data Dependency. In the actual implementation of the algorithm, due to data dependencies between updates of elements in queue A, mainly achieved by searching for a specific element in A to update it and pushing it into the queue if it doesn't exist, the decision whether to push the current element relies on the presence or absence of predecessor nodes in A. Through our analysis of the algorithm, this operation is primarily aimed at tracking the farthest wavefront. If we push all elements into the queue and incorporate a merging function for elements with the same key, we can achieve the same functionality by retaining the maximum value. By dissecting the original algorithm logic, we eliminate data dependencies and some branches, enabling this part of the algorithm to undergo vectorization optimization. Since this operation is a high-frequency task, vectorization optimization is necessary and expected to yield good results. Additionally, we expanded the number of bits for the LCP computation process, from the original 8 bits to 256 bits supported by the AVX2 instruction set, thus achieving vector optimization for this core computation.

Branch Masking Strategies. In the actual implementation of the algorithm, the expand process consumes more time compared to the extend process, which has a simpler structure and can be accelerated using bitwise operations. This is mainly because the expand process involves numerous branching instructions and loop operations to determine the current cell position. If the cell's position is in the center of the matrix, it requires updating the current wavefront. If the update fails, the cell needs to be pushed into queue B as a candidate for expansion. If the cell's position is on the edge of a node, it needs to traverse all neighboring nodes for expansion. If the position is at the end of the query sequence, the alignment might be nearing completion. Due to the excessive branching in this process, and with the majority of cases occurring within the matrix, we have vectorized the most time-consuming branches. We use masks to record the results of branch conditions and use the *blendv* operation to perform different calculations based on different conditions (Algorithm 1).

Algorithm 1. Handle branches with mask vector

Require: Loop bounds hi and lo, vector length $batchsize$, vertex length vl, query length ql
1: Initialize Queue Q, Q'
2: vec_k ← $\{k, k+1, \ldots, k + \text{batchsize} - 1\}$
3: vec_i ← $\{i, i+1, \ldots, i + \text{batchsize} - 1\}$
4: mask ← vec_and(vec_cmpgt($vl, k+1$), vec_cmpgt($ql, i+1$))
5: n_mask ← not mask
6: Q.push(vec_and(mask, elements))
7: Q'.push(vec_and(n_mask, elements))

---

3   Results

3.1   Evaluation Setup

We tested the performance of this vectorized algorithm on a local server. The configurations of the local server are shown in the Table 2.

In selecting the test data, we used parts of the human pangenome draft [16] to extract sequence graphs around the complement component 4 (C4) and major histocompatibility complex (MHC) loci in the human genome as the target sequence graphs for alignment. We used the sequences of HG002 and NA19240 as samples, and simulated data with varying sequence numbers and lengths using the pbsim tool [24]. The specific datasets are shown in Table 3. To ensure accurate and stable time recording, each dataset was repeatedly tested multiple times. During the testing process, our optimized version was configured with the same parameters as the comparison tools.

Table 2. Information about high performance computing platform.

Item	Local Server
CPU	Intel® Xeon Gold 6248R
Frequency	4.0 GHz
L1 Cache	1.5 MB
L2 Cache	24 MB
L3 Cache	35.75 MB
Cores	24
Sockets	2
Memory	384 GB
OS	Ubuntu 18.04
Node	–

Table 3. Table showing sequence datasets, their lengths, associated graphs, regions, and total lengths.

Sequence Dataset	Length (bp)	Graph	Region	Total Length (bp)
NA19240.real data	1905920	G1	C4	39926
NA19240.sim data	193577			
HG002.real data	2182962	G2	MHC	3293609
HG002.sim data	197856			

3.2   Performance Evaluation of Vectorization

For vectorization, we performed time statistics on the selected data. We ran both the pre-vectorization and post-vectorization test versions simultaneously. From the Fig. 1, we can find that vectorization has a noticeable speedup effect on some datasets, while for others, the optimization effect is not significant. This is mainly because the branch we optimized targets the majority of the cells inside the matrix. However, for nodes with smaller lengths or query sequences that frequently switch between nodes, there is less scope for optimization in this branch. For more continuous data, where most of the cells are within the matrix, our optimization is more effective. Thus, the vectorization is more recommended for graphs with long nodes and long data.

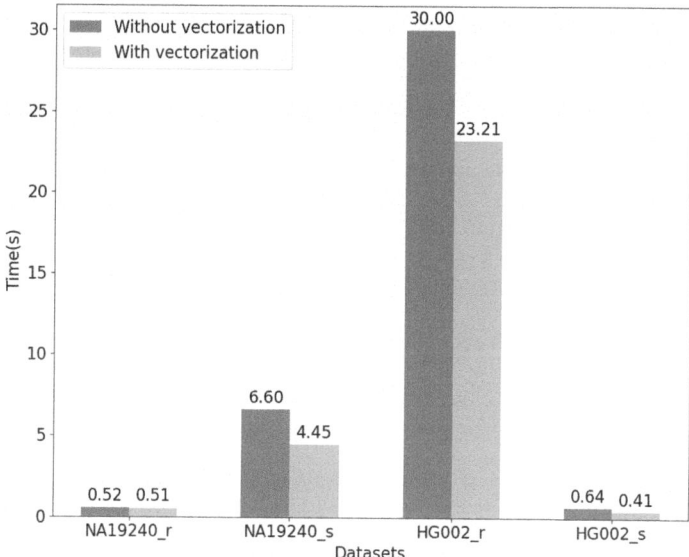

Fig. 1. The effect of vectorization obtained from tests using real and simulated datasets of different scales.

4   Conclusion

Sequence-to-graph alignment algorithms form the foundation of pangenome graph applications, with their accuracy and efficiency significantly impacting downstream research and applications. However, achieving more diverse and accurate alignments poses substantial computational challenges. To address these challenges, we propose an optimization algorithm, which accelerates sequence-to-graph alignment algorithm using vectorization and enables application across multiple platforms. Evaluation shows that, this vectorized algorithm achieves a speedup of about 1.5×, making it faster than many existing tools. Future research will explore more complex graphs and extend its application to other architectural platforms, further enhancing its speed and usability.

Acknowledgments. This work was supported by National Natural Science Foundation of China No. 62102427, Science and Technology Innovation Program of Hunan Province No. 2024RC3115 and Innovative Talent Program of National University of Defense Technology.

References

1. Cechova, M.: Probably correct: rescuing repeats with short and long reads. Genes **12**(1), 48 (2020)
2. Darby, C.A., Gaddipati, R., Schatz, M.C., Langmead, B.: Vargas: heuristic-free alignment for assessing linear and graph read aligners. Bioinformatics **36**(12), 3712–3718 (2020)

3. Ebler, J., et al.: Pangenome-based genome inference allows efficient and accurate genotyping across a wide spectrum of variant classes. Nat. Genet. **54**(4), 518–525 (2022)

4. Eizenga, J.M., et al.: Pangenome graphs. Ann. Rev. Genomics Hum. Genet. **21**, 139–162 (2020)

5. Gao, Y., et al.: A pangenome reference of 36 Chinese populations. Nature 1–10 (2023)

6. Garrison, E., et al.: Variation graph toolkit improves read mapping by representing genetic variation in the reference. Nat. Biotechnol. **36**(9), 875–879 (2018)

7. Gibbs, R.A.: The human genome project changed everything. Nat. Rev. Genet. **21**(10), 575–576 (2020)

8. Green, R.E., et al.: A draft sequence of the Neandertal genome. Science **328**(5979), 710–722 (2010)

9. Guo, Y., et al.: MTMap: a long-read alignment tool based on multi-core DSPs. In: 2023 IEEE International Conference on Bioinformatics and Biomedicine (BIBM), pp. 863–866. IEEE (2023)

10. Jain, C., Misra, S., Zhang, H., Dilthey, A., Aluru, S.: Accelerating sequence alignment to graphs. In: 2019 IEEE International Parallel and Distributed Processing Symposium (IPDPS), pp. 451–461. IEEE (2019)

11. Lander, E.S., et al.: C. International human genome sequencing. In: Initial Sequencing and Analysis of the Human Genome, pp. 860–921 (2001)

12. Lee, C., Grasso, C., Sharlow, M.F.: Multiple sequence alignment using partial order graphs. Bioinformatics **18**(3), 452–464 (2002)

13. Li, H.: Minimap and miniasm: fast mapping and de novo assembly for noisy long sequences. Bioinformatics **32**(14), 2103–2110 (2016)

14. Li, H.: Minimap2: pairwise alignment for nucleotide sequences. Bioinformatics **34**(18), 3094–3100 (2018)

15. Li, H., Feng, X., Chu, C.: The design and construction of reference pangenome graphs with minigraph. Genome Biol. **21**, 1–19 (2020)

16. Liao, W.-W., et al.: A draft human pangenome reference. Nature **617**(7960), 312–324 (2023)

17. Marco-Sola, S., Moure, J.C., Moreto, M., Espinosa, A.: Fast gap-affine pairwise alignment using the wavefront algorithm. Bioinformatics **37**(4), 456–463 (2021)

18. Marco-Sola, S., Eizenga, J.M., Guarracino, A., Paten, B., Garrison, E., Moreto, M.: Optimal gap-affine alignment in $O(s)$ space. Bioinformatics **39**(2), btad074 (2023)

19. Miga, K.H., Wang, T.: The need for a human pangenome reference sequence. Ann. Rev. Genomics Hum. Genet. **22**, 81–102 (2021)

20. Myers, E.W.: An $O(ND)$ difference algorithm and its variations. Algorithmica **1**(1–4), 251–266 (1986)

21. Navarro, G.: Improved approximate pattern matching on hypertext. Theor. Comput. Sci. **237**(1–2), 455–463 (2000)

22. Needleman, S.B., Wunsch, C.D.: A general method applicable to the search for similarities in the amino acid sequence of two proteins. J. Mol. Biol. **48**(3), 443–453 (1970)

23. Nurk, S., et al.: The complete sequence of a human genome. Science **376**(6588), 44–53 (2022)

24. Ono, Y., Asai, K., Hamada, M.: PBSIM2: a simulator for long-read sequencers with a novel generative model of quality scores. Bioinformatics **37**(5), 589–595 (2021)

25. Rautiainen, M., Marschall, T.: GraphAligner: rapid and versatile sequence-to-graph alignment. Genome Biol. **21**(1), 253 (2020)

26. Rautiainen, M., Mäkinen, V., Marschall, T.: Bit-parallel sequence-to-graph alignment. Bioinformatics **35**(19), 3599–3607 (2019)
27. Rhie, A., et al.: Towards complete and error-free genome assemblies of all vertebrate species. Nature **592**(7856), 737–746 (2021)
28. Sahlin, K., Baudeau, T., Cazaux, B., Marchet, C.: A survey of mapping algorithms in the long-reads era. Genome Biol. **24**(1), 1–23 (2023)
29. Schneider, V.A., et al.: Evaluation of GRCh38 and de novo haploid genome assemblies demonstrates the enduring quality of the reference assembly. Genome Res. **27**(5), 849–864 (2017)
30. Sherman, R.M., Salzberg, S.L.: Pan-genomics in the human genome era. Nat. Rev. Genet. **21**(4), 243–254 (2020)
31. Sirén, J., et al.: Pangenomics enables genotyping of known structural variants in 5202 diverse genomes. Science **374**(6574), abg8871 (2021)
32. Smith, T.F., Waterman, M.S., et al.: Identification of common molecular subsequences. J. Mol. Biol. **147**(1), 195–197 (1981)
33. Venter, J.C., et al.: The sequence of the human genome. Science **291**(5507), 1304–1351 (2001)
34. Wang, T., et al.: The human pangenome project: a global resource to map genomic diversity. Nature **604**(7906), 437–446 (2022)
35. Wenger, A.M., et al.: Accurate circular consensus long-read sequencing improves variant detection and assembly of a human genome. Nat. Biotechnol. **37**(10), 1155–1162 (2019)
36. Xia, Z., et al.: A review of parallel implementations for the Smith–Waterman algorithm. Interdisc. Sci. Comput. Life Sci. 1–14 (2021)
37. Xia, Z., et al.: Large-scale parallel alignment algorithm for SMRT reads. In: International Conference on Algorithms and Architectures for Parallel Processing, pp. 213–229. Springer (2021)
38. Zhang, H., Wu, S., Aluru, S., Li, H.: Fast sequence to graph alignment using the graph wavefront algorithm. arXiv preprint arXiv:2206.13574 (2022)
39. Zhang, Z., Schwartz, S., Wagner, L., Miller, W.: A greedy algorithm for aligning DNA sequences. J. Comput. Biol. **7**(1–2), 203–214 (2000)

Author Index

© The Editor(s) (if applicable) and The Author(s), under exclusive license
to Springer Nature Singapore Pte Ltd. 2025
T. Zhu et al. (Eds.): ICA3PP 2024, LNCS 15251, pp. 373–374, 2025.
https://doi.org/10.1007/978-981-96-1525-4

The manufacturer's authorised representative in the EU is Springer
Nature Customer Service Centre GmbH, Europaplatz 3, 69115 Heidelberg,
Germany. If you have any concerns regarding our products, please
contact ProductSafety@springernature.com

Printed and bound by CPI Group (UK) Ltd, Croydon, CR0 4YY

29/04/2026

02099546-0004